Hadoop 大数据分析实战

[美] 斯里达尔・奥拉　著

李　垚　译

清华大学出版社

北　京

内 容 简 介

本书详细阐述了与 Hadoop 3 大数据分析相关的基本解决方案，主要包括 Hadoop 简介、大数据分析概述、基于 MapReduce 的大数据处理、Python-Hadoop 科学计算和大数据分析、R-Hadoop 统计数据计算、Apache Spark 批处理分析、Apache Spark 实时数据分析、Apache Flink 批处理分析、Apache Flink 流式处理、大数据可视化技术、云计算简介、使用亚马逊 Web 服务等内容。此外，本书还提供了相应的示例、代码，以帮助读者进一步理解相关方案的实现过程。

本书适合作为高等院校计算机及相关专业的教材和教学参考书，也可作为相关开发人员的自学教材和参考手册。

图书在版编目（CIP）数据

Hadoop 大数据分析实战/（美）斯里达尔·奥拉（Sridhar Alla）著；李垚译. —北京：清华大学出版社，2019

书名原文：Big Data Analytics with Hadoop 3

ISBN 978-7-302-52789-3

Ⅰ. ①H… Ⅱ. ①斯… ②李… Ⅲ. ①数据处理软件-高等学校-教材 Ⅳ. ①TP274

中国版本图书馆 CIP 数据核字（2019）第 076915 号

责任编辑： 贾小红
封面设计： 刘 超
版式设计： 魏 远
责任校对： 马子杰
责任印制： 丛怀宇

出版发行： 清华大学出版社
网 址： http://www.tup.com.cn，http://www.wqbook.com
地 址： 北京清华大学学研大厦 A 座 **邮 编：** 100084
社 总 机： 010-62770175 **邮 购：** 010-62786544
投稿与读者服务： 010-62776969，c-service@tup.tsinghua.edu.cn
质 量 反 馈： 010-62772015，zhiliang@tup.tsinghua.edu.cn
印 装 者： 清华大学印刷厂
经 销： 全国新华书店
开 本： 185mm×230mm **印 张：** 24 **字 数：** 480 千字
版 次： 2019 年 5 月第 1 版 **印 次：** 2019 年 5 月第 1 次印刷
定 价： 129.00 元

产品编号：081453-01

译　者　序

Hadoop 是一个由 Apache 基金会所开发的分布式系统基础架构，它可以使用户在不了解分布式底层细节的情况下开发分布式程序，充分利用集群的威力进行高速运算和存储。Hadoop 解决了两大问题：大数据存储和大数据分析，也就是 Hadoop 的两大核心内容：HDFS 和 MapReduce。目前，Hadoop 已经成为业界大数据平台的首选方案之一，Hadoop 人才的需求量也是越来越大

本书旨在令读者具备 Hadoop 3 生态系统的分析能力，并能够构建强大的解决方案来执行大数据分析，同时毫不费力地从大数据分析结果中获得敏锐的洞察力。本书涉及 R 语言、Python 语言、Spark、Flink、Hadoop 的综合运用，同时实现了大数据分析的可视化结果。

在本书的翻译过程中，除李垚外，王辉、刘璋、刘晓雪、张博、刘祎、张华臻等人也参与了部分翻译工作，在此一并表示感谢。

由于译者水平有限，难免有疏漏和不妥之处，恳请广大读者批评指正。

译　者

前　　言

Apache Hadoop 是一类流行的大数据处理平台，并可与大多数大数据工具集成，以构建功能强大的数据分析方案。本书将围绕这一点对相关软件展开讨论，同时辅以大量的操作实例。

在本书阅读过程中，读者将会系统学习 HDFS、MapReduce、YARN 方面的知识，以及如何实现快速、高效的大数据处理方案。此外，本书还将 Hadoop 与其他开源工具集成，例如 Python 和 R 语言，进而分析和可视化数据，同时针对大数据进行统计计算。一旦读者掌握了这些内容，即可尝试在 Apache Spark 和 Apache Flink 的基础上应用 Hadoop，最终实现实时数据分析和流式处理。除此之外，本书还将讨论如何在云端和端到端管道上利用 Hadoop 构建数据分析方案，并通过操作实例执行大数据分析任务。

在阅读完本书后，读者将具备基于 Hadoop 生态系统的分析能力，同时可构建强大的解决方案执行大数据分析，并拥有自己的技术观点。

适用读者

如果读者希望使用 Hadoop 3 的强大功能为企业或业务构建高性能的分析解决方案，或者您是一名大数据分析新手，那么本书将十分适合于您。另外，本书需要读者具备 Java 编程方面的基础知识。

本书内容

第 1 章将介绍 Hadoop 环境及其核心组件，包括 HDFS 和 MapReduce。

第 2 章将讨论大型数据集的检测处理过程，从中发现数据的模式，生成相应的报告并采集有价值的内容。

第 3 章将讨论 MapReduce，这也是大多数计算/处理系统中的基本概念。

第 4 章探讨 Python 语言，并在此基础上通过 Hadoop 对大数据进行分析。

第 5 章介绍了 R 语言，同时阐述了如何使用 R 语言并借助于 Hadoop 执行大数据统计计算。

第 6 章将考查 Apache Spark，同时根据批处理模型使用 Spark 进行大数据分析。

第 7 章将对 Apache Spark 的流式处理模型进行分析，以及如何打造基于流式的实时分析应用程序。

第 8 章主要介绍 Apache Flink，及其基于批处理模型的、针对大数据分析的应用方式。

第 9 章讨论 DataStream API 和基于 Flink 的流处理。其中，Flink 用于接收和处理实时事件流，并在 Hadoop 集群中存储聚合和结果。

第 10 章考查数据可视化问题，并通过各种工具和技术实现这一功能，例如 Tableau。

第 11 章讲述云计算以及各种概念，例如 IaaS、PaaS 和 SaaS。除此之外，本章还将对云供应商加以简要介绍。

第 12 章介绍 AWS 和 AWS 中的各种服务，这些服务使用 Elastic MapReduce（EMR）在 AWS 云中建立 Hadoop 集群，这对执行大数据分析非常有用。

软件和硬件环境

本书示例是在 64 位 Linux 上使用 Scala、Java、R 和 Python 语言实现的。另外，还应在机器上安装下列内容（建议使用最新版本）：

- Spark 2.3.0（或更高版本）。
- Hadoop 3.1（或更高版本）。
- Flink 1.4。
- Java（JDK 和 JRE）1.8+。
- Scala 2.11.x（或更高版本）。
- Python 2.7+/3.4+。
- R 3.1+和 RStudio 1.0.143。
- Eclipse Mars 或 Idea IntelliJ（最新版本）。

关于操作系统，最好使用 Linux 发行版（包括 Debian、Ubuntu、Fedora、RHEL 和 CentOS）。具体来说，例如，对于 Ubuntu，建议使用完整的 14.04 （LTS） 64 位安装、VMWare player 12 或 Virtual box。此外，还可在 Windows（XP/7/8/10）或者 macOS X（10.4.7+）上运行代码。

关于硬件配置，可采用 Core i3、Core i5（推荐）～Core i7（获得最佳效果）。然而，多核处理将提供更快的数据处理以及较好的可伸缩性。另外，对于单系统模式，至少使用

8GB RAM（推荐）；单个 VM 至少使用 32GB RAM；对于集群，则至少使用 32GB RAM。足够的存储空间可运行繁重的任务（取决于将要处理的数据集大小），最好至少包含 50GB 的空闲磁盘存储空间（用于独立系统和 SQL 仓库）。

资源下载

读者可访问 http://www.packtpub.com 并通过个人账户下载示例代码文件。另外，http://www.packtpub.com/support，注册成功后，我们将以电子邮件的方式将相关文件发与读者。

读者可根据下列步骤下载代码文件：

（1）登录 www.packtpub.com 并注册我们的网站。

（2）选择 SUPPORT 选项卡。

（3）单击 Code Downloads & Errata。

（4）在 Search 文本框中输入书名并执行后续命令。

当文件下载完毕后，确保使用下列最新版本软件解压文件夹：

- Windows 系统下的 WinRAR/7-Zip。
- Mac 系统下的 Zipeg/iZip/UnRarX。
- Linux 系统下的 7-Zip/PeaZip。

另外，读者还可访问 GitHub 获取本书的代码包，对应网址为 https://github.com/PacktPublishing/Big-Data-Analytics-with-Hadoop-3。代码与 GitHub 存储库将实现同步更新。

此外，读者还可访问 https://github.com/PacktPublishing/以了解丰富的代码和视频资源。

除此之外，我们还提供了 PDF 文件，其中包含了本书所用截图/图表的彩色图像。读者访问 http://www.packtpub.com/sites/default/files/downloads/BigDataAnalyticswithHadoop3_ColorImages.pdf 进行下载。

本书约定

代码块则通过下列方式设置：

```
hdfs dfs -copyFromLocal temperatures.csv /user/normal
```

代码中的重点内容则采用黑体表示：

```
Map-Reduce Framework -- output average temperature per city name
    Map input records=35
    Map output records=33
    Map output bytes=208
    Map output materialized bytes=286
```

命令行输入或输出如下所示：

```
$ ssh-keygen -t rsa -P '' -f ~/.ssh/id_rsa
$ cat ~/.ssh/id_rsa.pub >> ~/.ssh/authorized_keys
$ chmod 0600 ~/.ssh/authorized_keys
```

图标表示较为重要的说明事项。

图标则表示提示信息和操作技巧。

读者反馈和客户支持

欢迎读者对本书的建议或意见予以反馈。

对此，读者可向 feedback@packtpub.com 发送邮件，并以书名作为邮件标题。若读者对本书有任何疑问，均可发送邮件至 questions@packtpub.com，我们将竭诚为您服务。

勘误表

尽管我们在最大程度上做到尽善尽美，但错误依然在所难免。如果读者发现谬误之处，无论是文字错误抑或是代码错误，还望不吝赐教。对此，读者可访问 http://www.packtpub.com/submit-errata，选取对应书籍，单击 Errata Submission Form 超链接，并输入相关问题的详细内容。

版权须知

一直以来，互联网上的版权问题从未间断，Packt 出版社对此类问题异常重视。若读者在互联网上发现本书任意形式的副本，请告知网络地址或网站名称，我们将对此予以处理。关于盗版问题，读者可发送邮件至 copyright@packtpub.com。

若读者针对某项技术具有专家级的见解，抑或计划撰写书籍或完善某部著作的出版工作，则可访问 www.packtpub.com/authors。

问题解答

若读者对本书有任何疑问，均可发送邮件至 questions@packtpub.com，我们将竭诚为您服务。

目　　录

第 1 章　Hadoop 简介

本章主要介绍 Hadoop 环境及其核心组件，即 Hadoop 分布式文件系统（HDFS）以及 MapReduce。本章首先探讨 Hadoop 3 版本中的变化内容以及最新特性。特别地，我们将介绍 HDFS 的新特性、YARN 和客户端应用程序方面的变化内容。进一步讲，本章将于本地安装 Hadoop 集群，同时展示某些新的特征，例如纠删码（EC）和时间轴服务。作为快速提示，第 10 章将向读者展示如何在 AWS 中创建 Hadoop 集群。

本章主要涉及以下内容：

- HDFS。
  - 高可用性。
  - Intra-DataNode 均衡器。
  - EC。
  - 端口映射。
- MapReduce。
  - 任务级优化。
- YARN。
  - 机会型容器。
  - 时间轴服务 v.2。
  - Docker 容器。
- 其他一些变化内容。
- Hadoop 3.1 的安装。
  - HDFS。
  - YARN。
  - EC。
  - 时间轴服务 v.2。

1.1　Hadoop 分布式文件系统

HDFS 是用 Java 实现的基于软件的文件系统，它位于本机文件系统之上。HDFS 背后的核心概念是，该系统将文件划分为块（一般为 128MB），而非整体处理文件。这将支持如分布、复制、故障恢复等诸多特性，更重要的是可使用多台机器对块进行分布式处理。具体来说，块尺寸可以是 64MB、128MB、256MB 或 512MB，且适用于多种用途。对于具有 128MB 块的 1GB 文件，将有 1024MB/128MB=8 个块。如果复制因子为 3，那么它就是 24 个块。HDFS 提供了基于容错和故障恢复的分布式存储系统，主要包含两个组件，即 NameNode 和 DataNode。其中，NameNode 涵盖了文件系统所有内容的元数据，涉及文件名、文件权限以及每个文件块的位置，因而也是 HDFS 中最为重要的组件。DataNode 负责连接 NameNode，并存储 HDFS 中的块，同时依赖 NameNode 获得关于文件系统中内容的所有元数据信息。如果 NameNode 未包含任何信息，那么，DataNode 则无法向读/取 HDFS 的客户端提供信息服务。

NameNode 和 DataNode 的处理过程可以在单机设备上运行；通常，HDFS 集群由运行 NameNode 进程的专用服务器和运行 DataNode 进程的数千台机器组成。为了能够访问存储于 NameNodeNameNode 中的内容信息，元数据整体结构将存储于内存中。当设备出现故障时，通过记录块的复制因子，可确保数据完整无损。鉴于这是一类单点故障，为了减少 NameNode 故障而导致的数据丢失的风险，可以使用一个二级 NameNode 生成 NameNode 主内存结构的快照。

DataNode 具有较强的存储能力，与 NameNode 不同，如果 DataNode 出现故障，HDFS 仍可正常运行。当 DataNode 出现故障时，NameNode 自动处理故障 DataNode 中所有数据块所减少的复制行为，同时确保该复制操作被构建。考虑到 NameNode 知晓所有复制块的位置，因而连接至集群的任意客户端都能够顺利地工作。

注意：

为了使每个数据块能够满足所需的最小复制因子，NameNode 将复制所丢失的数据块。

图 1.1 显示 NameNode 中文件与数据块间的映射关系、数据块的存储及其在 DataNode 中的复制行为。

在图 1.1 中，NameNode 在 Hadoop 开始阶段即为单点故障。

1.1.1　高可用性

在 Hadoop 1.x 和 Hadoop 2.x 中，NameNode 的丢失可导致集群崩溃。在 Hadoop 1.x 中，其恢复过程较为困难；而 Hadoop 2.x 则引入了高可用性（主动-被动设置），并可从 NameNode 故障中予以恢复。

图 1.2 显示了高可用性的工作方式。

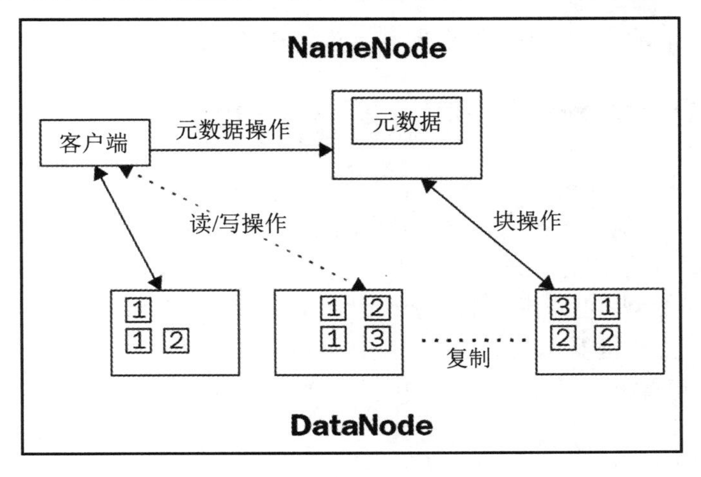

图 1.1

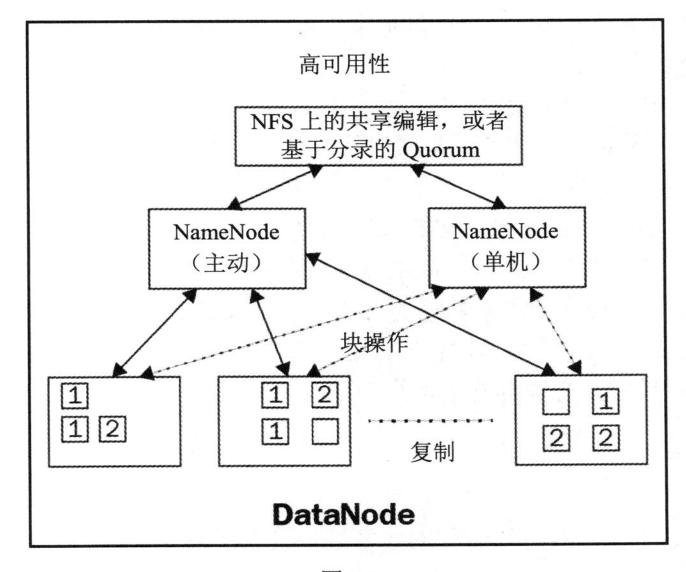

图 1.2

Hadoop 3.x 中包含了两个被动式 NameNode 和一个主动节点，以及 5 个 JournalNode，并以此协助从灾难性故障中予以恢复。

- NameNode 机器：运行主动和备用 NameNode 的机器，它们之间应拥有相同的硬件，以及在非 HA 集群中使用的硬件。
- JournalNode 机器：运行 JournalNode 的机器。JournalNode 守护进程则是相对轻

量级的，因此，这一类守护进程可与其他 Hadoop 守护进程合理地配置在机器上。

1.1.2 内部 DataNode 均衡器

HDFS 包含了一种方式，可均衡数据节点间的数据块，但是，在具有多个硬盘的相同数据节点中并不存在这样的平衡机制。因此，12 轴 DataNode 的物理磁盘可能处于非平衡状态。但是，为何这会对性能产生影响？对于非平衡磁盘，DataNode 级别的数据块可能与其他 DataNode 相同，但考虑到非平衡磁盘，读/写行为可能会产生倾斜。因此，Hadoop 3.x 引入了内部节点平衡器，以平衡每个数据节点内的物理磁盘，进而降低数据倾斜。

这增加了在集群上运行的任何进程执行的读/写速度，例如 mapper 或 reducer。

1.1.3 纠删码

自 Hadoop 诞生以来，HDFS 一直是基本组件。在 Hadoop 1.x 和 Hadoop 2.x 中，HDFS 安装所用的复制因子一般为 3。

与默认的复制因子（3）相比，EC 可能是 HDFS 多年来最大的变化。通过将复制因子从 3 降低到 1.4，从根本上增加了数据集的容量。下面将进一步考查 EC。

EC 可视为一种数据保护方法，在这种方法中，数据被分解为片段、扩展、用冗余数据片段进行编码，并存储在一组不同的位置或存储中。在此过程中的某个时间点处，如果由于数据损坏而导致丢失，那么可以使用存储在其他地方的信息进行重构。虽然 EC 具有 CPU 密集型特征，但对于大量数据的可靠存储（HDFS），可极大地减少所需的存储量。HDFS 通过复制行为提供可靠性存储，其代价也较为高昂——一般需要 3 个存储数据的副本，因而在存储空间方面其开销增至 200%。

1.1.4 端口号

在 Hadoop 3.x 中，针对各种服务的诸多端口均有所变化。

之前，多个 Hadoop 服务的默认端口位于 Linux 临时端口范围内（32768～61000）。这表明，在启动阶段，由于冲突问题，服务有时无法绑定至对应端口上。

此类冲突端口已从临时范围中被移除，同时也会对 NameNode、二级 NameNode 和 KMS 产生一定的影响。

具体变化如下所示。

- NameNode 端口：50470→9871、50070→9870 和 8020→9820。
- 二级 NameNode 端口：50091→9869 和 50090→9868。

- DataNode 端口：50020→9867、50010→9866、50475→9865 和 50075→9864。

1.2　MapReduce 框架

理解这个概念的一个简单方法是，想象你和你的朋友把成堆的水果分类装至盒子中。对此，你需要为每个人分配一项任务：首先将生水果装至一个篮子中（混在一起），然后把水果分装为不同的盒子。然后每个人都执行同样的任务，将这一篮子水果分成不同的种类。最后，可从你所有的朋友那里得到很多盒水果。随后，可将同类水果装箱，对其称重、封装并运输。单词计数则是另一个与 MapReduce 框架较为类似的示例。图 1.3 显示了处理输入数据的各个阶段。首先将输入划分至多个工作节点，随后生成输出结果，即单词计数结果。

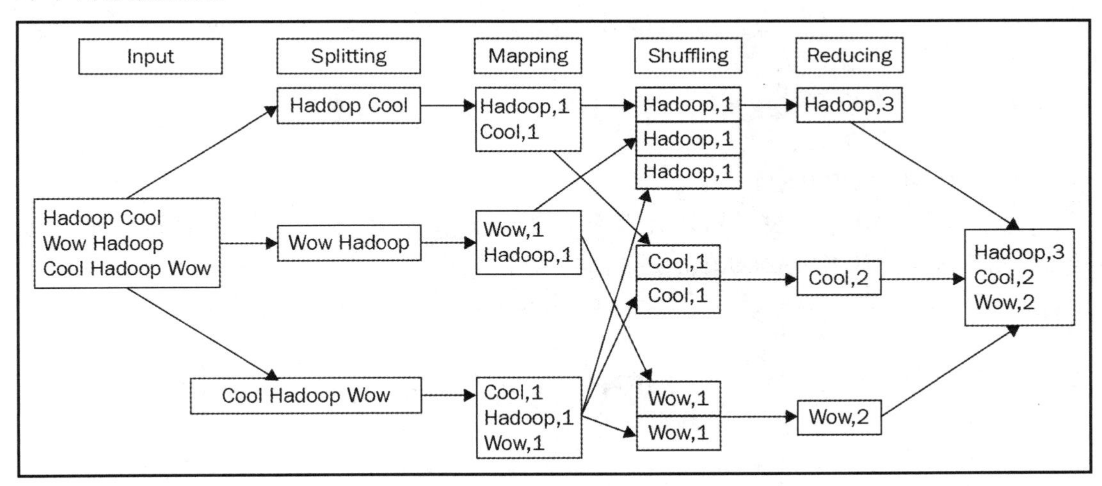

图 1.3

MapReduce 框架由单一的 ResourceManager 和多个 NodeManager 构成（通常，NodeManager 与 HDFS 的 DataNode 协同存在）。

MapReduce 增加了对映射输出收集器的本地实现的支持。这可以使性能提高大约 30%甚至更多，对于需要大量数据清洗工作的任务来说尤其如此。

本地库可通过 Pnative 自动构建。通过在任务配置中设置 mapreduce.job.map.output.collector.class=org.apache.hadoop.mapred.nativetask.NativeMapOutputCollectorDelegator，可在逐项任务的基础上选择新的采集器。

此处所涉及的基本理念是，可添加 NativeMapOutputCollector，进而处理映射器发出的键/值对。最终，sort、spill 和 IFile 序列化均可在本地代码中完成。经初步测试后（采用 Xeon E5410，jdk6u24），对应结果良好，如下所示：

- 与 Java 相比，sort 大约提升了 3～10 倍（仅支持二进制字符串比较）。
- IFile 比 Java 大约块 3 倍，即 500MB/秒。如果采用了 CRC32C 硬件，在每秒 1GB 或更高的范围内，速度还将得到进一步提升。
- 合并代码尚未完成，所以测试使用了 io.sort.mb 防止中途溢出。

1.3　YARN

当应用程序需要运行时，客户端启动 ApplicationMaster，然后它与 ResourceManager 协商以容器的形式获取集群中的资源。容器表示分配在单个节点上用于运行任务和进程的 CPU（内核）和内存。容器由 NodeManager 监测并由 ResourceManager 调度。

容器的示例如下所示：

- 单核和 4GB RAM。
- 双核和 6GB RAM。
- 4 核和 20MB RAM。

一些容器被指定为映射器，而某些容器则被指定为 reducer；所有这些都通过 ApplicationMaster 和 ResourceManager 协调。该框架称作 YARN，如图 1.4 所示。

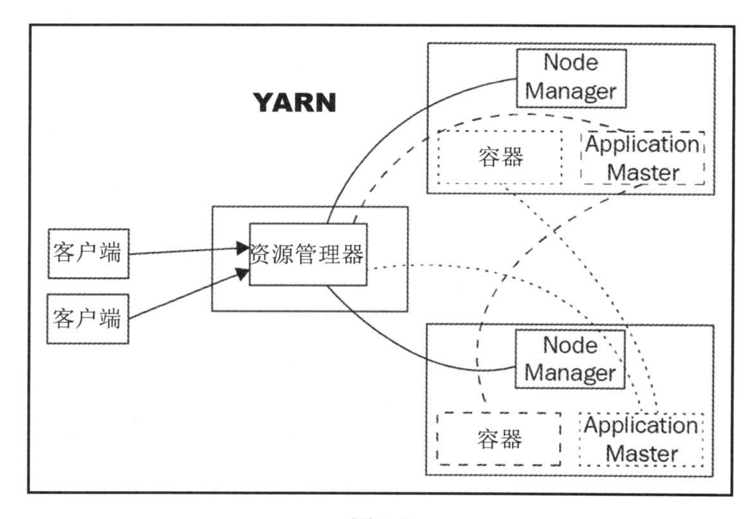

图 1.4

当采用 YARN 时，多个不同的应用程序可请求和执行容器中的任务，并可较好地共享集群资源。然而，随着集群尺寸的增加，以及各种应用程序和需求的变化，在一段时间内，资源利用效率也将随之降低。

1.3.1　机会型容器

机会型容器可传输至 NodeManager 中，即使它们在特定时间点不能立即开始执行。这与 YARN 容器不同，该容器是在节点中调度的，前提是存在未分配的资源。

在这一类应用中，机会型容器将在 NodeManager 中排队，直到所需的资源可用为止。这些容器的最终目标是提高集群资源利用率，进而提高任务吞吐量。

相应地，存在两种容器类型，具体如下。

- 保证型容器：此类容器与现有的 YARN 容器相对应，并由容量调度器分配。当且仅当相关资源可以立即开始执行时，它们才被传输到节点。
- 机会型容器：与保证型容器不同，在这种情况下，当容器被分派到一个节点时，无法保证存在可用资源以供执行。相反，它们将在 NodeManager 中排队，直到资源变得可用为止。

1.3.2　YARN 时间轴服务 v.2

YARN 时间轴服务 v.2 主要处理下列具有挑战性的问题：

- 增强时间轴服务的可伸缩性和可靠性。
- 通过引入流和聚合来提高可用性。

1．增强可伸缩性和可靠性

版本 2 采用了更具伸缩性的分布式 writer 体系结构和后端存储；而 v.1 使用了一个 writer/reader 体系结构和后端存储实例，因而其可伸缩性仅限于小型集群。

由于 Apache HBase 可以很好地扩展至更大的集群，同时可保持良好的读写响应时间，因而 v.2 选择 HBase 作为主后端存储。

2．可用性方面的改进

很多时候，用户更感兴趣的是在流级别，或 YARN 应用程序的逻辑分组中获得的信息。因此，可以更方便地启动一系列的 YARN 应用程序来实现一个逻辑工作流。

针对于此，v.2 支持流这一概念，以及流级别上的聚合度量结果。

3. 架构

YARN 时间轴服务使用一组收集器将数据写入后端存储。收集器与其专用的应用程序管理器一起分发和放置。除了资源管理器时间轴采集器之外，隶属于该应用程序的所有数据均被发送至应用程序级别的时间轴采集器中。

对于给定的应用程序，应用程序管理器可将应用程序数据写入共处的时间轴采集器中（这也是该版本中的 NM 辅助服务）。除此之外，运行应用程序容器的、其他节点的节点管理器，还将向运行应用程序管理器节点的时间轴采集器写入数据。

资源管理器还将维护自身的时间轴采集器，仅发送 YARN 通用生命周期事件，以保持其写入量的合理性。

时间轴读取器是独立于时间轴采集器的独立守护进程，且致力于通过 REST API 提供查询，如图 1.5 所示。

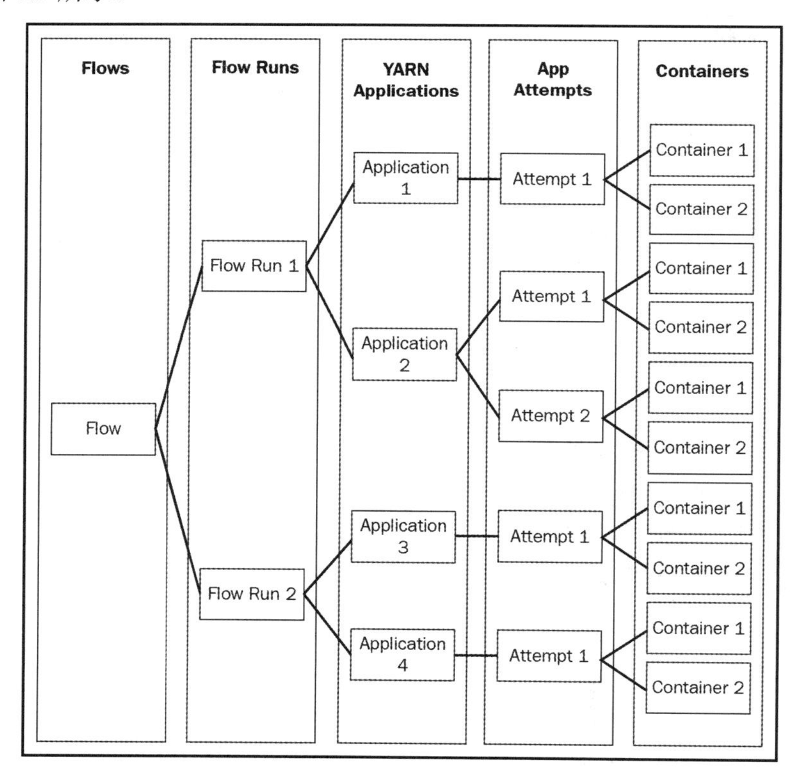

图 1.5

图 1.6 显示了高层设计结果。

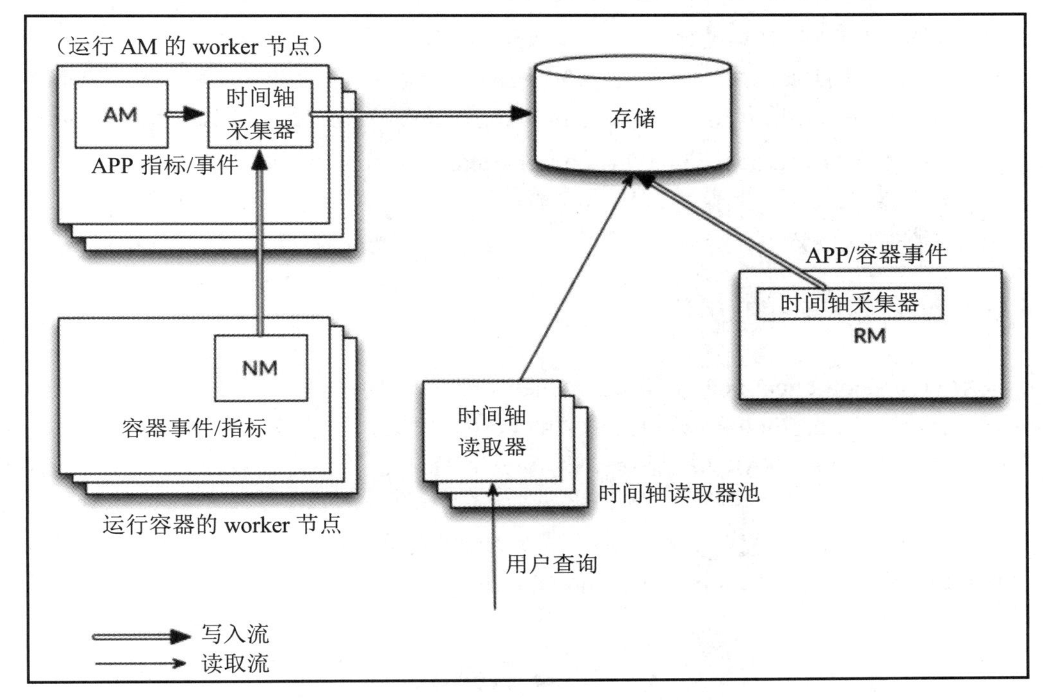

图 1.6

1.4　其他变化内容

Hadoop 3 中还包含了其他一些变化内容，主要涉及维护和操作方面的简化行为。特别地，命令行工具经修正后更加适用于团队操作的需求。

1.4.1　最低 Java 版本

全部 Hadoop JAR 均被编译为 Java 8 运行期版本。因此，使用 Java 7（或更低版本）的用户需要升级至 Java 8。

1.4.2　Shell 脚本重写

Hadoop Shell 脚本经重写后，修复了大量的 bug，同时也包含了一些最新特性。

版本更新文档中列出了相应的不兼容信息，读者可访问 https://issues.apache.org/jira/

browse/HADOOP-9902 对其进行查看。

除此之外，文档中还显示了更为详细的信息，读者可访问 https://hadoop.apache.org/docs/r3.0.0/hadoop-project-dist/hadoop-common/UnixShellGuide.html 以了解更多内容。https://hadoop.apache.org/docs/r3.0.0/hadoop-projectdist/hadoop-common/UnixShellAPI.html 中所展示的文档相信会引起某些高级用户的关注——其中描述了某些最新的功能，特别是与可伸缩性相关的内容。

1.4.3　覆盖客户端的 JAR

最新的 hadoop-client-api 和 hadoop-client-runtime 构建已被添加进来，读者可访问 https://issues.apache.org/jira/browse/HADOOP-11804 对其进行查看。这些构件将 Hadoop 的依赖关系隐藏到一个 JAR 中。因此，可避免将 Hadoop 的依赖关系泄漏到应用程序的类路径上。

作为兼容于 Hadoop 文件系统的一种备选方案，Hadoop 当前支持与 Microsoft Azure Data Lake 和 Aliyun Object Storage System 间的集成。

1.5　安装 Hadoop 3

本节将考查如何在本地机器上安装单节点 Hadoop 3 集群。对此，可访问 https://hadoop.apache.org/docs/current/hadoop-project-dist/hadoop-common/ingleCluster.html，并遵循其中的内容。

关于单节点设置的安装和配置问题，上述文档列出了详细的描述信息，据此，可通过 Hadoop MapReduce 和 HDFS 快速地执行某项操作。

1.5.1　准备条件

首先，需要针对 Hadoop 安装 Java 8。如果读者尚未在机器上安装 Java 8，可访问下载并安装 Java 8。

当在浏览器中打开下载链接后，对应结果如图 1.7 所示。

1.5.2　下载

通过链接 http://apache.spinellicreations.com/hadoop/common/hadoop-3.1.0/下载 Hadoop

3.1 版本。

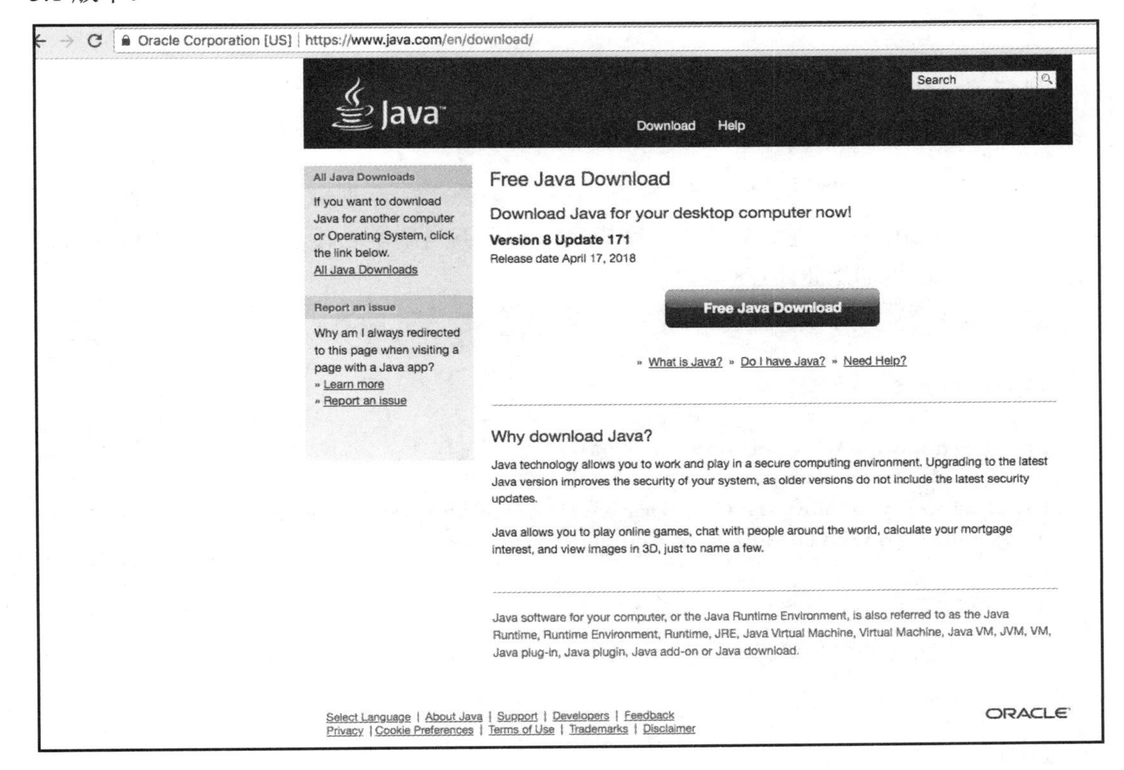

图 1.7

当打开下载链接后，对应页面如图 1.8 所示。

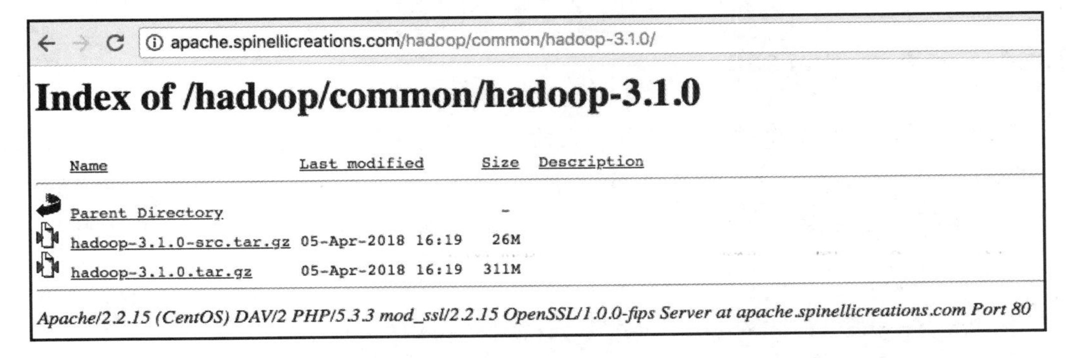

图 1.8

随后，将 hadoop-3.1.0.tar.gz 文件下载至本地机器上。

1.5.3　安装

执行下列各项步骤，并在机器上安装单节点 Hadoop 集群。

（1）利用下列命令解压下载后的文件。

```
tar -xvzf hadoop-3.1.0.tar.gz
```

（2）当 Hadoop 二进制文件解压完毕后，运行下列命令对其进行测试，以确保二进制文件可在本地机器上正常工作：

```
cd hadoop-3.1.0

mkdir input

cp etc/hadoop/*.xml input

bin/hadoop jar share/hadoop/mapreduce/hadoop-mapreduceexamples-3.1.0.jar
grep input output 'dfs[a-z.]+'

cat output/*
```

如果一切运行正常，将会看到一个输出目录，其中显示了一些输出结果，以及示例命令的工作效果。

提示：

缺失 Java 则是一类较为常见的错误。对此，需要检查是否在机器上安装了 Java，以及是否正确地设置了 JAVA_HOME 环境变量。

1.5.4　设置无密码 ssh

现在，通过运行一个简单的命令，来检查是否可以在不使用密码的情况下 ssh 到 localhost，如下所示：

```
$ ssh localhost
```

否则，可执行下列命令：

```
$ ssh-keygen -t rsa -P '' -f ~/.ssh/id_rsa
$ cat ~/.ssh/id_rsa.pub >> ~/.ssh/authorized_keys
$ chmod 0600 ~/.ssh/authorized_keys
```

1.5.5 设置 NameNode

下面针对 etc/hadoop/core-site.xml 配置文件进行如下修改：

```
<configuration>
    <property>
        <name>fs.defaultFS</name>
        <value>hdfs://localhost:9000</value>
    </property>
</configuration>
```

接下来，对 etc/hadoop/hdfs-site.xml 配置文件进行如下修改：

```
<configuration>
    <property>
        <name>dfs.replication</name>
        <value>1</value>
    </property>
        <name>dfs.name.dir</name>
        <value><YOURDIRECTORY>/hadoop-3.1.0/dfs/name</value>
    </property>
</configuration>
```

1.5.6 启动 HDFS

执行下列步骤以启动 HDFS（NameNode 和 DataNode）。

（1）格式化文件系统：

```
$ ./bin/hdfs namenode -format
```

（2）启动 NameNode 守护进程和 DataNode 守护进程：

```
$ ./sbin/start-dfs.sh
```

Hadoop 守护进程日志输出被写入$HADOOP_LOG_DIR 目录中（默认为$HADOOP_HOME/logs）。

（3）浏览 NameNode 的 Web 界面，默认为 http://localhost:9870/。

（4）令 HDFS 目录执行 MapReduce 任务：

```
$ ./bin/hdfs dfs -mkdir /user
$ ./bin/hdfs dfs -mkdir /user/<username>
```

（5）结束后，利用下列命令终止守护进程：

```
$ ./sbin/stop-dfs.sh
```

（6）打开浏览器检查本地 Hadoop，该操作可在 http://localhost:9870/处启动。图 1.9 显示了 HDFS 的安装结果。

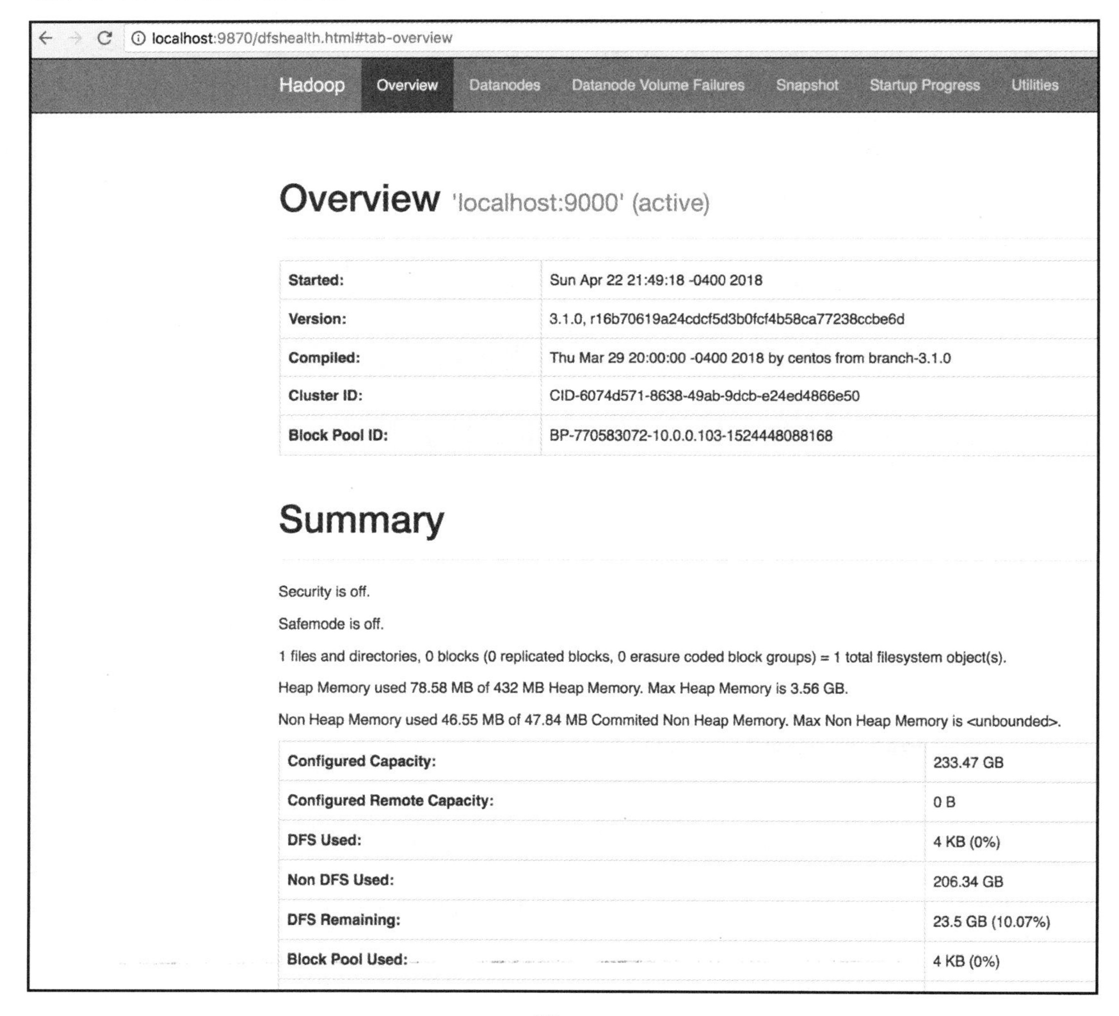

图 1.9

（7）选择 Datanodes 选项卡，对应节点如图 1.10 所示。

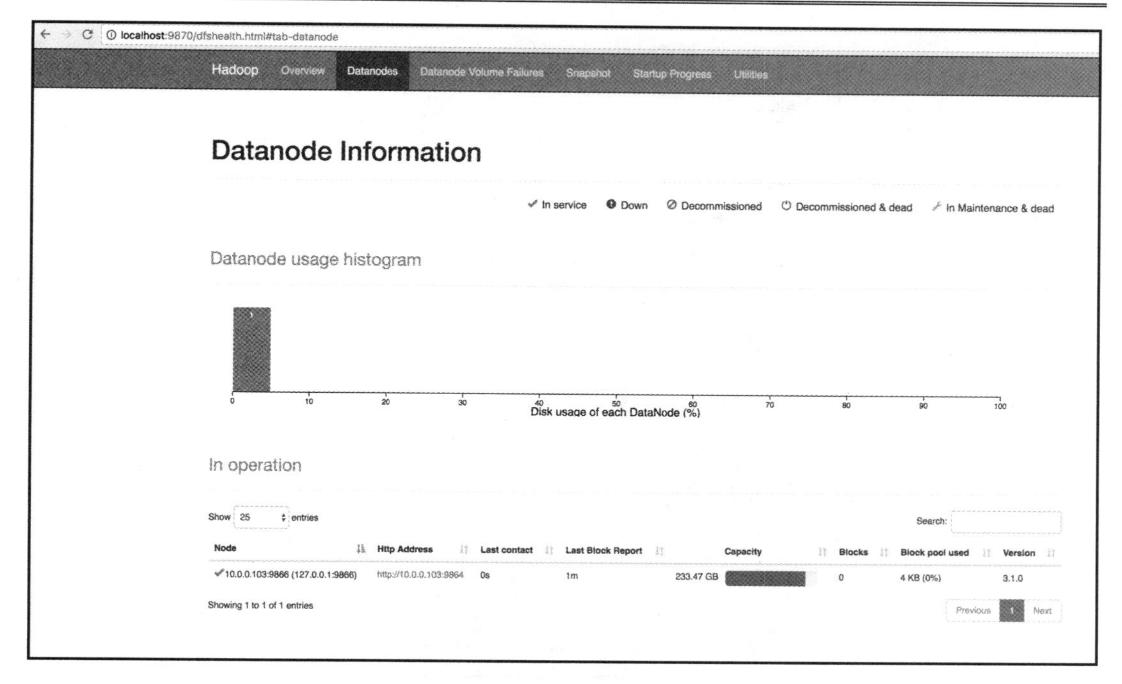

图 1.10

（8）单击 log，图 1.11 显示了集群中的各种日志。

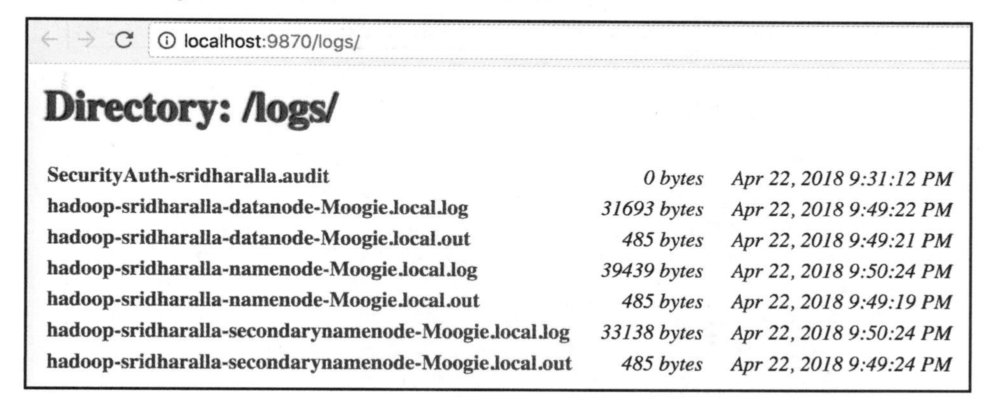

图 1.11

（9）图 1.12 显示了集群组件的 JVM 度量结果。

（10）如图 1.13 所示，还可检查当前配置状态，其中包括整体配置和默认设置内容。

localhost:9870/jmx


```
{
  "beans" : [ {
    "name" : "Hadoop:service=NameNode,name=JvmMetrics",
    "modelerType" : "JvmMetrics",
    "tag.Context" : "jvm",
    "tag.ProcessName" : "NameNode",
    "tag.SessionId" : null,
    "tag.Hostname" : "Moogie.local",
    "MemNonHeapUsedM" : 47.575027,
    "MemNonHeapCommittedM" : 49.148438,
    "MemNonHeapMaxM" : -1.0,
    "MemHeapUsedM" : 103.684074,
    "MemHeapCommittedM" : 432.0,
    "MemHeapMaxM" : 3641.0,
    "MemMaxM" : 3641.0,
    "GcCount" : 8,
    "GcTimeMillis" : 247,
    "GcNumWarnThresholdExceeded" : 0,
    "GcNumInfoThresholdExceeded" : 0,
    "GcTotalExtraSleepTime" : 435,
    "ThreadsNew" : 0,
    "ThreadsRunnable" : 10,
    "ThreadsBlocked" : 0,
    "ThreadsWaiting" : 6,
    "ThreadsTimedWaiting" : 36,
    "ThreadsTerminated" : 0,
    "LogFatal" : 0,
    "LogError" : 0,
    "LogWarn" : 8,
    "LogInfo" : 130
  }, {
    "name" : "JMImplementation:type=MBeanServerDelegate",
    "modelerType" : "javax.management.MBeanServerDelegate",
    "MBeanServerId" : "Moogie.local_1524448157872",
    "SpecificationName" : "Java Management Extensions",
    "SpecificationVersion" : "1.4",
```


图 1.12

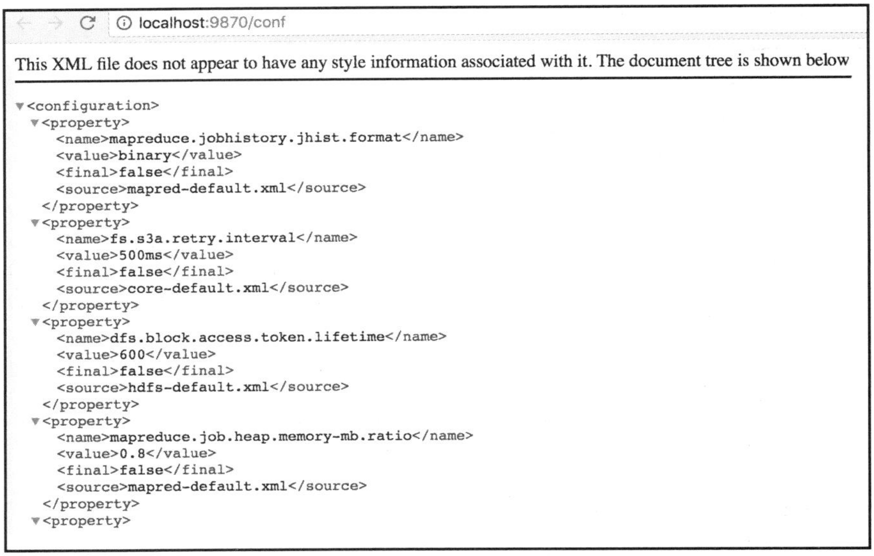

localhost:9870/conf

This XML file does not appear to have any style information associated with it. The document tree is shown below


```
▼<configuration>
  ▼<property>
     <name>mapreduce.jobhistory.jhist.format</name>
     <value>binary</value>
     <final>false</final>
     <source>mapred-default.xml</source>
   </property>
  ▼<property>
     <name>fs.s3a.retry.interval</name>
     <value>500ms</value>
     <final>false</final>
     <source>core-default.xml</source>
   </property>
  ▼<property>
     <name>dfs.block.access.token.lifetime</name>
     <value>600</value>
     <final>false</final>
     <source>hdfs-default.xml</source>
   </property>
  ▼<property>
     <name>mapreduce.job.heap.memory-mb.ratio</name>
     <value>0.8</value>
     <final>false</final>
     <source>mapred-default.xml</source>
   </property>
  ▼<property>
```


图 1.13

（11）此外，还可浏览最新安装的集群的文件系统，如图 1.14 所示。

图 1.14

至此，应可看到并使用基本的 HDFS 集群，但这仅是一个包含某些目录和文件的 HDFS 文件系统。除此之外，还需要一项作业/任务调度服务，并使用集群以满足计算需求，而非仅仅是存储。

1.5.7　设置 YARN 服务

本节将设置 YARN 服务、启动相关组件、运行和操控 YARN 集群。

（1）启动 ResourceManager 守护进程和 NodeManager 守护进程。

```
$ sbin/start-yarn.sh
```

（2）浏览 ResourceManager 的 Web 界面，默认状态下位于 http://localhost:8088/。

（3）运行 MapReduce 任务。

（4）结束后，利用下列命令终止守护进程：

```
$ sbin/stop-yarn.sh
```

图 1.15 显示了 YARN ResourceManager。对此，可在浏览器中输入 http://localhost:8088/ 进行查看。

图 1.16 所示视图展示了集群中的资源队列，以及处于运行状态下的应用程序。除此之外，还可于此处查看和监视运行的各项任务。

至此，我们应能够查看到运行 Hadoop 3.1 后，本地集群中处于运行状态下的 YARN 服务。稍后，还将继续考查 Hadoop 3.x 中的一些新特性。

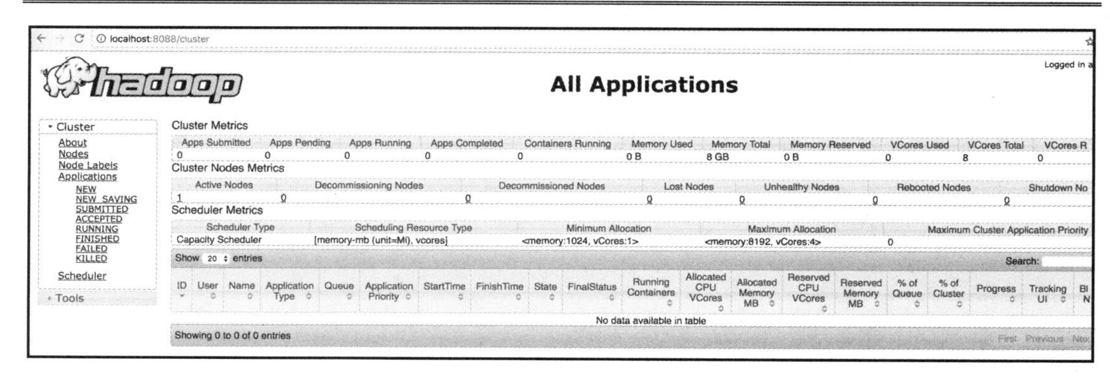

图 1.15

localhost:8088/cluster/scheduler

hadoop　Logged in as: dr.

NEW,NEW_SAVING,SUBMITTED,ACCEPTED,RUNNING Applications

Cluster　About　Nodes　Node Labels　Applications　NEW　NEW_SAVING　SUBMITTED　ACCEPTED　RUNNING　FINISHED　FAILED　KILLED　Scheduler　Tools

Cluster Metrics

Apps Submitted	Apps Pending	Apps Running	Apps Completed	Containers Running	Memory Used	Memory Total	Memory Reserved	VCores Used	VCores Total	VCores Reserve
0	0	0	0	0	0 B	8 GB	0 B	0	8	0

Cluster Nodes Metrics

Active Nodes	Decommissioning Nodes	Decommissioned Nodes	Lost Nodes	Unhealthy Nodes	Rebooted Nodes	Shutdown Nodes
1	0	0	0	0	0	0

Scheduler Metrics

Scheduler Type	Scheduling Resource Type	Minimum Allocation	Maximum Allocation	Maximum Cluster Application Priority
Capacity Scheduler	[memory-mb (unit=Mi), vcores]	<memory:1024, vCores:1>	<memory:8192, vCores:4>	0

Dump scheduler logs　1 min

Application Queues

Legend:　Capacity　Used　Used (over capacity)　Max Capacity　Users Requesting Resources　Auto Created Queues

- Queue: root　0.0% used
- + Queue: default　0.0% used

Show 20 entries　Search:

ID	User	Name	Application Type	Queue	Application Priority	StartTime	FinishTime	State	FinalStatus	Running Containers	Allocated CPU VCores	Allocated Memory MB	Reserved CPU VCores	Reserved Memory MB	% of Queue	% of Cluster	Progress	Tracking UI	Blacklist Nodes
No data available in table																			

Showing 0 to 0 of 0 entries　First　Previous　Next　Last

Aggregate scheduler counts

Total Container Allocations(count)	Total Container Releases(count)	Total Fulfilled Reservations(count)	Total Container Preemptions(count)
0	0	0	0

Last scheduler run

Time	Allocations(count - resources)	Reservations(count - resources)	Releases(count - resources)
Sun Apr 22 22:22:28 -0400 2018	0 - <memory:0, vCores:0>	0 - <memory:0, vCores:0>	0 - <memory:0, vCores:0>

图 1.16

1.5.8　纠删码

EC 是 Hadoop 3.x 中的一个重要变化，与之前的版本相比，旨在改进 HDFS 的应用效率。在之前的版本中，例如，复制因子设置为 3，对于各类数据来说，无论即将执行的任务有多重要，这都将导致集群文件系统的巨大浪费。

通过相关策略，并将对应策略分配至 HDFS 中的目录，可对 EC 进行适当设置。对此，HDFS 提供了 ec 子命令，进而执行与 EC 相关的管理命令：

```
hdfs ec [generic options]
    [-setPolicy -path <path> [-policy <policyName>] [-replicate]]
    [-getPolicy -path <path>]
    [-unsetPolicy -path <path>]
    [-listPolicies]
    [-addPolicies -policyFile <file>]
    [-listCodecs]
    [-enablePolicy -policy <policyName>]
    [-disablePolicy -policy <policyName>]
    [-help [cmd ...]]
```

每个命令的详细解释如下。

- [-setPolicy -path <path> [-policy <policyName>] [-replicate]]：在特定路径中，设置某个目录上的 EC 策略，其中包括以下内容。
  - path：HDFS 中的某个目录，这定义为一个强制型参数。此外，制定某项策略仅会对新创建的文件产生影响，而不会影响到现有的文件。
  - policyName：在当前目录下，文件所使用的 EC 策略。如果设置了 dfs.namenode.ec.system.default.policy，那么，该参数将被忽略。该路径的 EC 策略将采用配置中的默认值进行设置。
  - -replicate：在当前目录中，使用特定的 REPLICATION 策略，并强制当前目录采用 3x 复制方案。
  - -replicate and -policy <policyName>：定义为可选参数，且无法被同时指定。
- [-getPolicy -path <path>]：获取指定路径上文件或目录的 EC 策略的详细信息。
- [-unsetPolicy -path <path>]：取消某项 EC 策略（该策略通过之前在某个目录上调用 setPolicy 而被设置）。如果当前目录继承自上一级目录中的 EC 策略，unsetPolicy 将表示为一个空操作。在不包含显式策略集的目录中取消策略将不会返回错误。
- [-listPolicies]：列出所有在 HDFS 中注册的 EC 策略（包括已启用的、禁用的、被移除的策略）。其中，只有启用的策略适合与 setPolicy 命令一起使用。
- [-addPolicies -policyFile <file>]：添加 EC 策略列表。读者可访问 etc/hadoop/user_ecdfs.namenode.ec.policies.max.cellsize_policies.xml.template 以查看示例策略文件。此外，最大单元格尺寸定义于 dfs.namenode.ec.policies.max.cellsize 属性中，对应的默认值为 4MB。当前，HDFS 支持用户添加共计 64 项策略，被加入的策略 ID 位于 64～127。如果已经添加了 64 项策略，那么策略添加行为将失败。
- [-listCodecs]：获取系统中所支持的 EC 编解码器和编码器的列表。其中，编码器表示为编解码器的实现；编解码器可包含不同的实现，因而涉及不同的编码器。基于编解码器的编码器则是按照回退（fall back）顺序列出的。

- [-enablePolicy -policy <policyName>]：启用某项 EC 策略。
- [-disablePolicy -policy <policyName>]：禁用某项 EC 策略。

通过-listPolicies 命令，可列出集群中当前指定的所有 EC 策略，以及相关策略的状态（无论是否被 ENABLED 或 DISABLED），如图 1.17 所示。

```
Moogie:hadoop-3.1.0 sridharalla$ ./bin/hdfs ec -listPolicies
2018-04-22 22:55:59,587 WARN util.NativeCodeLoader: Unable to load native-hadoop library for your platform... using builtin-java classes
where applicable
Erasure Coding Policies:
ErasureCodingPolicy=[Name=RS-10-4-1024k, Schema=[ECSchema=[Codec=rs, numDataUnits=10, numParityUnits=4]], CellSize=1048576, Id=5], State=
DISABLED
ErasureCodingPolicy=[Name=RS-3-2-1024k, Schema=[ECSchema=[Codec=rs, numDataUnits=3, numParityUnits=2]], CellSize=1048576, Id=2], State=DI
SABLED
ErasureCodingPolicy=[Name=RS-6-3-1024k, Schema=[ECSchema=[Codec=rs, numDataUnits=6, numParityUnits=3]], CellSize=1048576, Id=1], State=EN
ABLED
ErasureCodingPolicy=[Name=RS-LEGACY-6-3-1024k, Schema=[ECSchema=[Codec=rs-legacy, numDataUnits=6, numParityUnits=3]], CellSize=1048576, I
d=3], State=DISABLED
ErasureCodingPolicy=[Name=XOR-2-1-1024k, Schema=[ECSchema=[Codec=xor, numDataUnits=2, numParityUnits=1]], CellSize=1048576, Id=4], State=
DISABLED
```

图 1.17

下面测试集群中的 EC。首先，在 HDFS 中的目录，如下所示：

```
./bin/hdfs dfs -mkdir /user/normal
./bin/hdfs dfs -mkdir /user/ec
```

在构建了两个目录后，可在任意路径上设置策略，如下所示：

```
./bin/hdfs ec -setPolicy -path /user/ec -policy RS-6-3-1024k
Set RS-6-3-1024k erasure coding policy on /user/ec
```

当前，将任意内容复制至/user/ec 文件夹中都将置入最新设置的 policy 中。

输入下列命令并执行测试：

```
./bin/hdfs dfs -copyFromLocal ~/Documents/OnlineRetail.csv /user/ec
```

图 1.18 显示了复制结果，正如预期的那样，系统会提示，本地系统中尚不存在足够的集群来实现 EC。但这也使我们了解到所欠缺的内容及其大致的样子。

```
Moogie:hadoop-3.1.0 sridharalla$ ./bin/hdfs dfs -copyFromLocal ~/Documents/OnlineRetail.csv /user/ec
2018-04-22 23:48:25,379 WARN util.NativeCodeLoader: Unable to load native-hadoop library for your platform... using builtin-java classes
where applicable
2018-04-22 23:48:26,284 WARN erasurecode.ErasureCodeNative: ISA-L support is not available in your platform... using builtin-java codec
here applicable
copyFromLocal: File /user/ec/OnlineRetail.csv._COPYING_ could only be written to 1 of the 6 required nodes for RS-6-3-1024k. There are 1
datanode(s) running and no node(s) are excluded in this operation.
```

图 1.18

1.5.9　内部 DataNode 平衡器

虽然 HDFS 一般具有在集群中的数据节点之间平衡数据的强大功能，但这通常会导致数据节点内的磁盘倾斜。例如，假设存在 4 个磁盘，其中两个磁盘可能会占用大部分数据，而另外两个磁盘可能没有得到充分利用。考虑到物理磁盘的读写速度较慢（例如 7200rpm 或 10000rpm），这种数据倾斜往往会导致较差的性能。使用节点内的平衡器，可以在磁盘之间重新平衡数据。

对此，运行下列命令，并在某个 DataNode 节点上调用磁盘平衡机制：

```
./bin/hdfs diskbalancer -plan 10.0.0.103
```

图 1.19 显示了磁盘平衡器命令的输出结果。

```
Moogie:hadoop-3.1.0 sridharalla$ ./bin/hdfs diskbalancer -plan 10.0.0.103
2018-04-23 00:00:14,789 WARN util.NativeCodeLoader: Unable to load native-hadoop library for your platform... using builtin-java classes
where applicable
2018-04-23 00:00:15,691 INFO planner.GreedyPlanner: Starting plan for Node : 10.0.0.103:9867
2018-04-23 00:00:15,691 INFO planner.GreedyPlanner: Compute Plan for Node : 10.0.0.103:9867 took 16 ms
2018-04-23 00:00:15,692 INFO command.Command: No plan generated. DiskBalancing not needed for node: 10.0.0.103 threshold used: 10.0
No plan generated. DiskBalancing not needed for node: 10.0.0.103 threshold used: 10.0
```

图 1.19

1.5.10　安装时间轴服务 v.2

在 1.3.2 节中曾讨论到，v.2 始终选择 Apache HBase 作为主备份存储，因为 Apache HBase 可以很好地扩展到更大的集群，同时继续保持良好的读/写响应时间。

关于时间轴服务 v.2 存储的准备工作，需要执行下列各项步骤：

（1）设置 HBase 集群。

（2）启用协处理器。

（3）创建时间轴服务 v.2 的模式。

下面对每个步骤加以详细解释。

1. 设置 HBase 集群

首先需要选择一个 Apache HBase 作为存储集群。支持时间轴服务 v.2 的 Apache HBase 版本为 1.2.6。注意，1.0.x 版本将不再支持时间轴服务 v.2；而后续 HBase 版本对此则未经测试。

如果打算为 Apache HBase 集群提供一个简单的部署配置文件，并实现轻量级的数据加载，但数据需要跨节点往复持久化，那么可以考虑使用 HDFS 部署模式的独立 HBase。

读者可访问 http://mirror.cogentco.com/pub/apache/hbase/1.2.6/下载 Apache HBase，如图 1.20 所示。

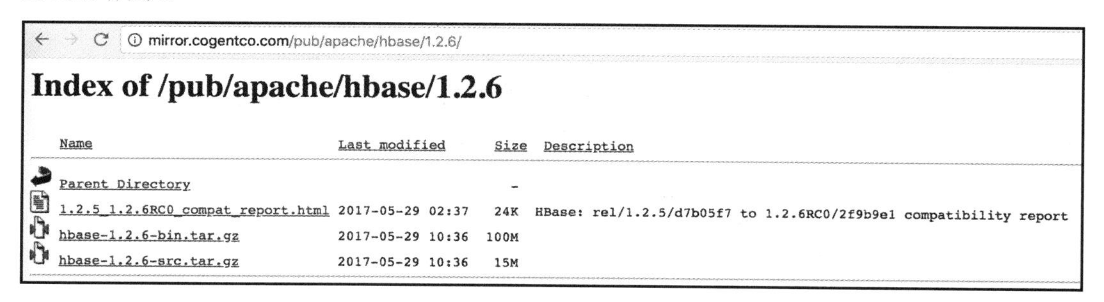

图 1.20

随后，可将 hbase-1.2.6-bin.tar.gz 下载至本地机器上，并利用下列命令解压 HBase 二进制文件：

```
tar -xvzf hbase-1.2.6-bin.tar.gz
```

图 1.21 显示了解压后 HBase 中的相关内容。

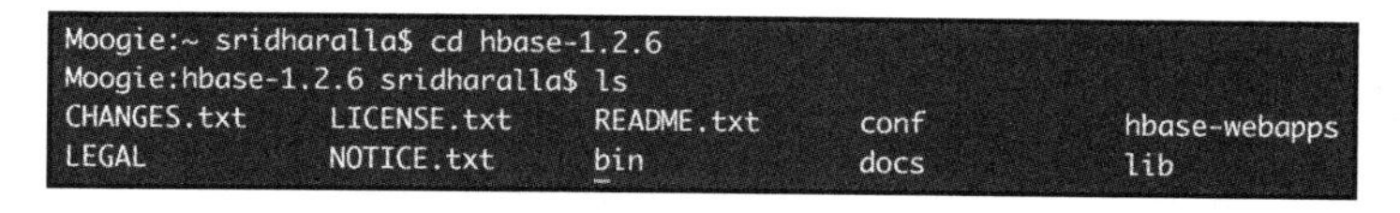

图 1.21

这表示为独立 HBase 设置上的一个变化版本，并将所有 HBase 守护进程运行在一个 JVM 中，而不是持久化到本地文件系统中（持久化至 HDFS 实例中）。写入 HDFS（数据于其中被复制）可确保数据在节点间实现往复持久化。当配置这一独立的变化版本时，可编辑 hbasesite.xml 文件，设置 hbase.rootdir 并指向 HDFS 实例中的某个目录，随后将 hbase.cluster.distributed 设置为 false。

hbase-site.xml 文件如下所示。作为一项属性，其中包含了针对之前安装的本地集群的 hdfs 端口 9000；否则，将不会安装 HBase 集群。

```
<configuration>
    <property>
        <name>hbase.rootdir</name>
        <value>hdfs://localhost:9000/hbase</value>
    </property>
    <property>
        <name>hbase.cluster.distributed</name>
```

```
        <value>false</value>
    </property>
</configuration>
```

下一步是启动 HBase。对此，可使用 start-hbase.sh 脚本，如下所示：

```
./bin/start-hbase.sh
```

图 1.22 显示了刚刚安装的 HBase 集群。

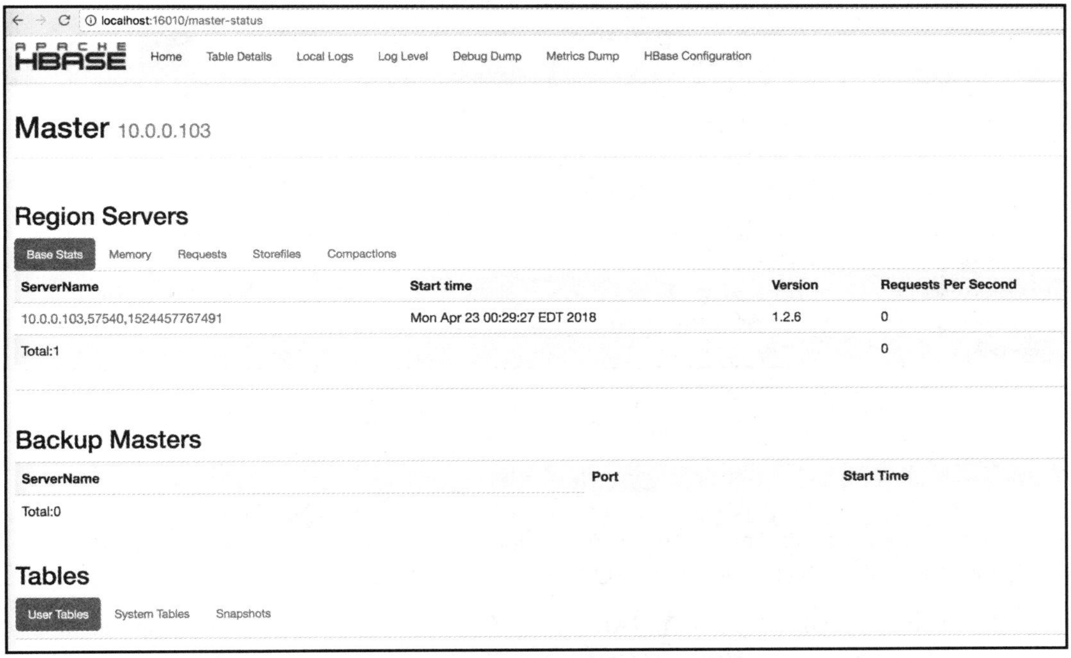

图 1.22

图 1.23 显示了 HBase 集群设置的更多属性，以及各种组件的版本。

一旦 Apache HBase 处于就绪状态，即可执行后续各项步骤，下面分别对此予以解释。

2. 启用协处理器

在当前版本中，协处理器将以动态方式被加载。

对此，将时间轴服务.jar 从 HBase 加载处复制至 HDFS 中。默认的 HDFS 位置为/hbase/coprocessor。

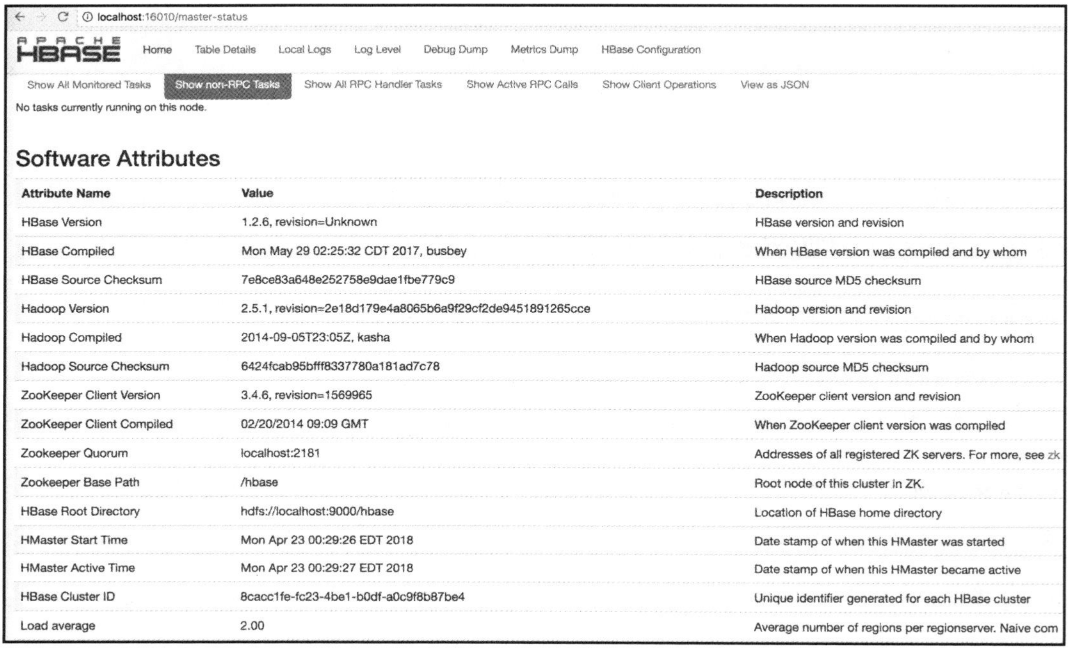

localhost:16010/master-status

APACHE HBASE　Home　Table Details　Local Logs　Log Level　Debug Dump　Metrics Dump　HBase Configuration

Show All Monitored Tasks　Show non-RPC Tasks　Show All RPC Handler Tasks　Show Active RPC Calls　Show Client Operations　View as JSON

No tasks currently running on this node.

Software Attributes

Attribute Name	Value	Description
HBase Version	1.2.6, revision=Unknown	HBase version and revision
HBase Compiled	Mon May 29 02:25:32 CDT 2017, busbey	When HBase version was compiled and by whom
HBase Source Checksum	7e8ce83a648e252758e9dae1fbe779c9	HBase source MD5 checksum
Hadoop Version	2.5.1, revision=2e18d179e4a8065b6a9f29cf2de9451891265cce	Hadoop version and revision
Hadoop Compiled	2014-09-05T23:05Z, kasha	When Hadoop version was compiled and by whom
Hadoop Source Checksum	6424fcab95bfff8337780a181ad7c78	Hadoop source MD5 checksum
ZooKeeper Client Version	3.4.6, revision=1569965	ZooKeeper client version and revision
ZooKeeper Client Compiled	02/20/2014 09:09 GMT	When ZooKeeper client version was compiled
Zookeeper Quorum	localhost:2181	Addresses of all registered ZK servers. For more, see zk
Zookeeper Base Path	/hbase	Root node of this cluster in ZK.
HBase Root Directory	hdfs://localhost:9000/hbase	Location of HBase home directory
HMaster Start Time	Mon Apr 23 00:29:26 EDT 2018	Date stamp of when this HMaster was started
HMaster Active Time	Mon Apr 23 00:29:27 EDT 2018	Date stamp of when this HMaster became active
HBase Cluster ID	8cacc1fe-fc23-4be1-b0df-a0c9f8b87be4	Unique identifier generated for each HBase cluster
Load average	2.00	Average number of regions per regionserver. Naive com

图 1.23

例如：

```
hadoop fs -mkdir /hbase/coprocessor hadoop fs -put hadoop-yarn-
servertimelineservice-hbase-3.0.0-alpha1-SNAPSHOT.jar /hbase/coprocessor/
hadoopyarn-server-timelineservice.jar
```

为了在不同的 HDFS 处放置 JAR，还存在一个称之为 yarn.timeline-service.hbase.coprocessor.jar.hdfs.location 的 YARN 配置设置项，如下所示：

```
<property>
  <name>yarn.timeline-service.hbase.coprocessor.jar.hdfs.location</name>
  <value>/custom/hdfs/path/jarName</value>
</property>
```

相应地，可使用模式生成器工具创建时间轴服务模式。要做到这一点，还需要确保所有的 jar 都被正确地找到，如下所示：

```
export
HADOOP_CLASSPATH=$HADOOP_CLASSPATH:/Users/sridharalla/hbase-1.2.6/lib/:/
Users/sridharalla/hadoop-3.1.0/share/hadoop/yarn/timelineservice/
```

一旦修正了类路径，即可使用一个简单的命令创建 HBase 模式/表，如下所示：

```
./bin/hadoop
org.apache.hadoop.yarn.server.timelineservice.storage.TimelineSchemaCreator
-create -skipExistingTable
```

图 1.24 显示了利用上述命令生成的 HBase 模式。

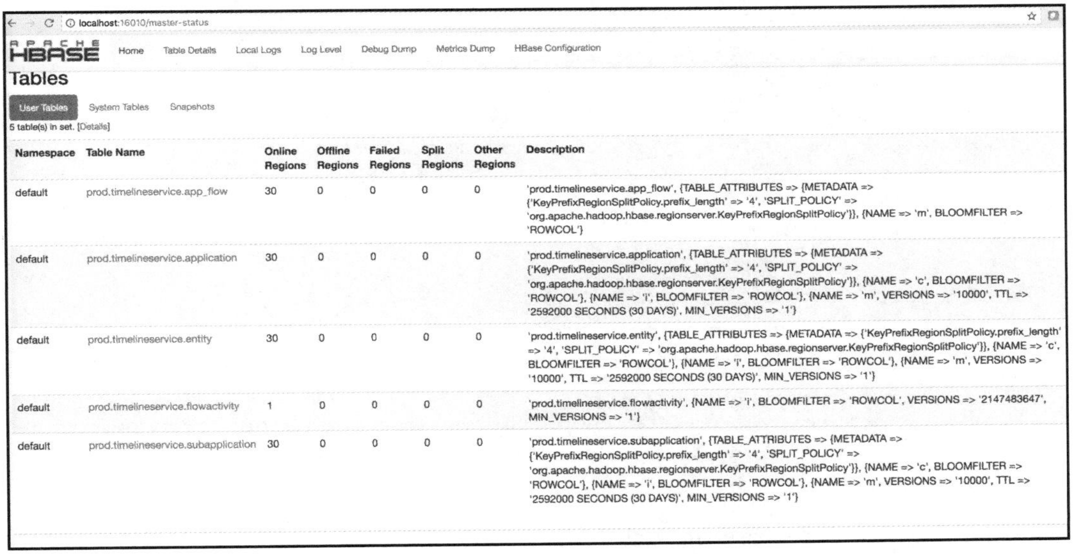

localhost:16010/master-status

APACHE HBASE Home Table Details Local Logs Log Level Debug Dump Metrics Dump HBase Configuration

Tables

User Tables System Tables Snapshots

5 table(s) in set. [Details]

Namespace	Table Name	Online Regions	Offline Regions	Failed Regions	Split Regions	Other Regions	Description
default	prod.timelineservice.app_flow	30	0	0	0	0	'prod.timelineservice.app_flow', {TABLE_ATTRIBUTES => {METADATA => {'KeyPrefixRegionSplitPolicy.prefix_length' => '4', 'SPLIT_POLICY' => 'org.apache.hadoop.hbase.regionserver.KeyPrefixRegionSplitPolicy'}}, {NAME => 'm', BLOOMFILTER => 'ROWCOL'}
default	prod.timelineservice.application	30	0	0	0	0	'prod.timelineservice.application', {TABLE_ATTRIBUTES => {METADATA => {'KeyPrefixRegionSplitPolicy.prefix_length' => '4', 'SPLIT_POLICY' => 'org.apache.hadoop.hbase.regionserver.KeyPrefixRegionSplitPolicy'}}, {NAME => 'c', BLOOMFILTER => 'ROWCOL'}, {NAME => 'i', BLOOMFILTER => 'ROWCOL'}, {NAME => 'm', VERSIONS => '10000', TTL => '2592000 SECONDS (30 DAYS)', MIN_VERSIONS => '1'}
default	prod.timelineservice.entity	30	0	0	0	0	'prod.timelineservice.entity', {TABLE_ATTRIBUTES => {METADATA => {'KeyPrefixRegionSplitPolicy.prefix_length' => '4', 'SPLIT_POLICY' => 'org.apache.hadoop.hbase.regionserver.KeyPrefixRegionSplitPolicy'}}, {NAME => 'c', BLOOMFILTER => 'ROWCOL'}, {NAME => 'i', BLOOMFILTER => 'ROWCOL'}, {NAME => 'm', VERSIONS => '10000', TTL => '2592000 SECONDS (30 DAYS)', MIN_VERSIONS => '1'}
default	prod.timelineservice.flowactivity	1	0	0	0	0	'prod.timelineservice.flowactivity', {NAME => 'i', BLOOMFILTER => 'ROWCOL', VERSIONS => '2147483647', MIN_VERSIONS => '1'}
default	prod.timelineservice.subapplication	30	0	0	0	0	'prod.timelineservice.subapplication', {TABLE_ATTRIBUTES => {METADATA => {'KeyPrefixRegionSplitPolicy.prefix_length' => '4', 'SPLIT_POLICY' => 'org.apache.hadoop.hbase.regionserver.KeyPrefixRegionSplitPolicy'}}, {NAME => 'c', BLOOMFILTER => 'ROWCOL'}, {NAME => 'i', BLOOMFILTER => 'ROWCOL'}, {NAME => 'm', VERSIONS => '10000', TTL => '2592000 SECONDS (30 DAYS)', MIN_VERSIONS => '1'}

图 1.24

3. 启用时间轴服务 v.2

下列内容显示了启动时间轴服务 v.2 的基本配置：

```
<property>
  <name>yarn.timeline-service.version</name>
  <value>2.0f</value>
</property>

<property>
  <name>yarn.timeline-service.enabled</name>
  <value>true</value>
</property>

<property>
  <name>yarn.nodemanager.aux-services</name>
```

```
  <value>mapreduce_shuffle,timeline_collector</value>
</property>

<property>
  <name>yarn.nodemanager.aux-services.timeline_collector.class</name>
<value>org.apache.hadoop.yarn.server.timelineservice.collector.PerNode
TimelineCollectorsAuxService</value>
</property>

<property>
  <description> This setting indicates if the yarn system metrics is
published by RM and NM by on the timeline service. </description>
  <name>yarn.system-metrics-publisher.enabled</name>
  <value>true</value>
</property>

<property>
  <description>This setting is to indicate if the yarn container events
are published by RM to the timeline service or not. This configuration is
for ATS V2. </description>
  <name>yarn.rm.system-metrics-publisher.emit-container-events</name>
  <value>true</value>
</property>
```

除此之外，还可将 hbase-site.xml 配置文件添加至客户端 Hadoop 集群配置中，以便将数据写入所用的 Apache HBase 集群中；或者，出于同样的目的（将数据写入 HBase 中），可将 yarn.timeline-service.hbase.configuration.file 设置为指向 hbase-site.xml 的 URL，如下所示：

```
<property>
  <description>This is an Optional URL to an hbase-site.xml configuration
file. It is to be used to connect to the timeline-service hbase cluster.
If it is empty or not specified, the HBase configuration will be loaded
from the classpath. Else, they will override those from the ones present
on the classpath. </description>
  <name>yarn.timeline-service.hbase.configuration.file</name>
  <value>file:/etc/hbase/hbase-site.xml</value>
</property>
```

（1）运行时间轴服务 v.2

重启 ResourceManager 和节点管理器，并加载新的配置内容。采集器将以嵌入的方式在资源管理器和节点管理器中启动。

相应地，时间轴读取器则是一个独立的 YARN 守护进程，并通过下列语法予以启动：

```
$ yarn-daemon.sh start timelinereader
```

（2）启动 MapReduce 并写入时间轴服务 v.2

为了将 MapReduce 框架数据写入时间轴服务 v.2，可启用 mapred-site.xml 中的下列配置：

```
<property>
  <name>mapreduce.job.emit-timeline-data</name>
  <value>true</value>
</property>
```

时间轴服务仍处于发展过程中，因而用户仅可尝试使用其中的各项功能，并对其进行测试，而不是在产品环境中加以使用。对此，我们可等待更为成熟的版本，相信这一天很快就会到来。

1.6　本章小结

本章讨论了 Hadoop 3.x 中的新特性，以及针对 Hadoop 2.x 的可靠性和性能方面的改进措施。除此之外，本章还考查了本地机器上独立 Hadoop 集群的安装过程。

第 2 章将深入讨论大数据分析方面的内容。

第 2 章　大数据分析概述

本章将讨论大数据分析，首先介绍一些普遍观点，随后将深入讨论数据分析所用的一些常见技术。本章将向读者介绍大型数据集的检测过程，以发现数据值的模式，进而获得有价值的观点。本章将会特别关注数据的 7 个“V”。除此之外，还将学习与数据分析和大数据相关的知识、所面临的挑战及其在分布式计算中的梳理方式。最后，本章还将通过 Hive 和 Tableau 展示一些较为常用的技术。

本章主要涉及以下主题：

- 数据分析简介。
- 大数据简介。
- 基于 Apache Hadoop 的分布式计算。
- MapReduce 框架。
- Hive。
- Apache Spark。

2.1　数据分析简介

数据分析是在检查数据时应用定性和定量技术的过程，其目的是提供有价值的见解。当使用各种技术和概念时，数据分析可以为探究性数据分析（EDA）提供手段，也可以为验证性数据分析（CDA）提供结论。EDA 和 CDA 可视为数据分析的基本概念，读者应理解二者间的差别。

EDA 涉及用于考查数据的方法、工具和技术，目的是在数据中找到相关模式，以及数据的各个元素之间的关系。CDA 则包含了相关的方法、工具和技术，同时基于各种假设和统计学技术，或者对数据简单的观察，进而针对特定问题提供某种简介或结论。

一旦数据被认为已处于就绪状态，数据科学家就可以使用 SAS 等统计方法对其进行分析和探索。数据管理也成为数据采集和数据保护过程中的一个重要因素。此外，另一个不太为人所知的角色是数据管理员。数据管理员专注于字节级别的数据，确切地说，数据源自哪里、数据发生的各种转换，以及基于数据列或字段的业务需求。

业务中的各种实体可能会采取不同的方式处理地址，例如：

```
123 N Main St vs 123 North Main Street.
```

然而，当前分析过程取决于获得正确的地址字段，否则这两个地址将视为不同，且分析结果的准确性无法保持一致。根据分析人员从数据仓库中获取的数据，分析处理过程始于数据采集过程，并收集各个部门（包括销售、市场营销、员工、工资和人力资源等）中的各类数据。这里，数据管理员和管理团队非常重要，以确保收集到正确的数据，以及任何被视为机密或私有的信息不会意外泄漏，即使当前终端用户均为公司内部雇员。此处，在分析中包含社会安全号码（SSN）或完整的地址并非是一类最佳方案，这会对组织机构带来大量的问题。

其间，需要对数据的质量进行把控，以保证所采集、设计的数据的准确性，进而满足数据科学家的要求。在这一阶段，主要的目标是发现和修复数据质量方面的问题，此类问题将影响到分析需求的准确性。其中，较为常见的技术是对数据进行分析和清理清洗，以确保数据集中的信息是一致的，同时删除任何错误和重复记录。

因此，数据分析应用程序可通过多种规则、团队和技能组予以实现。分析应用程序可全程生成相关报告，并自动触发业务操作。例如，我们可以简单地创建每天的销售报告，并在每天上午 8 点以电子邮件的方式发送至管理层。除此之外，还可与业务管理应用程序或某些定制的股票交易程序进行集成，进而执行相关操作，例如购入、卖出或者对股票市场中的某些操作予以警告。此外，还可尝试刊登一些新闻文章，或者社交媒体信息，并对决策制定产生进一步的影响。

数据可视化则是数据分析中的一个重要组成部分。通常，当人们看到大量的测算结果和计算过程时，往往难以理解数字的真正含义。相应地，人们越来越依赖商业智能（BI）工具，例如 Tableau 和 QlikView 等，进而考查并分析数据。当然，某些大规模的可视化行为，例如展示 Uber 在全国范围内的车辆分布，或者纽约城中自来水供应的热力图，仍然需要使用更加专业的定制应用程序或工具予以构建。

管理和分析数据一直是行业中不同规模的机构所面临的挑战。各家企业一直在努力寻找一种实用的方法来获取客户、产品和服务的信息。

对于一家公司来说，对小宗购物进行处理并非难事；但随着时间的推移，公司市场行为将呈现增长之势。此时，事态也变得越加复杂，例如品牌信息、社交媒体、互联网购物等。对此，我们需要制订不同的解决方案。对于 Web 开发、组织行为、价格机制、社交网络以及市场划分，当处理管理-组织行为时，以及尝试从数据中获取某种答案时，数据的多样化也使问题趋于复杂化。

2.2　大数据简介

Twitter、Facebook、Amazon、Verizon、Macy's 和 Whole Foods 等多家公司均采用了

数据分析运行期日常业务，并根据分析结果制定诸多决策。下面将对采集的数据类型、数据的采集量，以及数据的应用方式加以分析。

下面考查一个零售店示例。假设商店着手扩大其业务，并计划开设数百家分店，情况又当如何？不难想象，与一家店铺相比，业务数据的采集和存储规模必然比之前高出数百倍，且不再存在独立运作的业务行为。这其中将会涉及很多信息，例如本地新闻、Twitter 新闻、Yelp 评论、顾客投诉，调查活动、来自其他商店的竞争、当地人口或经济的变化，等等。所有这些额外的数据都有助于我们更好地理解客户行为和收益模型。

如果针对商店停车设施的负面消息越来越多，那么，我们可对此进行分析并采取适当措施，例如发放停车证；或者与城市公共交通部门进行协商以实现更多班次的火车和公交车。随着数据的种类和数量不断增加，在提供了较好的数据分析样本的同时，IT 部门也面临着巨大的挑战——需要存储、处理、分析全部数据。实际上，TB 级的数据目前已经十分常见。

每天，我们都会生成 2EB 字节的数据。据估计，仅最近几年就产生了 90%以上的数据。数据量单位关系如下所示：

1 KB = 1024 Bytes

1 MB = 1024 KB

1 GB = 1024 MB

1 TB = 1024 GB~1,000,000 MB

1 PB = 1024 TB~1,000,000 GB~1,000,000,000 MB

1 EB = 1024 PB~1,000,000 TB~1,000,000,000 GB~1,000,000,000,000 MB

自 20 世纪 90 年代以来，如此大量的数据以及对数据理解方面的需求，催生了大数据这一术语。

2001 年，当时在 Meta Group Inc 咨询公司（后被 Gartner 所收购）担任分析师的 Doug Laney 提出了多样性、速度和容量这 3 个概念（即 3 个“V”）。当前，在上述 3 个“V”的基础上又增加了数据的准确性，即 4 个“V”。

下列内容介绍了大数据的 4 个“V”，进而描述大数据的各种特性。

2.2.1　数据的多样性

数据可从多种资源处获得，例如气象传感器、车辆传感器、人口普查数据、Facebook 的更新数量、Tweet 的评论数量、交易数量、销售额以及市场份额，其中包含了结构化和非结构化的数据格式。另外，数据类型也不一而同，例如二进制、文本、JSON 和 XML。而多样化特征仅仅是数据的冰山一角。

2.2.2　数据的速度

数据可能来自数据仓库、批处理模式的存档文件、准实时更新数据，或者源自 Uber 打车软件的实时更新数据。这里，速度是指生成数据的增加速度，以及关系数据库处理、存储和分析数据时的增速。

2.2.3　数据的容量

数据可以收集并存储 1 小时、1 天、1 个月、1 年或 10 年。对于许多公司来说，数据的规模已经增长至数百 TB 这一级别。相应地，数据容量指的是数据规模，这也是大数据之所以庞大的部分原因。

2.2.4　数据的准确性

我们可以对数据进行分析并以此获得具体可行的解决方案。但考虑到数据源和数据类型的多样性，因而很难保证数据的正确性和准确性。

图 2.1 展示了大数据中的 4 个“V”。

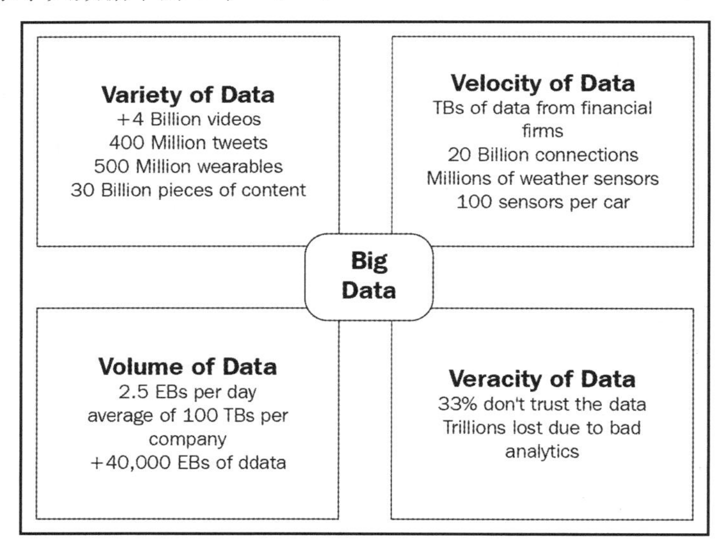

图 2.1

为了进一步对数据加以理解，并将数据分析过程应用至大数据中，我们需要扩展数据分析这一概念，并以更大的规模处理大数据中的 4 个“V”。其间，不仅数据分析的工具、技术和方法发生了变化，处理问题的方式也将有所不同。假设曾采用 SQL 数据库处理 1999 年业务数据。当今，对于同一业务，数据处理需要使用到分布式 SQL。此类数据库具有可伸缩性且适用于大数据环境。

之前所描述的 4 个“V”无法满足大数据分析的各项功能和需求，因此，现在更常听到的则是 7 个“V”。

2.2.5　数据的可变性

数据的可变性指的是含义不断变化的数据。很多时候，我们需要开发十分复杂的程序，以便能够理解它们的上下文，并解析它们的确切含义。

2.2.6　可视化

当数据处理完毕后需要以可读取、可访问的方式呈现时，可视化将以图像的方式解决此类问题。

2.2.7　数值

大数据的数量十分庞大，且每天都处于增长状态；同时，数据也处于杂乱无章、不断变化的状态，并以多种方式呈现于我们面前。因此，如果不执行数据分析和可视化操作，此类数据将无法投入使用。

2.3　使用 Apache Hadoop 的分布式计算

今天，智能冰箱、智能手表、手机、平板电脑、笔记本电脑、机场的自助终端、自动取款机等已十分常见，在这些设备的帮助下，我们的生活也发生了质的变化。此外，我们已经习惯了使用 Instagram、Snapchat、Gmail、Facebook、Twitter 和 Pinterest 等应用程序，这些应用程序已成为日常生活中不可缺少的内容；而云计算则进一步引入了以下概念：

- 作为服务的基础设施。
- 作为服务的平台。
- 作为服务的软件。

其背后的思想是可伸缩的分布式计算，这也使得存储和处理 PB 级别的数据成为可能：

- 1 EB = 1024 PB（相当于 5 千万部蓝光电影）。
- 1 PB = 1024 TB（相当于 50000 部蓝光电影）。

- 1 TB = 1024 GB（相当于 50 部蓝光电影）。

1 部蓝光电影的平均容量一般在 20GB 左右。

当今，分布式计算已不再是新鲜事物，几十年来，人们对分布式计算的研究从未停止，主要出现于一些科研机构和商业公司中。几十年前，大规模并行处理（MPP）即在海洋学、地震监测和空间探索等领域有所应用。多家公司（例如 Teradata）已实现了 MPP 平台，并发布了相关的商业产品和应用程序。

谷歌和亚马逊等科技公司将可扩展分布式计算的商业领域推向了一个新的发展阶段，最终，伯克利大学推出了 Apache Spark。随后，谷歌公司发表了一篇关于 MapReduce 和谷歌文件系统（GFS）的论文，将分布式计算的原理展现在我们每个人面前。当然，Doug Cutting 则更加令人敬佩，他实现了谷歌白皮书中的概念，并向全世界展示了他的作品——Hadoop。Apache Hadoop 框架是一个采用 Java 编写的开源软件框架，该框架所涉及的两个主要领域是存储和处理。对于存储而言，Apache Hadoop 框架采用了 Hadoop 分布式文件系统（HDFS），该系统基于 2003 年所发表的 GFS 论文。对于处理和计算来说，该框架依赖于 MapReduce，它是基于 2004 年 12 月发表的一篇关于 MapReduce 的谷歌论文。当前，MapReduce 框架已从 V1（基于 JobTracker 和 TaskTracker）发展至 V2（基于 YARN）。

2.4　MapReduce 框架

MapReduce 是一个框架，用于计算 Hadoop 集群中的大量数据。MapReduce 使用了 YARN 以及容器并作为任务调度 mapper 和 reducer。

图 2.2 展示了一项 MapReduce 任务，用于计算单词出现的频率。

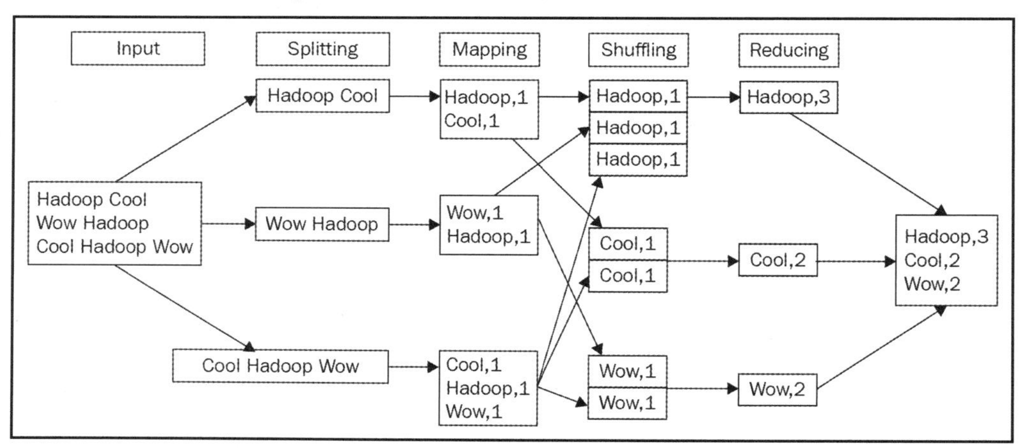

图 2.2

MapReduce 与 YARN 紧密合作，对作业以及作业中的各项任务进行规划，请求集群管理器（资源管理器）中的计算资源，调度集群计算资源上的任务执行，并于随后执行相关规划。当使用 MapReduce 时，可以读写不同格式、不同类型的文件，并以分布方式执行非常复杂的计算。我们将在第 3 章的 MapReduce 框架中看到更多这方面的内容。

2.5　Hive

Hive 在 MapReduce 框架上提供了一个 SQL 抽象层，同时包含了一些优化措施。鉴于 MapReduce 框架中编写代码的复杂度，因而这一行为不可或缺。例如，特定文件中简单的记录计数工作至少需要几十行代码，因而工作效率相对低下。通过将 SQL 语句中的逻辑封装至 MapReduce 框架代码中，Hive 实现了对 MapReduce 代码的抽象，并可在后端自动生成和执行。对于有效数据的处理，这节省了大量的时间，而无须针对每项任务、每次计算编写代码。图 2.3 显示了 Hive 的体系结构。

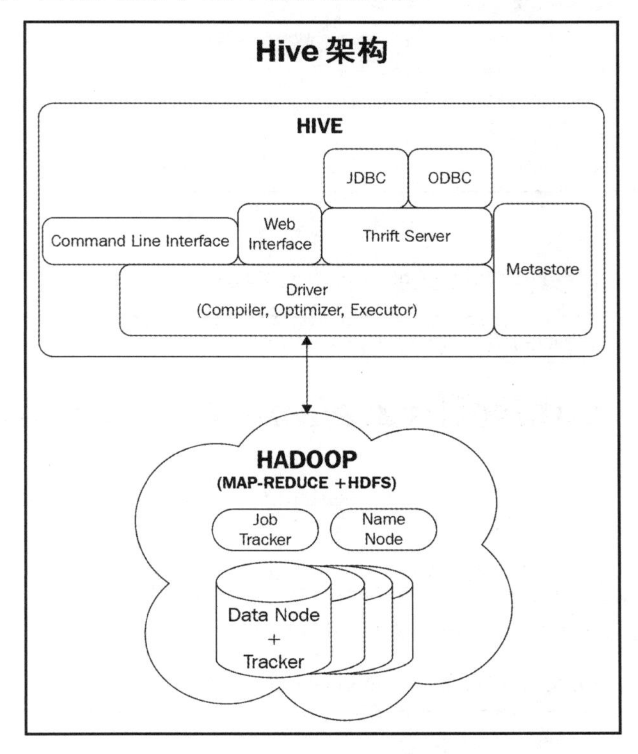

图 2.3

Hive 并不是为在线交易处理而设计的，也不提供实时查询和行级更新操作。

本节将考查 Hive 及其执行数据分析时的应用方式。读者可访问 https://hive.apache.org/downloads.html 下载 Hive，如图 2.4 所示。

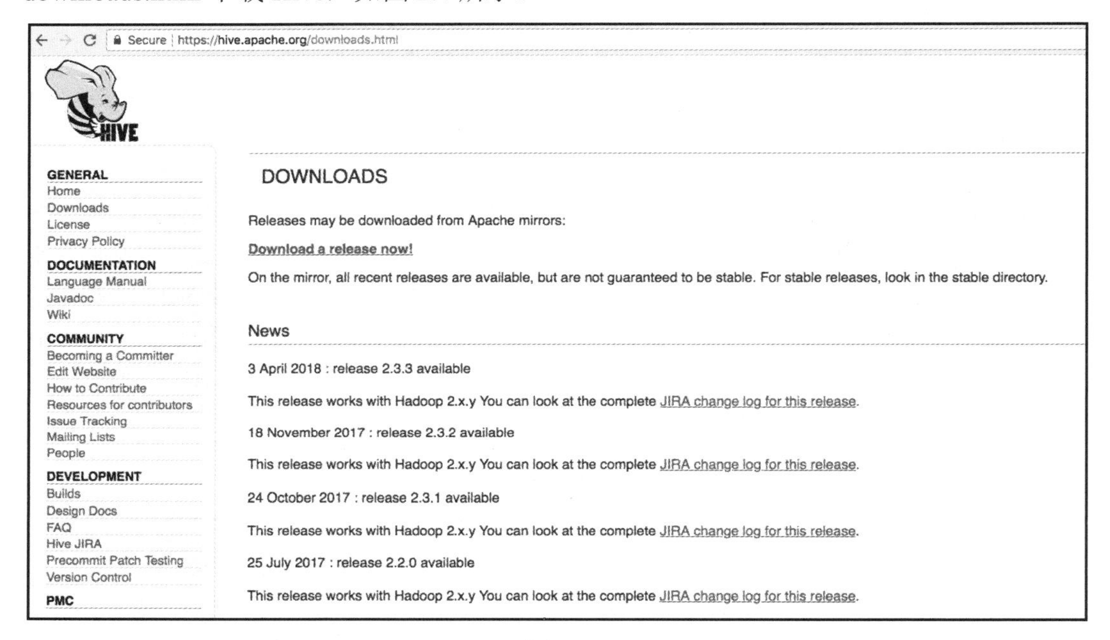

图 2.4

单击下载链接，并查看可供下载的文件，如图 2.5 所示。

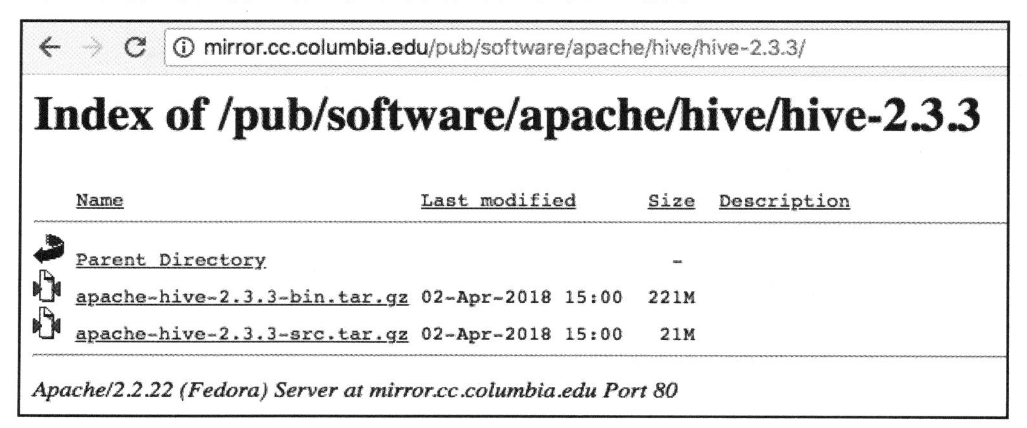

图 2.5

2.5.1　下载并解压 Hive 二进制文件

下面将解压下载后的二进制文件，并对其进行配置，如下所示：

```
tar -xvzf apache-hive-2.3.3-bin.tar.gz
```

随后，创建 hive-site.xml 文件，如下所示：

```
cd apache-hive-2.3.3-bin
vi conf/hive-site.xml
```

在属性列表上方，复制-粘贴下列内容：

```
<property>
 <name>system:java.io.tmpdir</name>
 <value>/tmp/hive/java</value>
</property>
```

在 hive-site.xml 文件底部，添加下列属性：

```
<property>
 <name>hive.metastore.local</name>
 <value>TRUE</value>

</property>
<property>
 <name>hive.metastore.warehouse.dir</name>
 <value>/usr/hive/warehouse</value>
 </property>
```

接下来，利用 Hadoop 命令生成 hive 所需的 HDFS 路径，如下所示：

```
cd hadoop-3.1.0
./bin/hadoop fs -mkdir -p /usr/hive/warehouse
./bin/hadoop fs -chmod g+w /usr/hive/warehouse
```

2.5.2　安装 Derby

Hive 在 MapReduce 框架的基础上工作，并使用表和模式为幕后运行的 MapReduce 作业创建 mapper 和 reducer。为了维护与数据相关的元数据，Hive 使用了一种较为简单的数据库，即 Derby。下面讨论 Derby 的安装过程及其在 Hive 中的应用方式。对此，首先可访问 https://db.apache.org/derby/derby_downloads.html 下载 Derby，如图 2.6 所示。

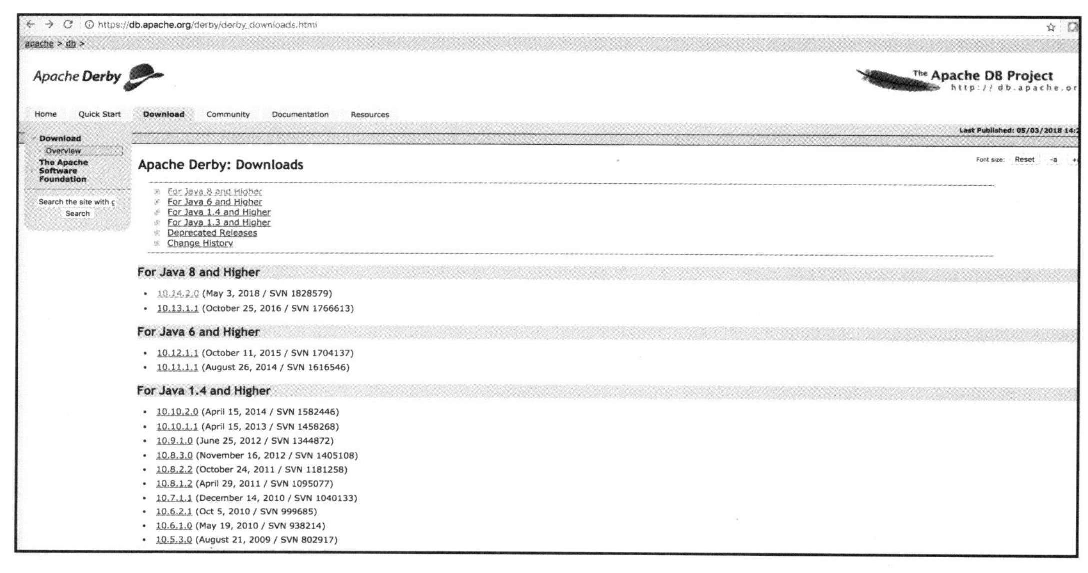

图 2.6

具体安装步骤解释如下：

（1）利用下列命令解压 Derby：

```
tar -xvzf db-derby-10.14.1.0-bin.tar.gz
```

（2）随后将目录修改为 derby，并创建一个名为 data 的目录。实际上，此处需要执行多个命令，如下所示：

```
export HIVE_HOME=<YOURDIRECTORY>/apache-hive-2.3.3-bin
export HADOOP_HOME=<YOURDIRECTORY>/hadoop-3.1.0
export DERBY_HOME=<YOURDIRECTORY>/db-derby-10.14.1.0-bin
export PATH=$PATH:$HADOOP_HOME/bin:$HIVE_HOME/bin:$DERBY_HOME/bin
mkdir $DERBY_HOME/data
cp $DERBY_HOME/lib/derbyclient.jar $HIVE_HOME/lib
cp $DERBY_HOME/lib/derbytools.jar $HIVE_HOME/lib
```

（3）利用下列命令启动 Derby 服务器：

```
nohup startNetworkServer -h 0.0.0.0
```

（4）生成并初始化 derby 实例，如下所示：

```
schematool -dbType derby -initSchema --verbose
```

（5）打开 hive 控制台，如下所示：

```
hive
```

对应结果如图 2.7 所示。

```
Moogie:apache-hive-2.3.3-bin sridharalla$ hive
SLF4J: Class path contains multiple SLF4J bindings.
SLF4J: Found binding in [jar:file:/Users/sridharalla/apache-hive-2.3.3-bin/lib/log4j-slf4j-impl-2.6.2.jar!/org/slf4j/impl/StaticLoggerBin
der.class]
SLF4J: Found binding in [jar:file:/Users/sridharalla/hbase-1.2.6/lib/slf4j-log4j12-1.7.5.jar!/org/slf4j/impl/StaticLoggerBinder.class]
SLF4J: Found binding in [jar:file:/Users/sridharalla/hadoop-3.1.0/share/hadoop/common/lib/slf4j-log4j12-1.7.25.jar!/org/slf4j/impl/Static
LoggerBinder.class]
SLF4J: See http://www.slf4j.org/codes.html#multiple_bindings for an explanation.
SLF4J: Actual binding is of type [org.apache.logging.slf4j.Log4jLoggerFactory]

Logging initialized using configuration in jar:file:/Users/sridharalla/apache-hive-2.3.3-bin/lib/hive-common-2.3.3.jar!/hive-log4j2.prope
rties Async: true
Hive-on-MR is deprecated in Hive 2 and may not be available in the future versions. Consider using a different execution engine (i.e. spa
rk, tez) or using Hive 1.X releases.
hive>
```

图 2.7

2.5.3　使用 Hive

与关系型数据仓库不同，内嵌数据模型包含了复杂类型，例如数组、映射和结构。利用 PARTITIONED BY 子句，可根据 1 列或多列对表进行划分。而且，表或划分结果可利用 CLUSTERED BY 子句进行分组，数据可通过 SORT BY 在桶中进行排序。具体如下。

- 表：类似于 RDBMS 并包含了行和表。
- 分区：Hive 表可包含多个分区，并映射至子目录和文件系统中。
- 桶：数据可划分至 Hive 中的桶，并可在底层文件系统中存储为分区中的文件。

Hive 查询语言提供了与 SQL 类型的基本操作。下列内容列出了 HQL 可执行的一些简单任务：

- 创建、管理表和分区。
- 支持多种关系、算术和逻辑运算符。
- 计算函数。
- 将表内容下载至本地目录中，或者将查询结果下载至 HDFS 目录中。

1．创建数据库

首先需要创建数据库，并装载 Hive 中生成的所有表。该步骤易于实现，且与大多数数据库中的操作类似，如下所示：

```
create database mydb;
```

图 2.8 显示了 hive 控制台中的查询过程。

在开始使用刚刚创建的数据库时，须生成数据库所需的表，如下所示：

```
use mydb;
```

图 2.9 显示了 hive 控制台中的查询过程。

```
hive> create database mydb;
OK
Time taken: 4.007 seconds
```

图 2.8

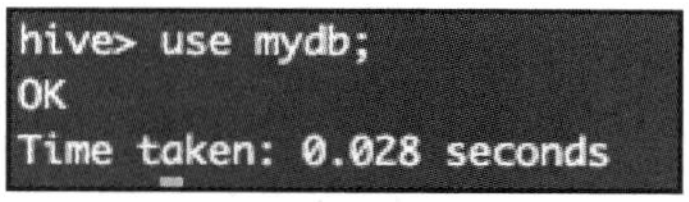

```
hive> use mydb;
OK
Time taken: 0.028 seconds
```

图 2.9

2．创建表

当数据库创建完毕后，下面将在该数据库中创建表。从语法上讲，表的创建操作与大多数 RDBMS（例如 Oracle、MySQL 等数据库）类似，如下所示：

```
create external table OnlineRetail (
 InvoiceNo string,
 StockCode string,
 Description string,
 Quantity integer,
 InvoiceDate string,
 UnitPrice float,
 CustomerID string,
 Country string
 ) ROW FORMAT DELIMITED
 FIELDS TERMINATED BY ','
 LOCATION '/user/normal';
```

图 2.10 显示了 hive 控制台中的输出结果。

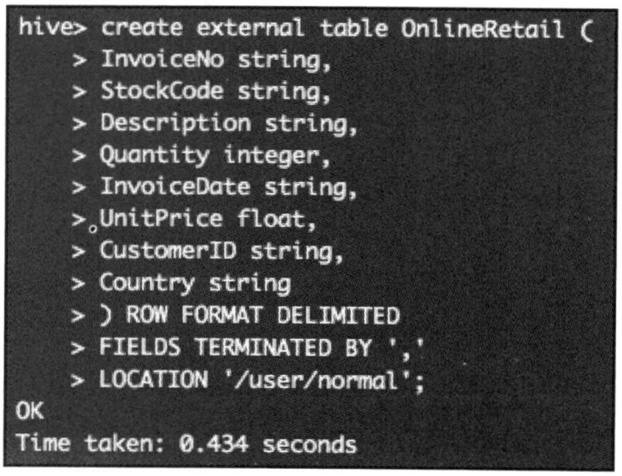

```
hive> create external table OnlineRetail (
    > InvoiceNo string,
    > StockCode string,
    > Description string,
    > Quantity integer,
    > InvoiceDate string,
    > UnitPrice float,
    > CustomerID string,
    > Country string
    > ) ROW FORMAT DELIMITED
    > FIELDS TERMINATED BY ','
    > LOCATION '/user/normal';
OK
Time taken: 0.434 seconds
```

图 2.10

我们的重点并不在于查询语句的语法知识，而是考查如何利用 Stinger Initiative 改善查询的性能，如下所示：

```
select count(*) from OnlineRetail;
```

图 2.11 显示了 hive 控制台中的查询过程。

```
hive> select count(*) from OnlineRetail;
WARNING: Hive-on-MR is deprecated in Hive 2 and may not be available in the future
(i.e. spark, tez) or using Hive 1.X releases.
Query ID = sridharalla_20180423173731_d68999d5-5618-4170-a3a8-42be21851d51
Total jobs = 1
Launching Job 1 out of 1
Number of reduce tasks determined at compile time: 1
In order to change the average load for a reducer (in bytes):
  set hive.exec.reducers.bytes.per.reducer=<number>
In order to limit the maximum number of reducers:
  set hive.exec.reducers.max=<number>
In order to set a constant number of reducers:
  set mapreduce.job.reduces=<number>
Job running in-process (local Hadoop)
2018-04-23 17:37:35,267 Stage-1 map = 100%,  reduce = 100%
Ended Job = job_local961179496_0001
MapReduce Jobs Launched:
Stage-Stage-1:  HDFS Read: 10714480 HDFS Write: 0 SUCCESS
Total MapReduce CPU Time Spent: 0 msec
OK
65500
Time taken: 3.518 seconds, Fetched: 1 row(s)
```

图 2.11

2.5.4　SELECT 语句的语法

在 Hive 中，SELECT 的语法如下所示：

```
SELECT [ALL | DISTINCT] select_expr, select_expr, ...
FROM table_reference
[WHERE where_condition]
[GROUP BY col_list]
[HAVING having_condition]
[CLUSTER BY col_list | [DISTRIBUTE BY col_list] [SORT BY col_list]]
[LIMIT number]
;
```

其中，SELECT 表示为 HiveSQL 中的投影（projection）运算符，具体如下：

- SELECT 扫描 FROM 子句所指定的表。
- WHERE 生成过滤条件。
- GROUP BY 生成包含多个列的列表，并指定记录的聚合方式。
- CLUSTER BY、DISTRIBUTE BY 和 SORT BY 分别指定了排序顺序和算法。
- LIMIT 定义了所检索的记录数量。

示例操作如下所示：

```
Select Description, count(*) as c from OnlineRetail group By Description
order by c DESC limit 5;
```

图 2.12 显示了 hive 控制台中的查询过程。

```
WHITE HANGING HEART T-LIGHT HOLDER      358
REGENCY CAKESTAND 3 TIER        278
HEART OF WICKER SMALL    224
HAND WARMER BABUSHKA DESIGN     213
SCOTTIE DOG HOT WATER BOTTLE    207
Time taken: 3.206 seconds, Fetched: 5 row(s)
```

图 2.12

又如：

```
select * from OnlineRetail limit 5;
```

图 2.13 显示了 hive 控制台中的查询过程。

```
hive> select * from OnlineRetail limit 5;
OK
InvoiceNo       StockCode       Description     NULL    InvoiceDate     NULL    CustomerID      Country
536365  85123A  WHITE HANGING HEART T-LIGHT HOLDER      6       12/1/10 8:26    2.55    17850   United Kingdom
536365  71053   WHITE METAL LANTERN     6       12/1/10 8:26    3.39    17850   United Kingdom
536365  84406B  CREAM CUPID HEARTS COAT HANGER  8       12/1/10 8:26    2.75    17850   United Kingdom
536365  84029G  KNITTED UNION FLAG HOT WATER BOTTLE     6       12/1/10 8:26    3.39    17850   United Kingdom
Time taken: 5.25 seconds, Fetched: 5 row(s)
```

图 2.13

图 2.14 显示了 hive 控制台中的查询过程。

```
select lower(description), quantity from OnlineRetail limit 5;
```

WHERE 子句用于过滤谓词运算符和逻辑运算符设置的结果，并借助于下列工具：

- 谓词运算符列表。
- 逻辑运算符列表。
- 函数列表。

```
hive> select lower(description), quantity from OnlineRetail limit 5;
OK
description     NULL
white hanging heart t-light holder      6
white metal lantern     6
cream cupid hearts coat hanger  8
knitted union flag hot water bottle     6
Time taken: 0.154 seconds, Fetched: 5 row(s)
```

图 2.14

WHERE 子句的应用示例如下所示：

```
select * from OnlineRetail where Description='WHITE METAL LANTERN' limit 5;
```

图 2.15 显示了 hive 控制台中的查询过程。

```
hive> select * from OnlineRetail where Description='WHITE METAL LANTERN' limit 5;
OK
536365  71053   WHITE METAL LANTERN     6       12/1/10 8:26    3.39    17850   United Kingdom
536373  71053   WHITE METAL LANTERN     6       12/1/10 9:02    3.39    17850   United Kingdom
536375  71053   WHITE METAL LANTERN     6       12/1/10 9:32    3.39    17850   United Kingdom
536396  71053   WHITE METAL LANTERN     6       12/1/10 10:51   3.39    17850   United Kingdom
536406  71053   WHITE METAL LANTERN     8       12/1/10 11:33   3.39    17850   United Kingdom
Time taken: 0.144 seconds, Fetched: 5 row(s)
```

图 2.15

下列查询操作显示了 group by 子句的使用方式：

```
select Description, count(*) from OnlineRetail group by Description limit
5;
```

图 2.16 显示了 hive 控制台中的查询过程。

```
        166
 4 PURPLE FLOCK DINNER CANDLES  4
 OVAL WALL MIRROR DIAMANTE      22
 SET 2 TEA TOWELS I LOVE LONDON         102
"ACRYLIC HANGING JEWEL  1
Time taken: 1.6 seconds, Fetched: 5 row(s)
```

图 2.16

下列查询操作展示了 group by 子句的应用示例，同时指定了相关条件，进而过滤 having 子句生成的结果。

```
select Description, count(*) as cnt from OnlineRetail group by Description
having cnt> 100 limit 5;
```

图 2.17 显示了 hive 控制台中的查询过程。

```
 SET 2 TEA TOWELS I LOVE LONDON         102
"KEY FOB         110
6 RIBBONS RUSTIC CHARM  121
60 TEATIME FAIRY CAKE CASES     108
Time taken: 1.551 seconds, Fetched: 5 row(s)
```

图 2.17

下列示例采用了 group by 子句，利用 having 子句过滤结果，并通过 order by 子句排序结果，此处使用了 DESC。

```
select Description, count(*) as cnt from OnlineRetail group by Description
having cnt> 100 order by cnt DESC limit 5;
```

图 2.18 显示了 hive 控制台中的查询过程。

```
WHITE HANGING HEART T-LIGHT HOLDER       358
REGENCY CAKESTAND 3 TIER         278
HEART OF WICKER SMALL    224
HAND WARMER BABUSHKA DESIGN      213
SCOTTIE DOG HOT WATER BOTTLE     207
Time taken: 3.045 seconds, Fetched: 5 row(s)
```

图 2.18

2.5.5 INSET 语句的语法

Hive 中的 INSERT 语句如下所示：

```
-- append new rows to tablename1
INSERT INTO TABLE tablename1 select_statement1 FROM from_statement;

-- replace contents of tablename1
INSERT OVERWRITE TABLE tablename1 select_statement1 FROM from_statement;

-- more complex example using WITH clause
WITH tablename1 AS (select_statement1 FROM from_statement) INSERT
[OVERWRITE/INTO] TABLE tablename2 select_statement2 FROM tablename1;
```

2.5.6 原始类型

类型与表中的列所关联。下面考查 Hive 所支持的类型，如表 2.1 所示。

表 2.1

类　　型	描　　述
整数	❑ TINYINT：单字节整数 ❑ SMALLINT：双字节整数 ❑ INT：4 字节整数 ❑ BIGINT：8 字节整数
布尔类型	BOOLEAN：TRUE 或 FALSE
浮点数	❑ FLOAT：单精度浮点数 ❑ DOUBLE：双精度浮点数
定点数	DECIMAL：定点值，用户负责定义缩放和精度
字符串类型	❑ STRING：特定字符集中的字符序列 ❑ VARCHAR：包含最大长度的特定字符集中的字符序列 ❑ CHAR：包含指定长度的特定字符集中的字符序列
日期和时间类型	❑ TIMESTAMP：特定的时间点，精确到纳秒 ❑ DATE：日期
二进制类型	BINARY：字节序列

2.5.7　复杂类型

借助于下列各项内容，还可利用原始类型和其他组合类型构建复杂类型。

- ❑ 结构：类型中的元素可通过“.”进行访问。
- ❑ 映射（键-值元组）：元素通过['element name']标识访问。
- ❑ 数组（索引表）：数组中的元素具有相同类型，并可通过[n]标识访问元素。其中，n 表示为数组的索引（以 0 开始）。

2.5.8　内建运算符和函数

下面列出的运算符和函数不一定是最新的（Hive 运算符和 UDF 中包含了更多的最新信息）。在 Beeline 或 Hive CLI 中，可使用以下命令显示最新的文档：

```
SHOW FUNCTIONS;
DESCRIBE FUNCTION <function_name>;
DESCRIBE FUNCTION EXTENDED <function_name>;
```

全部 Hive 关键字均为大小写敏感，其中包括 Hive 运算符和函数名称。

1．内建运算符

取决于操作数之间的比较是否成立，表 2.2 中的运算符对所传递的操作数进行比较，并生成 TRUE 或 FALSE 值。

表 2.2

运　算　符	类　　型	描　　述
A = B	全部原始类型	如果表达式 A 等于表达式 B，则结果为 TRUE，否则为 FALSE
A != B	全部原始类型	如果表达式 A 不等于表达式 B，则结果为 TRUE，否则为 FALSE
A < B	全部原始类型	如果表达式 A 小于表达式 B，则结果为 TRUE，否则为 FALSE
A <= B	全部原始类型	如果表达式 A 小于或等于表达式 B，则结果为 TRUE，否则为 FALSE
A > B	全部原始类型	如果表达式 A 大于表达式 B，则结果为 TRUE，否则为 FALSE
A >= B	全部原始类型	如果表达式 A 大于或等于表达式 B，则结果为 TRUE，否则为 FALSE
A IS NULL	所有类型	如果表达式 A 计算为 NULL，则结果为 TRUE，否则为 FALSE
A IS NOT NULL	所有类型	如果表达式 A 计算为 NULL，则结果为 FALSE，否则为 TRUE
A LIKE B	字符串	如果字符串 A 与 SQL 正则表达式 B 匹配，则结果为 TRUE，否则为 FALSE。其中，比较过程以逐字符方式执行。B 中的_字符与 A 中的任意字符匹配（类似于 POSIX 正则表达式中的“.”）；B 中的%字符匹配 A 中的任意数量的字符（类似于 POSIX 正则表达式中的“.*”）。例如，foobar LIKE foo 计算为 FALSE；而 foobar LIKE foo___则计算为 TRUE；'foobar' LIKE foo%的计算结果也为 TRUE。当对%进行转义时，可使用\（%匹配于一个%字符）。如果数据中包含了一个分号，且需要对此进行搜索，那么，该字符需要被转义，即 columnValue LIKE a\;b
A RLIKE B	字符串	如果 A 和 B 均为 NULL，则结果为 NULL。如果 A 中的任意子字符串（可能为空串）匹配于 Java 正则表达式 B（读者可参考 Java 正则表达式的语法知识），则结果为 TRUE；否则返回 FALSE。例如，'foobar' rlike 'foo'和'foobar' rlike '^f.*r$'的计算结果均为 TRUE
A REGEXP B	字符串	等同于 RLIKE

表 2.3 中的运算符支持各种常见的算术运算，且返回结果均为数字类型。

表 2.3

运　算　符	类　　型	描　　述
A + B	全部数字类型	生成 A+B 的计算结果。相应地，结果类型为操作数的公共父类型（在类型层次结构中）。例如，每个整数均为一个浮点数。因此，浮点数是整数的包含类型。因此，浮点数和整数上的+运算符将生成一个浮点数
A − B	全部数字类型	生成 A−B 的计算结果。结果类型表示为操作数类型的公共副类型（在类型层次结构中）
A * B	全部数字类型	生成 A×B 的计算结果。结果类型为操作数类型的公共父类型（在类型层次结构中）。注意，如果乘法计算上溢，需要将其中一个操作符强制转换为类型层次结构中更高的类型
A / B	全部数字类型	生成 A 除以 B 的计算结果。结果类型为操作数类型的公共类型（在类型层次结构中）。如果操作数为整数类型，最终结果为除法的商
A % B	全部数字类型	生成 A 除以 B 的余数。结果类型为操作数类型的公共类型（在类型层次结构中）
A & B	全部数字类型	A 和 B 位运算 AND 的计算结果。结果类型为操作数类型的公共类型（在类型层次结构中）
A | B	全部数字类型	A 和 B 位运算 OR 的计算结果。结果类型为操作数类型的公共类型（在类型层次结构中）
A ^ B	全部数字类型	A 和 B 位运算 XOR 的计算结果。结果类型为操作数类型的公共类型（在类型层次结构中）
~A	全部数字类型	A 的位运算 NOT 的计算结果。结果类型等同于 A 的类型

表 2.4 显示了逻辑表达式的构建过程。取决于操作数的布尔值，全部结果将返回 TRUE 或 FALSE。

表 2.4

运　算　符	类　　型	描　　述
A AND B	布尔型	如果 A 和 B 均为 TRUE，则结果为 TRUE
A && B	布尔型	等同于 A AND B
A OR B	布尔型	如果 A 或 B 为 TRUE，或者二者均为 TRUE，则结果为 TRUE；否则为 FALSE
A || B	布尔型	等同于 A OR B
NOT A	布尔型	如果 A 为 FALSE，则结果为 TRUE，否则为 FALSE
!A	布尔型	等同于 NOT A

表 2.5 提供了复杂类型中元素的访问机制。

表 2.5

运 算 符	类 型	描 述
A[n]	A 表示为一个数组，n 表示为一个 int	返回数组 A 中的第 n 个元素。其中，第 1 个元素的索引为 0。例如，如果数组由['foo','bar']构成，那么，A[0]将返回'foo'，A[1]则返回'bar'
M[key]	M 表示为一个 Map <K, V>，key 包含了类型 K	返回与映射中 key 对应的数值。例如，如果 M 定义为由('f' -> 'foo', 'b' -> 'bar', 'all' -> 'foobar')构成的映射，那么，M['all']将返回'foobar'
S.x	S 表示为一个结构	返回 S 的 x 字段。例如，对于 struct foobar (int foo, int bar)，foobar.foo 将返回 struct 中 foo 字段的整数值

2．内建函数

表 2.6 列出了 Hive 所支持的内建函数。

表 2.6

数据类型	函 数	描 述
BIGINT	round(double a)	返回双精度浮点数舍入后的 BIGINT 值
BIGINT	floor(double a)	返回小于或等于当前双精度浮点数的最大 BIGINT 值
BIGINT	ceil(double a)	返回大于或等于当前双精度浮点数的最小 BIGINT 值
double	rand(), rand(int seed)	返回一个随机数（在行间变化）。另外，指定 seed 可确保所生成的随机数具有确定性
string	concat(string A, string B,...)	返回在 A 后连接 B 所产生的字符串。例如，concat('foo', 'bar')的结果为'foobar'。该函数接收任意数量的参数，并返回所有的连接结果
string	substr(string A, int start)	返回 A 的子字符串（起始于 start 位置，终止于字符串 A 的结尾位置）。例如，substr('foobar', 4)将返回'bar'
string	substr(string A, int start, int length)	返回包含既定长度的 A 的子字符串（始于 start 位置）。例如，substr('foobar', 4, 2)将返回'ba'
string	upper(string A)	返回大写字符的字符串。例如，upper('fOoBaR')将返回'FOOBAR'
string	ucase(string A)	等同于 upper(string A)
string	lower(string A)	返回小写字符的字符串。例如，lower('fOoBaR')将返回'foobar'
string	lcase(string A)	等同于 lower(string A)

续表

数 据 类 型	函　　数	描　　述
string	trim(string A)	移除 A 两侧空格后的字符串。例如，trim('foobar ')将返回'foobar'
string	ltrim(string A)	移除 A 左侧空格后的字符串。例如，ltrim(' foobar ')将返回'foobar '
string	rtrim(string A)	移除 A 右侧空格后的字符串。例如，rtrim(' foobar ') 将返回'foobar'
string	regexp_replace(string A, string B, string C)	利用 C 替换 B 中的全部子字符串，并返回结果字符串（与 Java 正则表达式相匹配，参见 Java 正则表达式语法）。例如，regexp_replace('foobar', 'oo|ar',)将返回'fb'
int	size(Map<K.V>)	返回映射类型中元素的数量
int	size(Array<T>)	返回数组类型中元素的数量
value of <type>	cast(<expr> as <type>)	将表达式 expr 的结果转换为<type>。例如，cast('1' as BIGINT)将字符串'1'转换为其整数表达形式。如果转换失败，函数将返回 NULL
string	from_unixtime(int unixtime)	将 UNIX 时间（自 1970-01-01 00:00:00 UTC 起）的秒数转换为当前系统时区中该时刻时间戳的字符串，格式为 1970-01-01 00:00:00
string	to_date(string timestamp)	返回时间戳中的日期部分，例如 string: to_date("1970-01-01 00:00:00") ="1970-01-01"
int	year(string date)	返回日期或时间戳中的年份，例如 string: year("1970-01-01 00:00:00") =1970, year("1970-01-01") = 1970
int	month(string date)	返回日期或时间戳中的月份，例如 string:month("1970-11-01 00:00:00") = 11, month("1970-11-01") = 11
int	day(string date)	返回日期或时间戳中的天数，例如 string: day("1970-11-01 00:00:00") =1, day("1970-11-01") = 1
string	get_json_object(string json_string, string path)	根据所指定的 path 路径，从 json 字符串中获取 json 对象，并返回析取后的.json 对象的 json 字符串。如果输入的 json 字符串无效，函数将返回 null

表 2.7 显示了 Hive 中所支持的内建聚合函数。

表 2.7

数据类型	函　数	描　述
BIGINT	count(*), count(expr), count(DISTINCT expr[, expr_.])	count(*)返回检索的行数，包括包含 NULL 值的行 count(expr)返回输入表达式为非 NULL 的行数 count(DISTINCT expr[, expr])返回输入表达式唯一且非 NULL 的行数
DOUBLE	sum(col), sum(DISTINCT col)	返回分组中元素的和，或分组中列的不同值的和
DOUBLE	avg(col), avg(DISTINCT col)	返回分组中元素的平均值，或分组中列的不同值的平均值
DOUBLE	min(col)	返回分组中列的最小值
DOUBLE	max(col)	返回分组中列的最大值

2.5.9　语言的功能

Hive 的 SQL 提供了以下基本的 SQL 操作，并可在表或分区上工作：

- ❑ 利用 WHERE 子句，过滤表中的行。
- ❑ 利用 SELECT 子句，选择表中的特定列。
- ❑ 执行两个表间的等值连接。
- ❑ 针对存储于某个表中的数据，计算多个 group by 列上的聚合结果。
- ❑ 将查询结果存储于另一个表中。
- ❑ 将某个表中的内容下载至本地目录中（例如 nfs）。
- ❑ 将查询结果存储于某个 hadoop dfs 目录中。
- ❑ 管理表和分区（创建、移除和修改）。
- ❑ 对于自定义 map/reduce 作业，向所选语言中插入自定义脚本。

对于某些常用函数，表 2.8 显示了信息的检索方式。

表 2.8

函　数	Hive
检索信息（通用方式）	SELECT from_columns FROM table WHERE conditions;
检索全部值	SELECT * FROM table;
检索部分值	SELECT * FROM table WHERE rec_name = "value";
根据多项条件进行检索	SELECT * FROM TABLE WHERE rec1 = "value1" AND rec2 = "value2";
检索特定列	SELECT column_name FROM table;

续表

函　　数	Hive
检索唯一的输出结果	SELECT DISTINCT column_name FROM table;
排序	SELECT col1, col2 FROM table ORDER BY col2;
逆向排序	SELECT col1, col2 FROM table ORDER BY col2 DESC;
行计算	SELECT COUNT(*) FROM table;
利用计算结果进行分组	SELECT owner, COUNT(*) FROM table GROUP BY owner;
最大值	SELECT MAX(col_name) AS label FROM table;
在多个表中进行选择（利用别名 w/"AS"连接相同的表）	SELECT pet.name, comment FROM pet JOIN event ON (pet.name = event.name)

2.6　Apache Spark

Apache Spark 是跨不同工作负载和平台的统一分布式计算引擎。Spark 可以连接至不同的平台，并使用各种范例（如 Spark Streaming、Spark ML、Spark SQL 和 Spark Graphx）处理不同的数据工作负载。

Apache Spark 是一个快速的内存数据处理引擎，具有优雅而富有表现力的开发 API，允许数据工作人员高效地执行流式机器学习或 SQL 工作负载，这一类操作需要对数据集进行快速的交互访问。

在核心之上构建的其他库还支持流式工作负载、SQL、图形处理和机器学习。例如，SparkML 是为数据科学设计的，其抽象机制使得数据科学变得更加容易。

Spark 提供了实时流式处理、查询、机器学习和图形处理等功能。在 Apache Spark 出现之前，我们必须为不同类型的工作负载使用不同的技术，且分别用于批量分析、交互式查询、实时流处理以及机器学习算法。然而，仅通过 Apache Spark，就可以完成所有这些任务，而非“零散”的多种技术。

使用 Apache Spark，可以处理所有类型的工作负载。另外，Spark 还支持 Scala、Java、R 和 Python 作为编写客户机程序的各种手段。

Apache Spark 是一个开源的分布式计算引擎，与 MapReduce 相比具有以下优点：

- 尽可能地使用内存处理。
- 可对批量、实时工作负载使用通用引擎。
- 兼容于 YARN 和 Mesos。
- 可与 HBase、Cassandra、MongoDB、HDFS、Amazon S3 以及其他文件系统和数据源实现较好的集成。

Spark 于 2009 年在伯克利发布，同时也是构建 Mesos 项目过程中的一项成果；Mesos 则是一个支持各种集群计算系统的集群管理框架。

Hadoop 和 Apache Spark 都是十分流行的大数据框架，但它们的用途并不相同。Hadoop 提供了分布式存储和 MapReduce 分布式计算框架；而 Spark 则是一个数据处理框架，并对其他技术提供的分布式数据存储进行操作。

注意：

鉴于 Spark 的数据处理方式，其速度明显快于 MapReduce。MapReduce 通过磁盘进行操作；而 Spark 比 MapReduce 更加高效，其性能提高的主要原因是内存处理中的堆外处理，而不是仅仅依赖于基于磁盘的计算。

如果数据操作以及反馈机制多为静态，同时可采用批处理方式，那么，MapReduce 处理已然足够。如果需要对流数据进行分析，或者对多级处理逻辑中的处理需求进行分析，则可使用 Spark。

图 2.19 显示了 Apache Spark 栈。

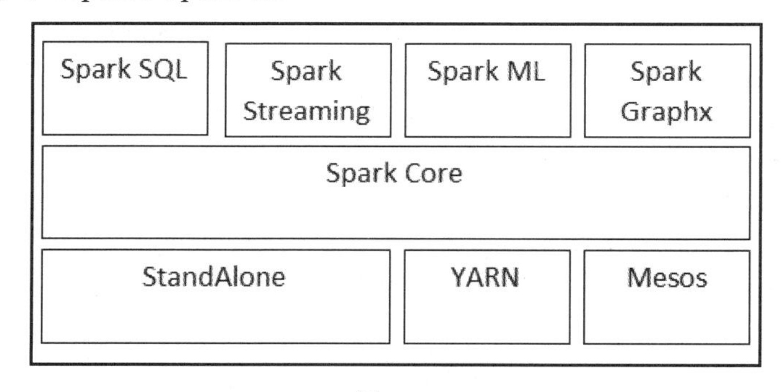

图 2.19

2.7 基于 Tableau 的可视化操作

无论我们使用哪一种方法执行大数据分布式计算，如果缺少 Tableau 这样的工具，将很难理解数据的具体含义——此类工具可以提供一个易于理解的数据可视化结果。

可以通过多种工具实现可视化结果，例如 Cognos、Tableau、Zoom data、KineticaDB、Python Matplotlib、R + Shine、JavaScript 等。第 10 章将对此加以详细讨论。

图 2.20 显示了一幅简单的 Tableau 横条图。

图 2.21 则显示了 Tableau 中的一幅地理空间数据视图。

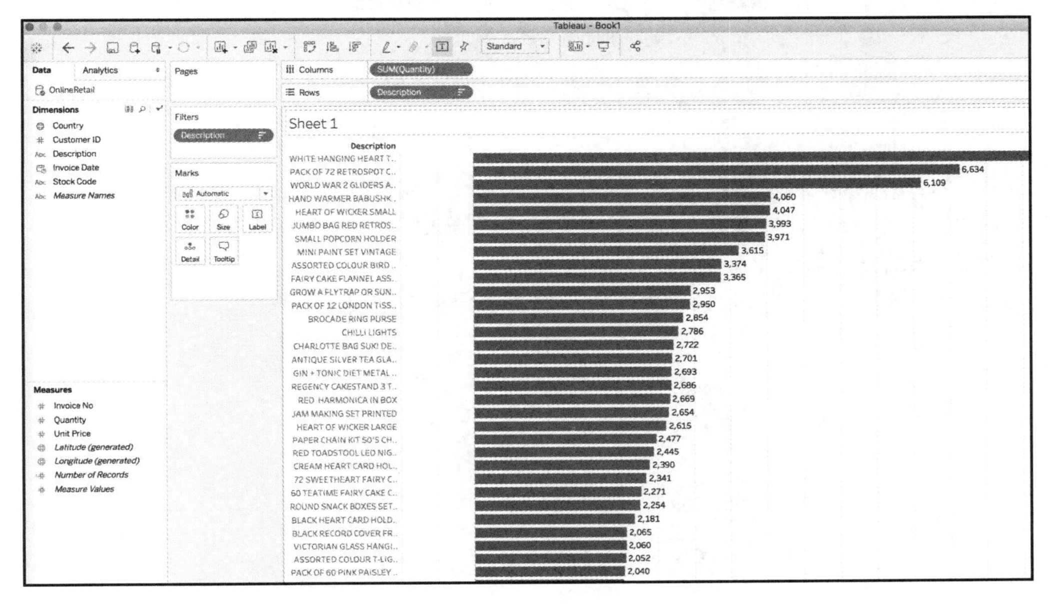
Tableau - Book1
Standard
Data
Analytics
OnlineRetail
Dimensions
Country
Customer ID
Description
Invoice Date
Stock Code
Measure Names
Measures
Invoice No
Quantity
Unit Price
Latitude (generated)
Longitude (generated)
Number of Records
Measure Values
Pages
Filters
Description
Marks
Automatic
Color
Size
Label
Detail
Tooltip
Columns
SUM(Quantity)
Rows
Description
Sheet 1
Description
WHITE HANGING HEART T..
PACK OF 72 RETROSPOT C..
WORLD WAR 2 GLIDERS A..
HAND WARMER BABUSHK..
HEART OF WICKER SMALL
JUMBO BAG RED RETROS..
SMALL POPCORN HOLDER
MINI PAINT SET VINTAGE
ASSORTED COLOUR BIRD ..
FAIRY CAKE FLANNEL ASS..
GROW A FLYTRAP OR SUN..
PACK OF 12 LONDON TISS..
BROCADE RING PURSE
CHILLI LIGHTS
CHARLOTTE BAG SUKI DE..
ANTIQUE SILVER TEA GLA..
GIN + TONIC DIET METAL ..
REGENCY CAKESTAND 3 T..
RED HARMONICA IN BOX
JAM MAKING SET PRINTED
HEART OF WICKER LARGE
PAPER CHAIN KIT 50'S CH..
RED TOADSTOOL LED NIG..
CREAM HEART CARD HOL..
72 SWEETHEART FAIRY C..
60 TEATIME FAIRY CAKE C..
ROUND SNACK BOXES SET..
BLACK HEART CARD HOLD..
BLACK RECORD COVER FR..
VICTORIAN GLASS HANGI..
ASSORTED COLOUR T-LIG..
PACK OF 60 PINK PAISLEY ..
6,634
6,109
4,060
4,047
3,993
3,971
3,615
3,374
3,365
2,953
2,950
2,854
2,786
2,722
2,701
2,693
2,686
2,669
2,654
2,615
2,477
2,445
2,390
2,341
2,271
2,254
2,181
2,065
2,060
2,052
2,040

图 2.20

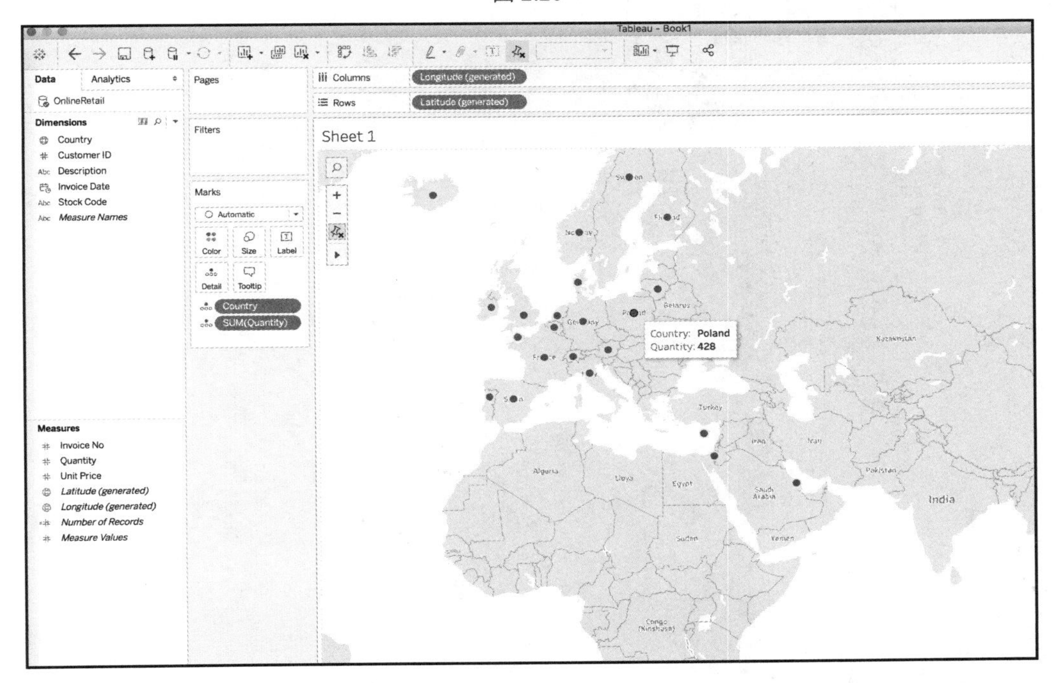
Tableau - Book1
Data
Analytics
OnlineRetail
Dimensions
Country
Customer ID
Description
Invoice Date
Stock Code
Measure Names
Measures
Invoice No
Quantity
Unit Price
Latitude (generated)
Longitude (generated)
Number of Records
Measure Values
Pages
Filters
Marks
Automatic
Color
Size
Label
Detail
Tooltip
Country
SUM(Quantity)
Columns
Longitude (generated)
Rows
Latitude (generated)
Sheet 1
Country: Poland
Quantity: 428
Kazakhstan
Turkey
Iran
Algeria
Libya
Egypt
Saudi Arabia
Pakistan
India
Sudan
Yemen

图 2.21

2.8 本章小结

本章讨论了大数据分析及其各种概念，以及与大数据相关的 7 个“V”，即容量、速度、多样性、准确性、可变性、数值、可视化。除此之外，本章还介绍了某些技术，以辅助数据分析的执行过程。

第 3 章将讨论 MapReduce，以及分布式计算中较为常用的模式。

第 3 章　基于 MapReduce 的大数据处理

本章将前述章节所学知识整合至实践用例中，并构建端到端管线，进而执行大数据分析。

本章主要涉及以下内容：

- MapReduce 框架。
- MapReduce 作业类型。
  - 单 mapper 作业。
  - 单 mapper-reducer 作业。
  - 多 mapper-reducer 作业。
- MapReduce 模式。
  - 聚合模式。
  - 过滤模式。
  - 连接模式。

3.1　MapReduce 框架

MapReduce 框架用于计算 Hadoop 集群中的大量数据。MapReduce 利用 YARN 并作为任务对 mapper 和 reduce 进行调度，同时还使用了容器。MapReduce 支持编写分布式应用程序，并可处理源自文件系统中的大量数据，例如 Hadoop 分布式文件系统（HDFS），其可靠性和容错性也极大地得到保证。当采用 MapReduce 框架处理数据时，需要创建运行于该框架上的作业以执行所需任务。通常，MapReduce 作业将输入数据划分为多个 worker 节点，并以并行方式运行 mapper 任务。

此处，任何 HDFS 级别或 mapper 任务中的故障均会被自动加以处理，以实现相应的容错机制。待 mapper 结束后，对应结果将通过网络被复制至运行 reduce 任务的其他机器上。

图 3.1 显示了一项 MapReduce 作业示例，并计算单词的出现频率。

MapReduce 使用 YARN 作为资源管理器，如图 3.2 所示。

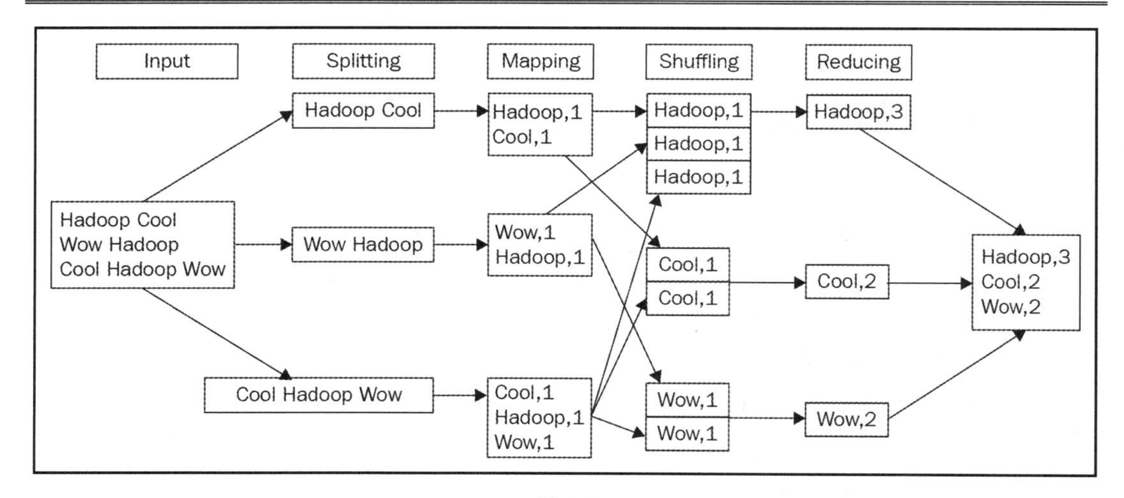

图 3.1

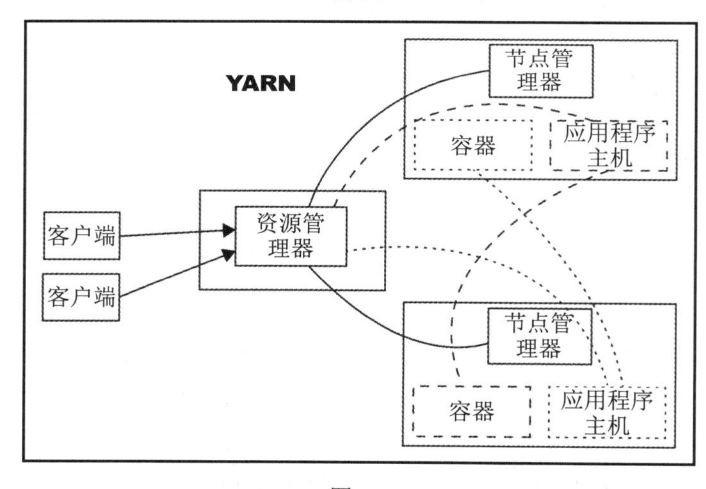

图 3.2

实际上，术语 MapReduce 涉及两个 Hadoop 程序执行的独立和不同的任务。第一项任务是映射作业，并接收一个数据集并将其转换为另一个数据集。其中，单个元素被分解为元组（即键/值对）。

reduce 作业接收来自 map 的输出结果作为输入内容，并将此类数据元组组合为较小的元组集。正如 MapReduce 名称所体现的那样，reduce 作业通常在 map 作业之后执行。

MapReduce 的输入内容表示为数据存储中的一组文件，且分布于 HDFS 中。在 Hadoop 中，此类文件利用某种输入格式被划分，进而定义了如何将某个文件分隔于输入划分结果中。输入划分表示为文件的块字节视图，并通过 map 任务加载。Hadoop 中的每个 map 任务被划分为以下阶段：记录读取器、映射器、组合器和分区器。其中，map 任务的输出结果（称作中间键值）被发送至 reducer 中。reduce 任务则被划分为以下几个阶段：混洗（shuffle）、排序、reducer 和输出格式化。运行 map 任务的节点处于数据所处的最优节点上。通过这一方式，数据一般无须在网络上移动，并可在本地机器上被计算。

本章将考查不同的应用示例，以及如何使用 MapReduce 作业输出所期望的输出结果。对此，我们将使用简单的数据集。

3.1.1　数据集

第 1 个数据集表示为一个城市表，其中包含了城市 ID 以及 City 名称，如下所示：

```
ID,City
1,Boston
2,New York
3,Chicago
4,Philadelphia
5,San Francisco
7,Las Vegas
```

上述 cities.csv 文件可通过下列命令移至 hdfs 中：

```
hdfs dfs -copyFromLocal cities.csv /user/normal
```

第 2 个数据集表示为某个城市的温度，其中包含了测量的 Date、城市 ID，以及某个城市某天的 Temperature，如下所示：

```
Date,ID,Temperature
2018-01-01,1,21
2018-01-01,2,22
2018-01-01,3,23
2018-01-01,4,24
2018-01-01,5,25
2018-01-01,6,22
2018-01-02,1,23
2018-01-02,2,24
2018-01-02,3,25
```

通过运行下列命令，可将上述 temperatures.csv 文件移至 hdfs 中：

```
hdfs dfs -copyFromLocal temperatures.csv /user/normal
```

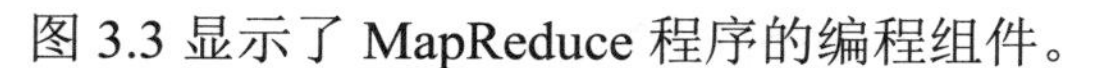
图 3.3 显示了 MapReduce 程序的编程组件。

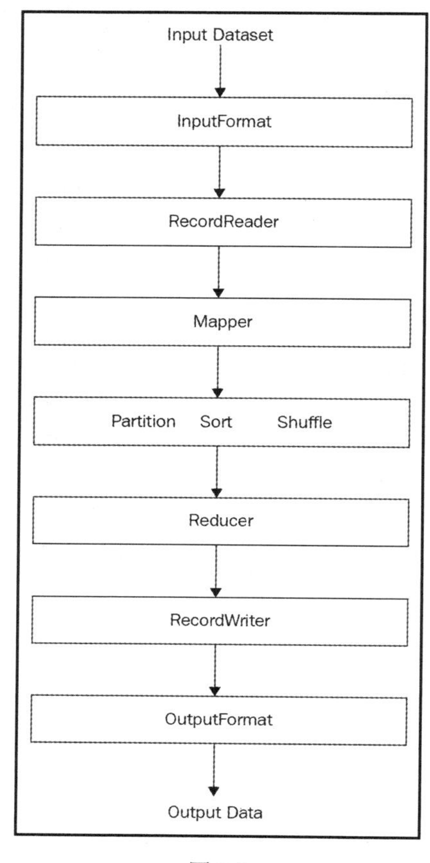

图 3.3

3.1.2　记录读取器

输入读取器将输入分为相应尺寸的划分结果（在实际操作过程中一般为 64MB～128MB）；同时，该框架将某个划分结果分配至每个 map 函数中。另外，输入读取器从稳定存储器（一般是分布式文件系统）中读取数据，并生成键/值对。

提示：

一个常见的示例是读取一个文本文件目录，并将每一行作为记录返回。

记录读取器将输入格式生成的输入划分结果转换为记录。这里，记录读取器的功能是将数据解析为记录，而不是解析记录自身；随后将该数据以键/值对形式传递至 mapper 中。通常，这一上下文中的键表示为位置信息，值则表示为构成记录的数据块。自定义记录读取器的详细内容则超出了本书的讨论范围，这里假设读者已针对数据具备了相应的记录读取器。具体来说，LineRecordReader 即为 TextInputFormat 提供的默认 RecordReader，并将输入文件中的各行视作新值；所关联的键则表示为字节偏移。除了首个划分结果之外，LineRecordReader 一般会跳过划分内容（或其中的部分内容）的第 1 行，并在结尾分割线之后读取一行（如果数据有效，且不为最后一个划分结果）。

3.1.3　映射

map 函数接收一系列的键/值对，经逐一处理后生成 0 个或多个输出的键/值对。map 的输入和输出类型间可彼此不同（且通常如此）。

如果应用程序执行单词计数操作，map 函数将每一行分解为多个单词，并针对每个单词输出键/值对。相应地，每个输出对将包含单词作为键，行中该单词的实例数量作为值。

在 mapper 中，代码将在源自记录读取器的每个键/值对上被执行，并生成 0 个或多个新的键/值对，称作 mapper 的中间输出（通常也由键/值对构成）。这里，将确定每条记录的键/值与 MapReduce 作业间的直接关联方式。其中，键表示为分组的数据，而值则表示为 reducer 中所用的部分数据，进而生成相应的输出结果。模式中讨论的关键内容之一即是不同类型的用例如何确定特定的键/值逻辑。实际上，这种逻辑的语义也是 MapReduce 设计模式之间的一个关键区别。

3.1.4　组合器

如果每个 mapper 的各项输出直接传递至每个 reducer，这将占用大量的资源和时间。对此，组合器（可选的本地 reducer）可以在 map 阶段对数据进行分组，同时接收源自 mapper 的中间键，并将用户提供的方法应用于该映射器的小范围内的聚合值。例如，由于聚合计数表示为各部分的求和结果，因而可生成中间计数值，随后对此类中间计数求和并生成最终结果。在大多数场合下，这可显著地降低在网络间移动的数据量。具体来说，当考查城市和温度的数据集时，通过网络发送(Boston, 66)所需的字节要比发送(Boston, 20)、(Boston, 25)、(Boston, 21)少 3 倍。因此，合成器通常可有效地改善性能问题且不存在任何副作用。

稍后，我们将指出哪些模式可从组合器中收益，而哪些模式则无法使用组合器。鉴于无法保证组合器总是能够被执行，因而不可将其视为整体算法中的一部分内容。

3.1.5　分区器

分区器接收源自 mapper 的中间键/值对（或者可用的合成器），并将其划分为多个分片，且每个 reducer 对应 1 个分片。

针对于分片，每个 map 函数通过应用程序的 partition 函数被分配至特定的 reducer。partition 函数接收键和 reducer 的数量，并返回期望 reducer 的索引。

一般的做法是，对键进行哈希计算，并利用哈希值对 reducer 的数量求模，如下所示：

```
partitionId = hash(key) % R, where R is number of Reducers
```

对于均衡负载来说，重要的是选取一个 partition 函数，以实现每个分片数据的均匀分布；否则，MapReduce 操作将会等待较慢的 reducer（也就是说，reducer 持有较大比例的倾斜数据）。

在 map 和 reduce 阶段，为了将数据从生成节点处移至分片（数据将于其中执行 reduce 操作）中，数据将被混洗（即并行排序以及随后的节点间的互换）。取决于网络带宽、CPU 速度、所生成的数据，以及 map 和 reduce 计算所占用的时间，混洗操作所占用的时间有时会超过计算时间。

默认状态下，分区器计算每个对象的哈希码，通常是一个 md5 校验和。随后，将键空间均匀地分布在 reducer 上，同时确保不同 mapper 中具有相同值的键最终位于同一个 reducer 上。利用诸如排序这一类操作，可自定义分区器的默认行为。针对每项映射任务，分区后的数据将被写入本地文件系统中，并等待对应的 reducer 读取。

3.1.6　混洗和排序

当 mapper 结束了输入数据的处理工作后（主要是数据的划分和键/值对的生成），输出结果将在集群中予以分布以启动 reduce 任务。因此，reduce 任务起始于混洗和排序步骤，即接收全部 mapper 和后续分区器写入的输出文件，并将其下载至 reducer 任务所运行的本地机器中。这一类数据个体随后通过键存储至一个较大的键/值对列表中。这里，排序的主要目的是将等价的键整合至一起，以便在 reduce 任务中可以轻松地对其进行遍历。利用自定义代码控制键的存储和分组方式，框架可自动处理一切事物。

3.1.7　reducer 任务

reducer 接收分组后的数据作为输入，并针对每个键分组运行一次 reduce 函数。该函数接收一个键和迭代器，进而遍历与该键关联的所有值。在许多模式中都会看到，该函数将执行大范围的处理操作。其中，数据将通过多种方式被聚合、过滤和组合。待 reduce 函数执行完毕后，将向最后一个步骤中传递 0 个或多个键/值对，即之前所谈到的输出格式。类似于 map 函数，作为解决方案中的核心逻辑，reduce 将在不同的作业间发生变化。reducer 可能会包含大量的自定义操作，包括 HDFS 的写入操作、Elasticsearch 索引的输出，以及 RDBMS 或 NoSQL 的输出，例如 Cassandra、HBase 等。

3.1.8　输出格式

输出格式负责转换源自 reduce 函数中的最终键/值对，并通过记录写入器将其写入某个文件中。默认状态下，将使用一个标签分隔键和值，并利用换行符分隔记录。通常，这可通过自定义方式提供更为丰富的输出格式。最终，无论格式如何，数据都将被写入 HDFS 中。其间，不仅是默认状态下所支持的 HDFS 写入操作，还包括输出至 Elasticsearch 索引，以及 RDBMS 或 NoSQL 的输出行为，例如 Cassandra、HBase 等。

3.2　MapReduce 作业类型

MapReduce 作业可通过多种方式编写，这取决于所期望的结果。MapReduce 作业的基本结构如下所示：

```
import java.io.IOException;
import java.util.StringTokenizer;
import java.util.Map;
import java.util.HashMap;
import org.apache.hadoop.conf.Configuration;
import org.apache.hadoop.fs.Path;
import org.apache.hadoop.io.IntWritable;
import org.apache.hadoop.io.Text;
import org.apache.hadoop.mapreduce.Job;
import org.apache.hadoop.mapreduce.Mapper;
import org.apache.hadoop.mapreduce.Reducer;
import org.apache.hadoop.mapreduce.lib.input.FileInputFormat;
import org.apache.hadoop.mapreduce.lib.output.FileOutputFormat;
import org.apache.hadoop.util.GenericOptionsParser;
import org.apache.commons.lang.StringEscapeUtils;

public class EnglishWordCounter {
public static class WordMapper
extends Mapper<Object, Text, Text, IntWritable> {
```

```
...
}
public static class CountReducer
extends Reducer<Text, IntWritable, Text, IntWritable> {
...
}

public static void main(String[] args) throws Exception {
Configuration conf = new Configuration();
Job job = new Job(conf, "English Word Counter");
job.setJarByClass(EnglishWordCounter.class);
job.setMapperClass(WordMapper.class);
job.setCombinerClass(CountReducer.class);
job.setReducerClass(CountReducer.class);
job.setOutputKeyClass(Text.class);
job.setOutputValueClass(IntWritable.class);
FileInputFormat.addInputPath(job, new Path(args[0]));
FileOutputFormat.setOutputPath(job, new Path(args[1]));
System.exit(job.waitForCompletion(true) ? 0 : 1);
}
}
```

驱动程序的目的在于对作业进行编排。main 函数中的前几行代码用于解析命令行参数，并开始设置作业对象——通知所使用的计算类、输入和输出路径。

下面考查 Mapper 代码。代码将对输入字符串进行简单的标记，并作为 mapper 的输出写入每个单词中，如下所示：

```
public static class WordMapper
extends Mapper<Object, Text, Text, IntWritable> {
private final static IntWritable one = new IntWritable(1);
private Text word = new Text();
public void map(Object key, Text value, Context context)
throws IOException, InterruptedException {
// Grab the "Text" field, since that is what we are counting over
String txt = value.toString()
StringTokenizer itr = new StringTokenizer(txt);
while (itr.hasMoreTokens()) {
word.set(itr.nextToken());
context.write(word, one);
}
}
}
```

reducer 代码则相对简单。相应地，reduce 函数针对每个键分组调用一次，此处为每个单词。随后将遍历数字值，并获取相应的求和结果，如下所示：

```
public static class CountReducer
extends Reducer<Text, IntWritable, Text, IntWritable> {
private IntWritable result = new IntWritable();
public void reduce(Text key, Iterable<IntWritable> values,
Context context) throws IOException, InterruptedException {
int sum = 0;
for (IntWritable val : values) {
sum += val.get();
}
result.set(sum);
context.write(key, result);
}
}
```

后续内容将进一步讨论基本的 MapReduce 作业类型。

3.2.1　SingleMapper 作业

单一 mapper 作业一般用于转换用例中。如果我们只想改变数据的格式，例如某种转换行为，那么就可以使用这种模式，如图 3.4 所示。

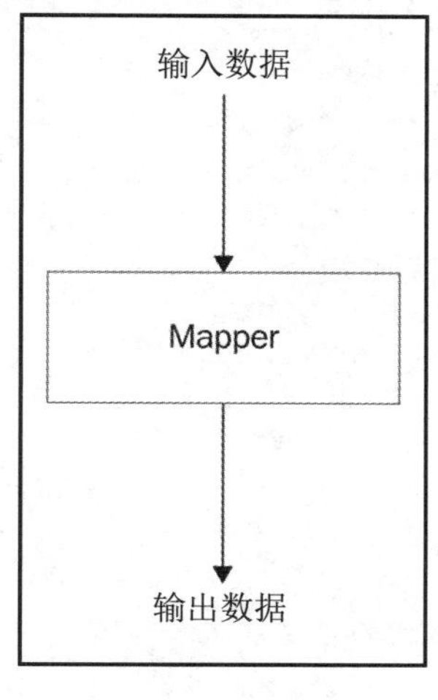

图 3.4

表 3.1 显示了相应的应用场景。

表 3.1

场　　景	某些城市采用简称表示，例如 BOS、NYC 等
Map(Key, Value)	❑ Key：城市名称 ❑ Value：简称。具体来说，如果城市为 Boston/boston，则转换为 BOS；如果城市为 New York/new york，则转换为 NYC

下面考查完整的单一 mapper 作业示例。对此，可输出前述 temperature.csv 文件中的 cityID 和 temperature。

对应代码如下所示：

```
package io.somethinglikethis;

import org.apache.hadoop.conf.Configuration;
import org.apache.hadoop.fs.Path;
import org.apache.hadoop.io.IntWritable;
import org.apache.hadoop.io.Text;
import org.apache.hadoop.mapreduce.Job;
import org.apache.hadoop.mapreduce.Mapper;
import org.apache.hadoop.mapreduce.Reducer;
import org.apache.hadoop.mapreduce.lib.input.FileInputFormat;
import org.apache.hadoop.mapreduce.lib.output.FileOutputFormat;
import java.io.IOException;

public class SingleMapper
{
    public static void main(String[] args) throws Exception {
        Configuration conf = new Configuration();
        Job job = new Job(conf, "City Temperature Job");
        job.setMapperClass(TemperatureMapper.class);
        job.setOutputKeyClass(Text.class);
        job.setOutputValueClass(IntWritable.class);

        FileInputFormat.addInputPath(job, new Path(args[0]));
        FileOutputFormat.setOutputPath(job, new Path(args[1]));

        System.exit(job.waitForCompletion(true) ? 0 : 1);
    }

    /*
    Date,Id,Temperature
    2018-01-01,1,21
```

```
    2018-01-01,2,22
    */
    private static class TemperatureMapper
            extends Mapper<Object, Text, Text, IntWritable> {

        public void map(Object key, Text value, Context context)
                throws IOException, InterruptedException {
            String txt = value.toString();
            String[] tokens = txt.split(",");
            String date = tokens[0];
            String id = tokens[1].trim();
            String temperature = tokens[2].trim();
            if (temperature.compareTo("Temperature") != 0)
                context.write(new Text(id), new
IntWritable(Integer.parseInt(temperature)));
        }
    }

}
```

当执行上述作业时，需要利用编辑器创建一个 Maven 项目，并编辑 pom.xml 文件，如下所示：

```
<?xml version="1.0" encoding="UTF-8"?>

<project xmlns="http://maven.apache.org/POM/4.0.0"
xmlns:xsi="http://www.w3.org/2001/XMLSchema-instance"
  xsi:schemaLocation="http://maven.apache.org/POM/4.0.0
http://maven.apache.org/xsd/maven-4.0.0.xsd">
  <modelVersion>4.0.0</modelVersion>
  <packaging>jar</packaging>
  <groupId>io.somethinglikethis</groupId>
  <artifactId>mapreduce</artifactId>
  <version>1.0-SNAPSHOT</version>

  <name>mapreduce</name>
  <!-- FIXME change it to the project's website -->
  <url>http://somethinglikethis.io</url>

  <properties>
    <project.build.sourceEncoding>UTF-8</project.build.sourceEncoding>
    <maven.compiler.source>1.7</maven.compiler.source>
    <maven.compiler.target>1.7</maven.compiler.target>
```

```
  </properties>

  <dependencies>
    <dependency>
      <groupId>junit</groupId>
      <artifactId>junit</artifactId>
      <version>4.11</version>
      <scope>test</scope>
    </dependency>
    <dependency>
      <groupId>org.apache.hadoop</groupId>
      <artifactId>hadoop-mapreduce-client-core</artifactId>
      <version>3.1.0</version>
    </dependency>
    <dependency>
      <groupId>org.apache.hadoop</groupId>
      <artifactId>hadoop-client</artifactId>
      <version>3.1.0</version>
    </dependency>
  </dependencies>
  <build>
      <plugins>
        <plugin>
          <groupId>org.apache.maven.plugins</groupId>
          <artifactId>maven-shade-plugin</artifactId>
          <version>3.1.1</version>
          <executions>
              <execution>
                  <phase>package</phase>
                  <goals>
                      <goal>shade</goal>
                  </goals>
              </execution>
          </executions>
            <configuration>
                <finalName>uber-${project.artifactId}-
${project.version}</finalName>
                <transformers>
                    <transformer
implementation="org.apache.maven.plugins.shade.resource.ServicesResour
ceTransformer"/>
                </transformers>
```

```
                <filters>
                    <filter>
                        <artifact>*:*</artifact>
                        <excludes>
                            <exclude>META-INF/*.SF</exclude>
                            <exclude>META-INF/*.DSA</exclude>
                            <exclude>META-INF/*.RSA</exclude>
                            <exclude>META-INF/LICENSE*</exclude>
                            <exclude>license/*</exclude>
                        </excludes>
                    </filter>
                </filters>
            </configuration>
        </plugin>
      </plugins>
  </build>
</project>
```

据此，可以使用 Maven 构建 shaded/fat .jar，如下所示：

```
Moogie:mapreduce sridharalla$ mvn clean compile package
[INFO] Scanning for projects...
[INFO]
[INFO]
------------------------------------------------------------------------
----
[INFO] Building mapreduce 1.0-SNAPSHOT
[INFO]
------------------------------------------------------------------------
----
[INFO]
[INFO] --- maven-clean-plugin:2.5:clean (default-clean) @ mapreduce ---
[INFO] Deleting /Users/sridharalla/git/mapreduce/target
......
...........
```

随后，在目标目录中，应可看到 uber-mapreduce-1.0-SNAPSHOT.jar。下面开始执行当前作业。

提示：

确保本地 Hadoop 集群已处于启动状态（参见第 1 章），并可在浏览器中访问 http://localhost:9870。

当执行作业时，将使用到 Hadoop 二进制文件，以及之前创建的 fat .jar 文件，如下所示：

```
export PATH=$PATH:/Users/sridharalla/hadoop-3.1.0/bin
hdfs dfs -chmod -R 777 /user/normal
```

运行下列命令：

```
hadoop jar target/uber-mapreduce-1.0-SNAPSHOT.jar
io.somethinglikethis.SingleMapper /user/normal/temperatures.csv
/user/normal/output/SingleMapper
```

作业运行后，将输出下列结果：

```
Moogie:target sridharalla$ hadoop jar uber-mapreduce-1.0-SNAPSHOT.jar
io.somethinglikethis.SingleMapper /user/normal/temperatures.csv
/user/normal/output/SingleMapper
2018-05-20  18:38:01,399  WARN  util.NativeCodeLoader:  Unable  to  load
nativehadoop library for your platform... using builtin-java classes where
applicable
2018-05-20 18:38:02,248 INFO impl.MetricsConfig: loaded properties from
hadoop-metrics2.properties
......
```

这里，需要特别注意以下输出计数器：

```
Map-Reduce Framework
 Map input records=28
 Map output records=27
 Map output bytes=162
 Map output materialized bytes=222
 Input split bytes=115
 Combine input records=0
 Combine output records=0
 Reduce input groups=6
 Reduce shuffle bytes=222
 Reduce input records=27
 Reduce output records=27
 Spilled Records=54
 Shuffled Maps =1
 Failed Shuffles=0
 Merged Map outputs=1
 GC time elapsed (ms)=13
 Total committed heap usage (bytes)=1084227584
```

其中，mapper 输出了 27 条记录，且不存在 reducer 操作，所有输入记录都按 1:1 的比例输出。通过 http://localhost:9870 并转至/user/normal/output 下的输出目录，读者将能够使用 HDFS 浏览器对此进行检查，如图 3.5 所示。

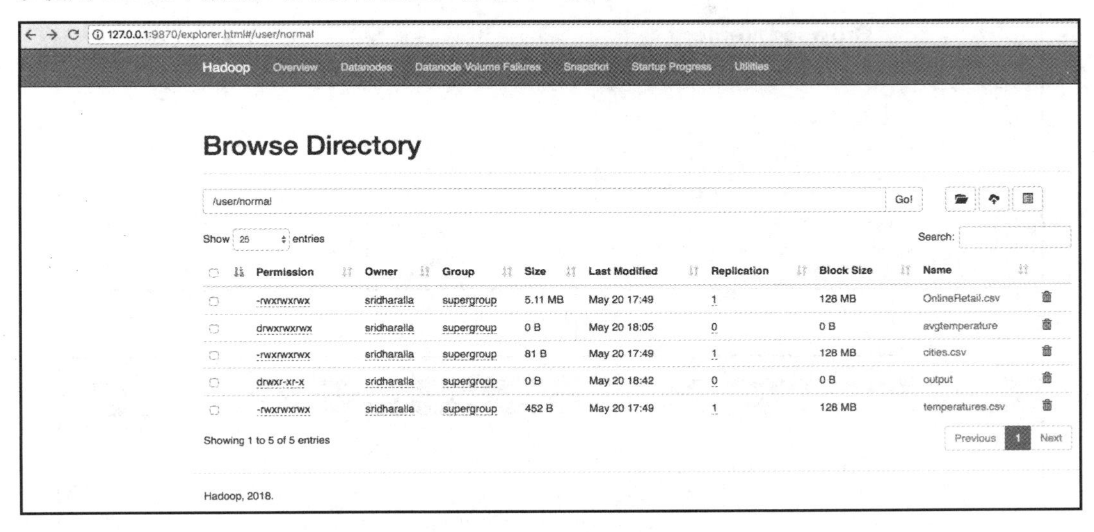

图 3.5

下面访问 SingleMapper 文件夹，如图 3.6 所示。

图 3.6

接下来，访问其中的 SingleMapper 文件夹，如图 3.7 所示。

最后，单击图中的 part-r-00000 文件，如图 3.8 所示。

对应的文件属性如图 3.9 所示。

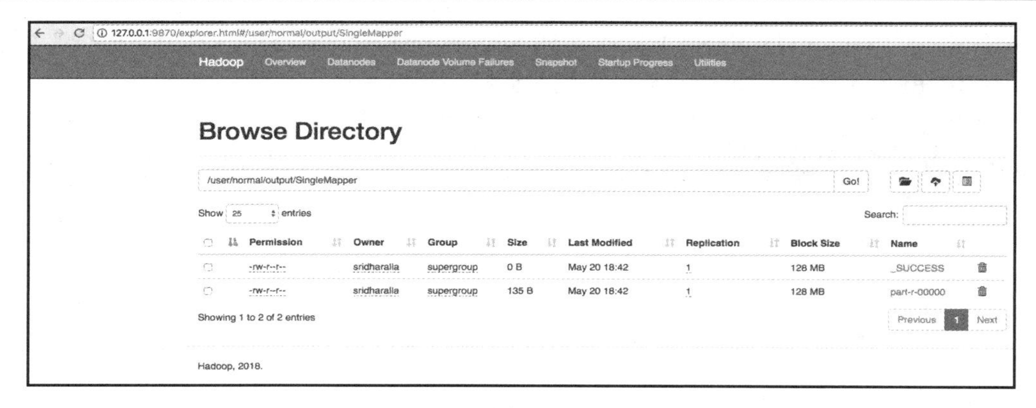
127.0.0.1:9870/explorer.html#/user/normal/output/SingleMapper
Hadoop
Overview
Datanodes
Datanode Volume Failures
Snapshot
Startup Progress
Utilities
Browse Directory
/user/normal/output/SingleMapper
Go!
Show 25 entries
Search:
Permission Owner Group Size Last Modified Replication Block Size Name
-rw-r--r-- sridharalla supergroup 0 B May 20 18:42 1 128 MB _SUCCESS
-rw-r--r-- sridharalla supergroup 135 B May 20 18:42 1 128 MB part-r-00000
Showing 1 to 2 of 2 entries
Previous 1 Next
Hadoop, 2018.

图 3.7

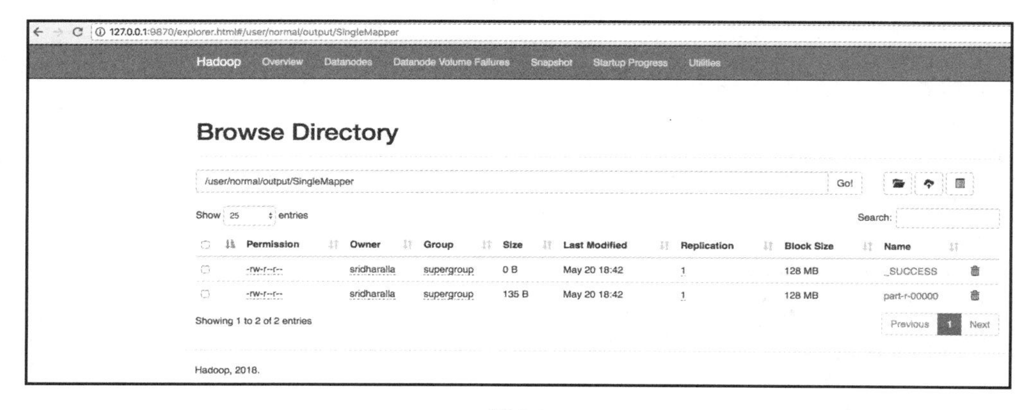
127.0.0.1:9870/explorer.html#/user/normal/output/SingleMapper
Hadoop
Overview
Datanodes
Datanode Volume Failures
Snapshot
Startup Progress
Utilities
Browse Directory
/user/normal/output/SingleMapper
Go!
Show 25 entries
Search:
Permission Owner Group Size Last Modified Replication Block Size Name
-rw-r--r-- sridharalla supergroup 0 B May 20 18:42 1 128 MB _SUCCESS
-rw-r--r-- sridharalla supergroup 135 B May 20 18:42 1 128 MB part-r-00000
Showing 1 to 2 of 2 entries
Previous 1 Next
Hadoop, 2018.

图 3.8

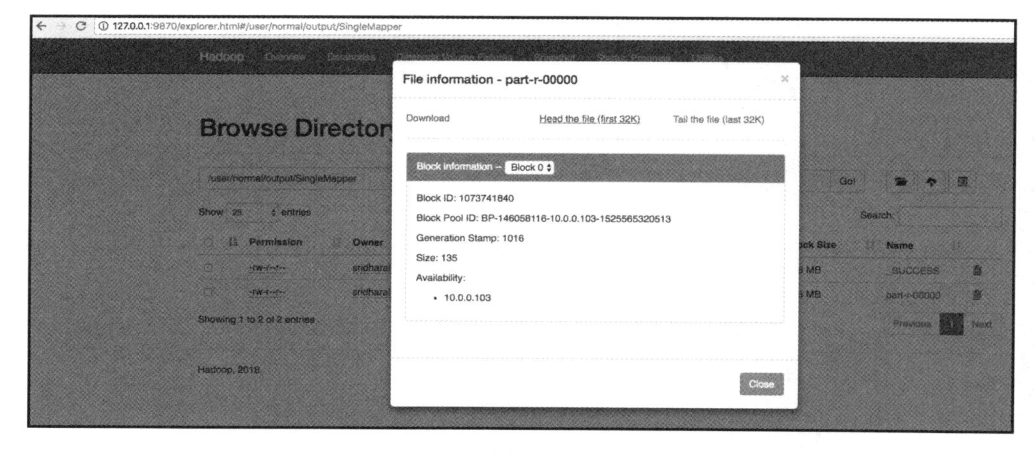
127.0.0.1:9870/explorer.html#/user/normal/output/SingleMapper
File information - part-r-00000
Download
Head the file (first 32K)
Tail the file (last 32K)
Block information -- Block 0
Block ID: 1073741840
Block Pool ID: BP-146058116-10.0.0.103-1525565320513
Generation Stamp: 1016
Size: 135
Availability:
10.0.0.103
Close

图 3.9

当使用图 3.9 中的 head/tail 选项时，即可查看到当前文件的内容，如图 3.10 所示。

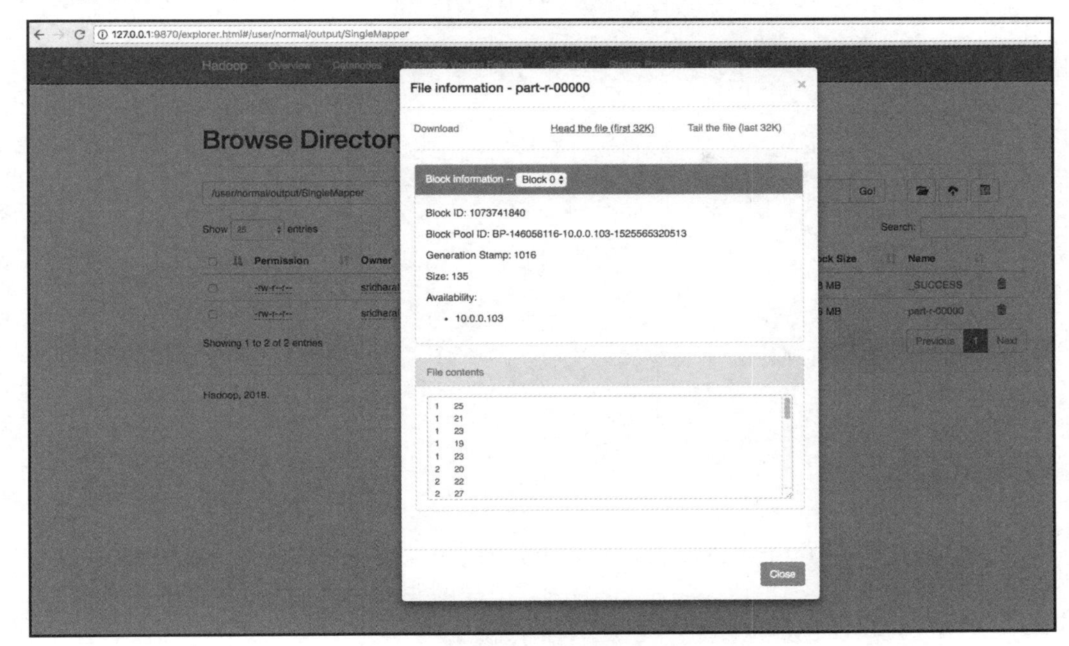

图 3.10

这显示了 SingleMapper 作业的输出结果，即简单地写入每行的 cityID 和 temperature，且不涉及任何计算。

提示：

除此之外，还可使用命令行查看输出内容，如下所示：

```
hdfs dfs -cat /user/normal/output/SingleMapper/part-r-00000
```

文件的输出内容如下所示：

```
1 25
1 21
1 23
1 19
1 23
2 20
2 22
2 27
2 24
```

```
2 26
3 21
3 25
3 22
3 25
3 23
4 21
4 26
4 23
4 24
4 22
5 18
5 24
5 22
5 25
5 24
6 22
6 22
```

至此，我们得到了 SingleMapper 作业的输出结果，且与预期结果一致。

3.2.2　SingleMapperReducer 作业

单一 mapper-reducer 作业用于聚合用例中。如果希望通过键执行诸如计数这一类聚合操作，则可使用该模式，如图 3.11 所示。

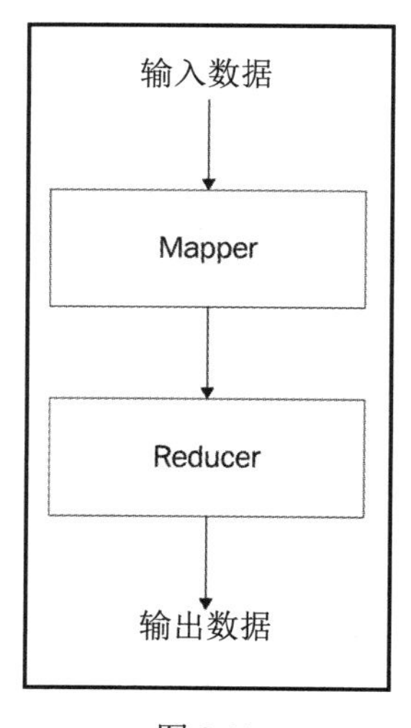

图 3.11

表 3.2 列出了相关场景。

表 3.2

场　　景	计算所有城市的全部/平均温度
Map (Key, Value)	❑ Key：城市 ❑ Value：城市的温度
Reduce	通过城市进行分组，并计算每所城市的平均温度

下面考查单一 mapper-reducer 作业的完整示例。对此，我们仅是简单地输出 temperature.csv 文件中的 cityID 和 temperature。

对应代码如下所示：

```
package io.somethinglikethis;

import org.apache.hadoop.conf.Configuration;
import org.apache.hadoop.fs.Path;
import org.apache.hadoop.io.IntWritable;
import org.apache.hadoop.io.Text;
import org.apache.hadoop.mapreduce.Job;
import org.apache.hadoop.mapreduce.Mapper;
import org.apache.hadoop.mapreduce.Reducer;
import org.apache.hadoop.mapreduce.lib.input.FileInputFormat;
import org.apache.hadoop.mapreduce.lib.output.FileOutputFormat;

import java.io.IOException;

public class SingleMapperReducer
{
    public static void main(String[] args) throws Exception {
        Configuration conf = new Configuration();
        Job job = new Job(conf, "City Temperature Job");
        job.setMapperClass(TemperatureMapper.class);
        job.setReducerClass(TemperatureReducer.class);
        job.setOutputKeyClass(Text.class);
        job.setOutputValueClass(IntWritable.class);

        FileInputFormat.addInputPath(job, new Path(args[0]));
        FileOutputFormat.setOutputPath(job, new Path(args[1]));

        System.exit(job.waitForCompletion(true) ? 0 : 1);
    }
```

```
    /*
    Date,Id,Temperature
    2018-01-01,1,21
    2018-01-01,2,22
    */
    private static class TemperatureMapper
            extends Mapper<Object, Text, Text, IntWritable> {

        public void map(Object key, Text value, Context context)
                throws IOException, InterruptedException {
            String txt = value.toString();
            String[] tokens = txt.split(",");
            String date = tokens[0];
            String id = tokens[1].trim();
            String temperature = tokens[2].trim();
            if (temperature.compareTo("Temperature") != 0)
                context.write(new Text(id), new
IntWritable(Integer.parseInt(temperature)));
        }
    }

    private static class TemperatureReducer
            extends Reducer<Text, IntWritable, Text, IntWritable> {
        private IntWritable result = new IntWritable();
        public void reduce(Text key, Iterable<IntWritable> values,
                           Context context) throws IOException,
InterruptedException {
            int sum = 0;
            int n = 0;
            for (IntWritable val : values) {
                sum += val.get();
                n +=1;
            }
            result.set(sum/n);
            context.write(key, result);
        }
    }
}
```

下面运行下列命令：

```
hadoop jar target/uber-mapreduce-1.0-SNAPSHOT.jar
io.somethinglikethis.SingleMapperReducer/user/normal/temperatures.csv/
user/normal/output/SingleMapperReducer
```

当前作业运行后，将输出下列计数器结果：

```
Map-Reduce Framework
    Map input records=28
    Map output records=27
    Map output bytes=162
    Map output materialized bytes=222
    Input split bytes=115
    Combine input records=0
    Combine output records=0
    Reduce input groups=6
    Reduce shuffle bytes=222
    Reduce input records=27
    Reduce output records=6
    Spilled Records=54
    Shuffled Maps =1
    Failed Shuffles=0
    Merged Map outputs=1
    GC time elapsed (ms)=12
    Total committed heap usage (bytes)=1080557568
```

其中，mapper 输出了 27 条记录，reducer 输出了 6 条记录。读者可通过 HDFS 浏览器进行检测，即使用 http://localhost:9870，并转至/user/normal/output 下的输出目录，如图 3.12 所示。

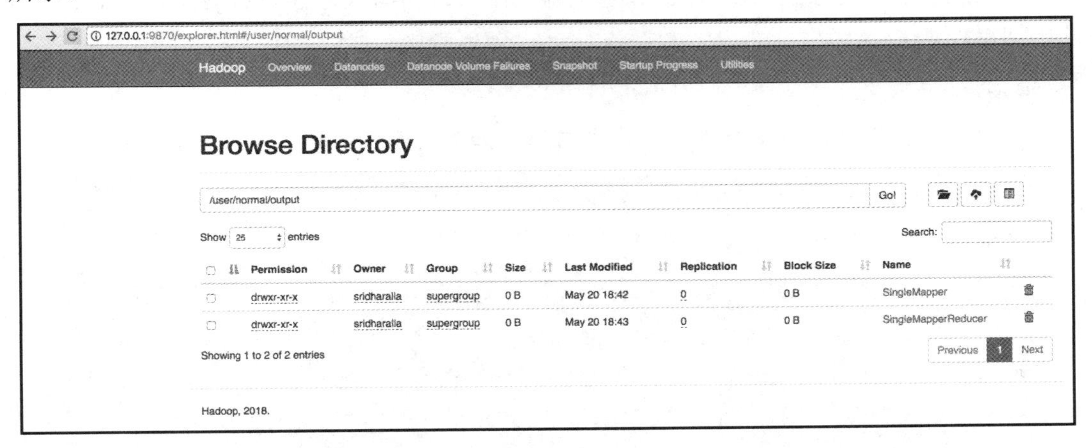

图 3.12

接下来，访问 SingleMapperReducer 文件夹，采用与之前 SingleMapper 作业相同的操作。图 3.13 显示了文件内容。

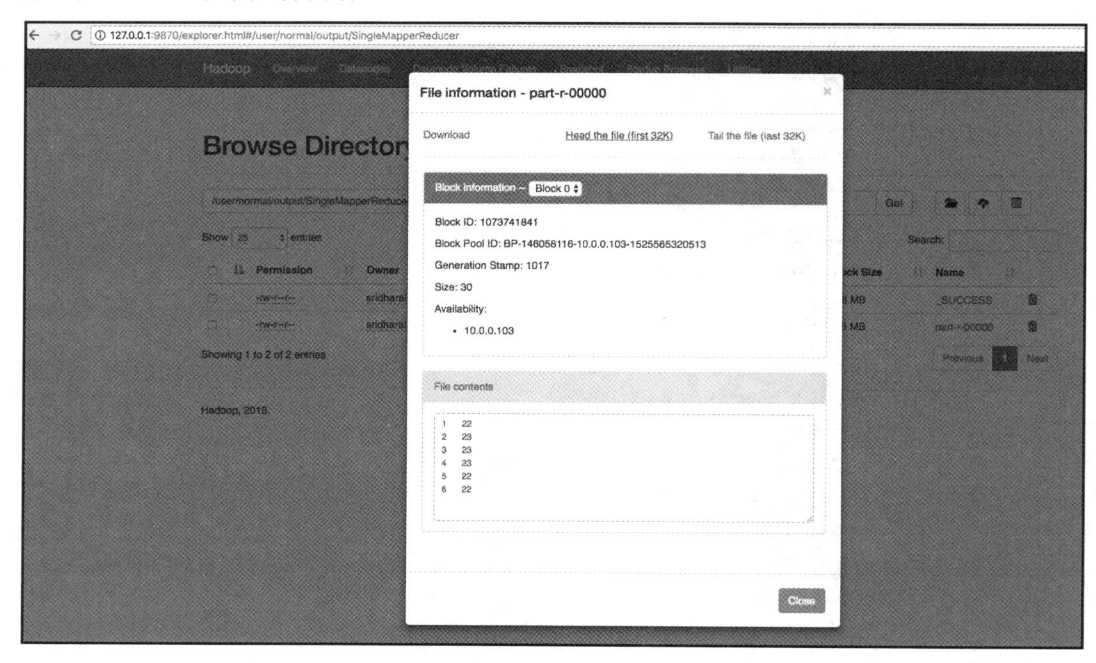

图 3.13

这显示了 SingleMapperReducer 作业的输出结果，即写入每行的 cityID 和每个 cityID 的平均温度。

提示：

除此之外，还可使用命令行查看输出内容，如下所示：

```
hdfs dfs -cat /user/normal/output/SingleMapperReducer/part-r-00000
```

文件的输出内容如下所示：

```
1 22
2 23
3 23
4 23
5 22
6 22
```

至此，我们得到了作业的输出结果，且与期望结果保持一致。

3.2.3　MultipleMappersReducer 作业

MultipleMappersReducer 作业用于连接用例中。在该设计模式中，输入源自多个输入文件，并生成连接/聚合的输出结果，如图 3.14 所示。

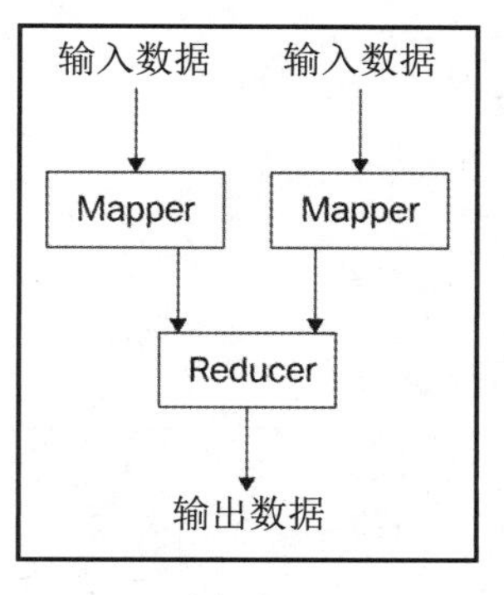

图 3.14

对应场景如表 3.3 所示。

表 3.3

场　景	计算城市的平均温度值，但持有两个包含不同模式的文件，分别对应于城市和温度数据。具体来说，输入文件 1 包含了 cityID 和 Name；输入文件 2 包含了每所城市每天的温度
Map (Key,Value)	❑ Map1（针对输入 1）：需要编写一个程序以划分 cityID 和 Name，并根据 cityID 写入 Name。随后，准备相应的键/值对（cityID 和 Name） ❑ Map2（针对输入 2）：需要编写一个程序以划分 date、cityID 和 temperature，并根据 cityID 写入 temperature。随后，准备相应的键值对（cityID 和 temperature）
Reduce	根据 cityID 进行分组，并根据每个城市名称计算平均温度值

下面考查 MultipleMappersReducer 作业的完整示例。对此，将输出 temperature.csv 文件中的 cityID 和平均温度值，如下所示：

```
package io.somethinglikethis;

import org.apache.hadoop.conf.Configuration;
import org.apache.hadoop.fs.Path;
import org.apache.hadoop.io.IntWritable;
import org.apache.hadoop.io.Text;
import org.apache.hadoop.mapreduce.Job;
```

```
import org.apache.hadoop.mapreduce.Mapper;
import org.apache.hadoop.mapreduce.Reducer;
import org.apache.hadoop.mapreduce.lib.input.FileInputFormat;
import org.apache.hadoop.mapreduce.lib.input.MultipleInputs;
import org.apache.hadoop.mapreduce.lib.input.TextInputFormat;
import org.apache.hadoop.mapreduce.lib.output.FileOutputFormat;

import java.io.IOException;

public class MultipleMappersReducer
{
    public static void main(String[] args) throws Exception {
        Configuration conf = new Configuration();
        Job job = new Job(conf, "City Temperature Job");
        job.setMapperClass(TemperatureMapper.class);
        MultipleInputs.addInputPath(job, new Path(args[0]),
TextInputFormat.class, CityMapper.class);
        MultipleInputs.addInputPath(job, new Path(args[1]),
TextInputFormat.class, TemperatureMapper.class);

        job.setMapOutputKeyClass(Text.class);
        job.setMapOutputValueClass(Text.class);
        job.setReducerClass(TemperatureReducer.class);
        job.setOutputKeyClass(Text.class);
        job.setOutputValueClass(IntWritable.class);

        FileOutputFormat.setOutputPath(job, new Path(args[2]));

        System.exit(job.waitForCompletion(true) ? 0 : 1);
    }

    /*
    Id,City
    1,Boston
    2,New York
    */
    private static class CityMapper

            extends Mapper<Object, Text, Text, Text> {

        public void map(Object key, Text value, Context context)
                throws IOException, InterruptedException {
            String txt = value.toString();
```

```
            String[] tokens = txt.split(",");
            String id = tokens[0].trim();
            String name = tokens[1].trim();
            if (name.compareTo("City") != 0)
                context.write(new Text(id), new Text(name));
        }
    }

    /*
    Date,Id,Temperature
    2018-01-01,1,21
    2018-01-01,2,22
    */
    private static class TemperatureMapper
            extends Mapper<Object, Text, Text, Text> {

        public void map(Object key, Text value, Context context)
                throws IOException, InterruptedException {
            String txt = value.toString();
            String[] tokens = txt.split(",");
            String date = tokens[0];
            String id = tokens[1].trim();
            String temperature = tokens[2].trim();
            if (temperature.compareTo("Temperature") != 0)
                context.write(new Text(id), new Text(temperature));
        }
    }

    private static class TemperatureReducer
            extends Reducer<Text, Text, Text, IntWritable> {
        private IntWritable result = new IntWritable();
        private Text cityName = new Text("Unknown");
        public void reduce(Text key, Iterable<Text> values,
                           Context context) throws IOException,
InterruptedException {
            int sum = 0;
            int n = 0;

            cityName = new Text("city-"+key.toString());

            for (Text val : values) {
                String strVal = val.toString();
                if (strVal.length() <=3)
```

```
                {
                    sum += Integer.parseInt(strVal);
                    n +=1;
                } else {
                    cityName = new Text(strVal);
                }
            }
            if (n==0) n = 1;
            result.set(sum/n);
            context.write(cityName, result);
        }
    }
}
```

随后，运行下列命令：

```
hadoop jar target/uber-mapreduce-1.0-SNAPSHOT.jar
io.somethinglikethis.MultipleMappersReducer/user/normal/cities.csv/
user/normal/temperatures.csv /user/normal/output/MultipleMappersReducer
```

当前作业运行后，将输出下列计数器结果：

```
Map-Reduce Framework -- mapper for temperature.csv
    Map input records=28
    Map output records=27
    Map output bytes=135
    Map output materialized bytes=195
    Input split bytes=286
    Combine input records=0
    Spilled Records=27
    Failed Shuffles=0
    Merged Map outputs=0
    GC time elapsed (ms)=0
    Total committed heap usage (bytes)=430964736

Map-Reduce Framework. -- mapper for cities.csv
    Map input records=7
    Map output records=6
    Map output bytes=73
    Map output materialized bytes=91
    Input split bytes=273
    Combine input records=0
    Spilled Records=6
    Failed Shuffles=0
```

```
    Merged Map outputs=0
    GC time elapsed (ms)=10
    Total committed heap usage (bytes)=657457152

Map-Reduce Framework -- output average temperature per city name
    Map input records=35
    Map output records=33
    Map output bytes=208
    Map output materialized bytes=286
    Input split bytes=559
    Combine input records=0
    Combine output records=0
    Reduce input groups=7
    Reduce shuffle bytes=286
    Reduce input records=33
    Reduce output records=7
    Spilled Records=66
    Shuffled Maps =2
    Failed Shuffles=0
    Merged Map outputs=2
    GC time elapsed (ms)=10
    Total committed heap usage (bytes)=1745879040
```

其中，一个 mapper 输出了 27 条记录，mapper2 输出了 6 条记录，reducer 输出了 7 条记录。利用 HDFS 浏览器，我们可对此予以检测，即访问 http://localhost:9870，并访问 /user/normal/output 下的输出目录，如图 3.15 所示。

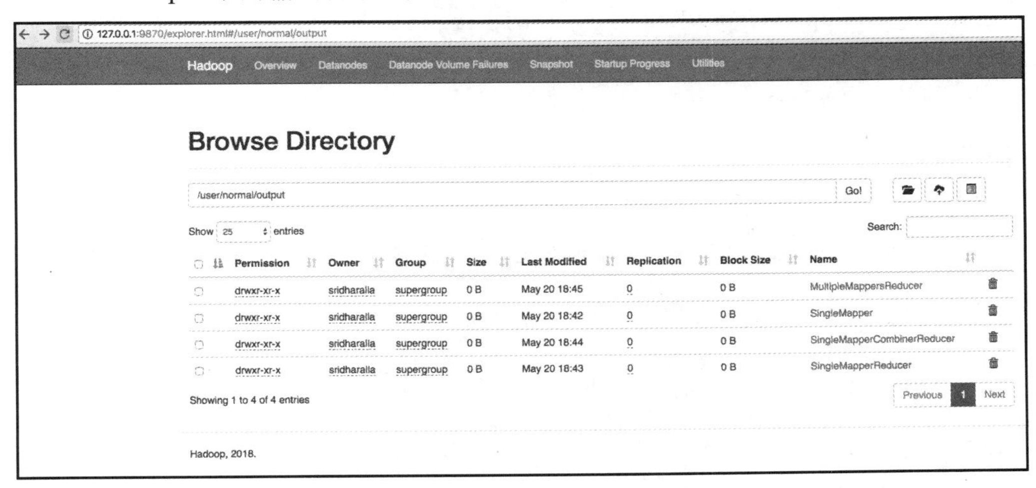

图 3.15

访问 MultipleMappersReducer 文件夹，并执行与 SingleMapper 作业任务相同的操作。随后，选择 head/tail 选项，并查看文件内容，如图 3.16 所示。

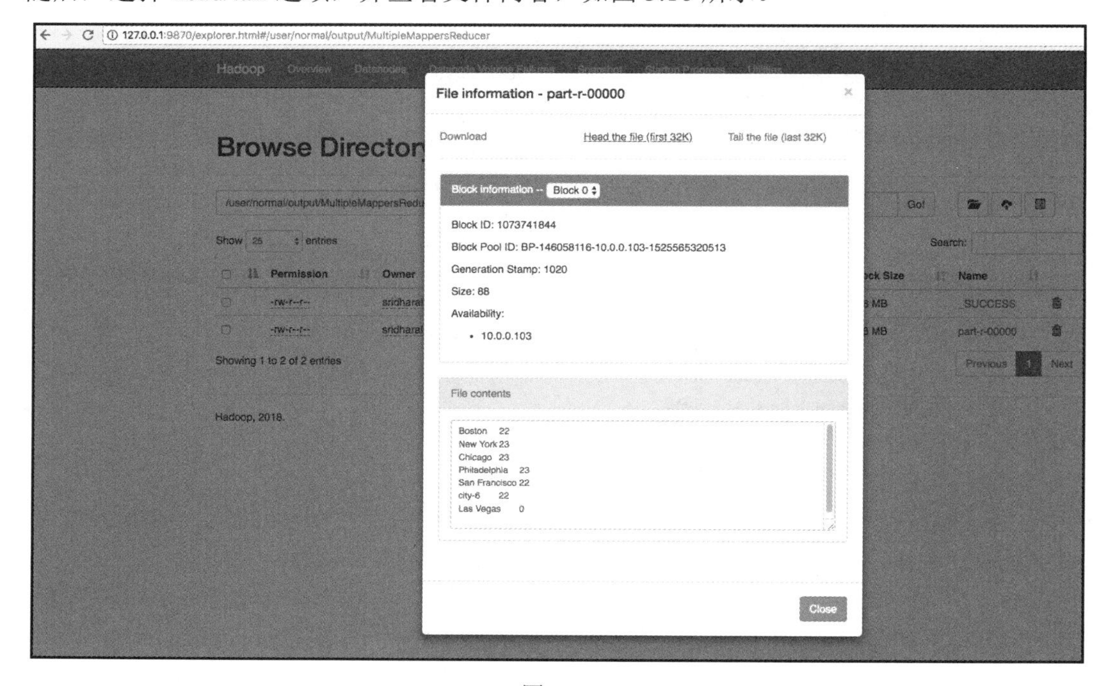

图 3.16

这显示了 MultipleMappersReducer 作业的输出结果，即城市名和每所城市的平均温度值。如果 cityID 未包含 temperature.csv 中的对应记录，那么，平均值将显示为 0。类似地，如果 cityID 未包含 cities.csv 中的名称，城市名将显示为 city-N。

提示：

除此之外，还可使用命令行查看输出内容，如下所示：

```
hdfs dfs -cat/user/normal/output/MultipleMappersReducer/partr-00000
```

输出后的文件内容如下所示：

```
Boston 22
New York 23
Chicago 23
Philadelphia 23
San Francisco 22
city-6 22 //city ID 6 has no name in cities.csv only temperature
```

```
measurements
Las Vegas 0 // city of Las vegas has no temperature measurements in
temperature.csv
```

至此，我们得到了 MultipleMappersReducer 作业的执行结果，且与预期结果保持一致。

3.2.4　SingleMapperReducer 作业

SingleMapperReducer 作业用于聚合用例中。其中，组合器（也称作 semi-reducer）定义为一个可选类，并接收源自 map 类中的输入内容，随后将输出的键/值对传递至 reducer 类中。这里，组合器的用途在于降低 reducer 的工作负载，如图 3.17 所示。

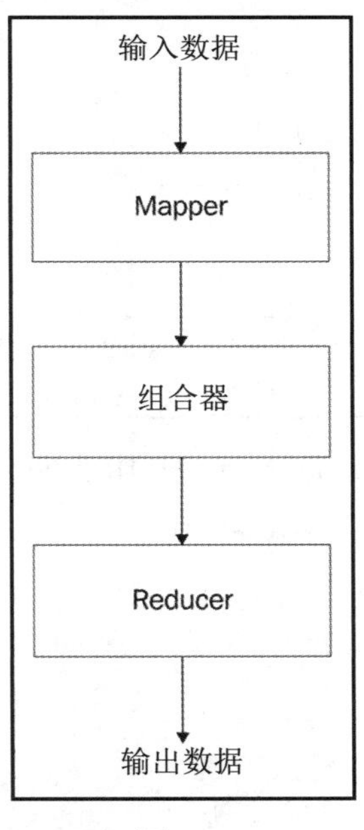

图 3.17

在 MapReduce 程序中，25%的工作量在 map 阶段完成，该阶段也称作数据准备阶段，并以并行方式工作。同时，75%的工作量在 reduce 阶段完成，也称作计算阶段，这一过程并未采用并行方式，因而与 map 阶段相比，其速度相对较慢。为了进一步节省时间，一些 reduce 阶段的工作可在组合器阶段完成。

例如，如果设置了一个组合器，那么，我们将从 mapper 发送(Boston, 66)，mapper 将(Boston, 22)、(Boston, 24)、(Boston, 20)视为输入记录，而不是通过网络发送 3 个单独的键/值对记录。

3.2.5 应用场景

当前场景涉及多所城市，同时包含了针对每所城市的每日温度，进而计算城市的平均温度值。然而，存在多种规则可计算平均值。在针对每所城市计算了总值后，即可计算其平均值，如表 3.4 所示。

表 3.4

输入文件 （多个文件）	Map(,Value= Name) （并行）	组合器（并行）	Reducer（非并行）	输　出
City 1	1<10,20,25,45,15,45,25,20> 2 <10,30,20,25,35>	1 <250,20> 2 <120,10>	1 Boston, < 250,20,155, 10,90,90,30> 2 New York, <120,10,175,10,135, 10,110,10,130,10>	Boston <645> New York <720>
City 2	1<Boston> 2 <New York>	1 <Boston> 2 <New York>		

下面考查 SingleMapperCombinerReducer 作业的完整示例。对此，可简单地输出 temperature.csv 文件中的 cityID 和平均温度值。

对应 diamante 如下所示：

```
package io.somethinglikethis;

import org.apache.hadoop.conf.Configuration;
import org.apache.hadoop.fs.Path;
import org.apache.hadoop.io.IntWritable;
import org.apache.hadoop.io.Text;
import org.apache.hadoop.mapreduce.Job;
import org.apache.hadoop.mapreduce.Mapper;
import org.apache.hadoop.mapreduce.Reducer;
import org.apache.hadoop.mapreduce.lib.input.FileInputFormat;
import org.apache.hadoop.mapreduce.lib.output.FileOutputFormat;

import java.io.IOException;

public class SingleMapperCombinerReducer
```

```
{
    public static void main(String[] args) throws Exception {
        Configuration conf = new Configuration();
        Job job = new Job(conf, "City Temperature Job");
        job.setMapperClass(TemperatureMapper.class);
        job.setCombinerClass(TemperatureReducer.class);
        job.setReducerClass(TemperatureReducer.class);
        job.setOutputKeyClass(Text.class);
        job.setOutputValueClass(IntWritable.class);

        FileInputFormat.addInputPath(job, new Path(args[0]));
        FileOutputFormat.setOutputPath(job, new Path(args[1]));

        System.exit(job.waitForCompletion(true) ? 0 : 1);
    }

    /*
    Date,Id,Temperature
    2018-01-01,1,21
    2018-01-01,2,22
    */
    private static class TemperatureMapper
            extends Mapper<Object, Text, Text, IntWritable> {

        public void map(Object key, Text value, Context context)
                throws IOException, InterruptedException {
            String txt = value.toString();
            String[] tokens = txt.split(",");
            String date = tokens[0];
            String id = tokens[1].trim();
            String temperature = tokens[2].trim();
            if (temperature.compareTo("Temperature") != 0)
                context.write(new Text(id), new
IntWritable(Integer.parseInt(temperature)));
        }
    }

    private static class TemperatureReducer
            extends Reducer<Text, IntWritable, Text, IntWritable> {
        private IntWritable result = new IntWritable();
        public void reduce(Text key, Iterable<IntWritable> values,
                            Context context) throws IOException,
InterruptedException {
```

```
            int sum = 0;
            int n = 0;
            for (IntWritable val : values) {
                sum += val.get();
                n +=1;
            }
            result.set(sum/n);
            context.write(key, result);
        }
    }
}
```

接下来，运行下列命令：

```
hadoop jar target/uber-mapreduce-1.0-SNAPSHOT.jar
io.somethinglikethis.SingleMapperCombinerReducer
/user/normal/temperatures.csv
/user/normal/output/SingleMapperCombinerReducer
```

运行上述作业后，将会看到下列计数器输出结果：

```
Map-Reduce Framework
    Map input records=28
    Map output records=27
    Map output bytes=162
    Map output materialized bytes=54
    Input split bytes=115
    Combine input records=27
    Combine output records=6
    Reduce input groups=6
    Reduce shuffle bytes=54
    Reduce input records=6
    Reduce output records=6
    Spilled Records=12
    Shuffled Maps =1
    Failed Shuffles=0
    Merged Map outputs=1
    GC time elapsed (ms)=11
    Total committed heap usage (bytes)=1077936128
```

其中，mapper 输出了 27 条记录，reducer 则输出了 6 条记录。需要注意的是，当前设置了一个组合器，该组合器接收 27 个输入记录，并输出了 6 条记录，通过减少 mapper 和 reducer 间的混洗记录，性能得到了显著的改善。利用 HDFS 浏览器可对此进行检测，即访问 http://localhost:9870，并转至/user/normal/output 下的输出目录，如图 3.18 所示。

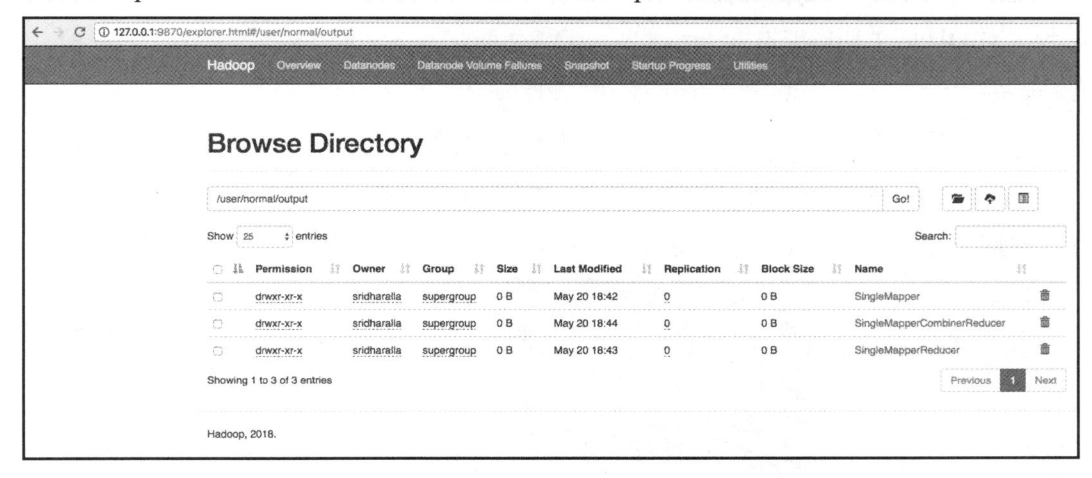

图 3.18

访问 SingleMapperCombinerReducer 文件夹，并执行与 SingleMapper 相同的操作，随后通过 head/tail 选项查看文件内容，如图 3.19 所示。

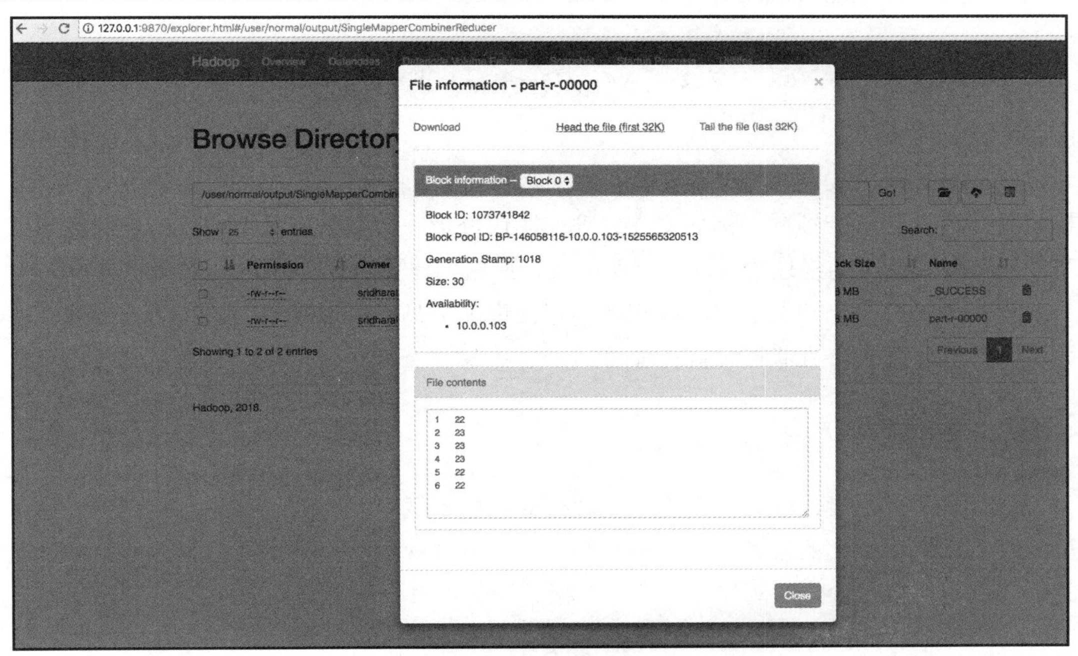

图 3.19

这显示了 SingleMapperCombinerReducer 作业的输出结果，即写入每行的 cityID，以及每个 cityID 的平均温度值。

提示：

除此之外，还可使用命令行查看输出内容，如下所示：

```
dfs - cat/user/normal/output/SingleMapperCombinerReducer/partr-00000
```

对应的输出文件内容如下所示：

```
1 22
2 23
3 23
4 23
5 22
6 22
```

至此，我们得到了 SingleMapperCombinerReducer 作业的执行结果，且与期望结果保持一致。

下面将考查 MapReduce 作业中与模式相关的更多信息。

3.3　MapReduce 模式

MapReduce 模式表示为一个模板，并以此处理常见的和一般的数据处理问题。这里，模式并非特定于某个领域，例如文本处理或图像分析，而是处理某个问题的通用解决方案。设计模式是采用经过验证的、真实的设计原则构建更为优异的软件系统。

设计模式可简化开发过程，相应地，存在许多工具并可以复用、通用的方式处理问题。据此，开发人员可节省大量的时间。

3.3.1　聚合模式

本节集中讨论设计模式，进而生成数据的顶层、概括性视图。据此，我们可以通过单独查看一组本地化的记录集得出某些结论。聚合（或汇总）分析是将相似的数据分组到一起，然后执行操作，例如计算统计数据、构建索引或简单的计数。

这里所讨论的模式主要包括数字汇总、倒排索引以及计数器计数，如图 3.20 所示。

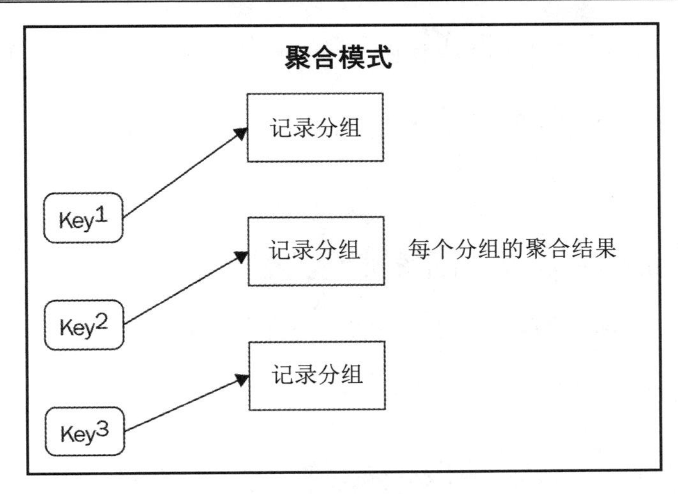

图 3.20

聚合模式是一类通用模式，用于计算数据中的聚合统计数值，其间需要恰当地使用组合器，并在编写代码前深入理解所执行的计算过程。基本上讲，其中的逻辑是通过键字段对记录进行分组，进而计算每个分组的数值聚合结果。

当满足下列条件时，可采用聚合或数值汇总模式：

- 处理数值数据或计数行为。
- 数据可通过特定字段进行分组。

1. 城市的平均温度

应用程序将输出每条记录的城市名（作为键）及其温度值（作为值），因而可通过城市进行分组。随后，reduce 阶段对全部整数求和，并输出包含平均温度值的城市。

2. 记录计数

一种较为常见的汇总是根据键获取记录计数，并将其分解为每日、每周和每月的计数。

3. 最小值/最大值/计数

这一分析过程将确定最小值、最大值以及特定事件的计数结果，例如某所城市第一次被采样、某所城市被最后一次采样，以及温度在此时间段被采样的次数。如果用户仅对某项内容感兴趣，则无须采集上述 3 种聚合数据（或者此处列出的任何其他数据）。

4. 平均值/中值/标准偏差

这与最小值/最大值/计数有些相似，但其实现过程并不直观——这一类操作缺乏关联性。这里，可针对这 3 种情形使用组合器，但与复用 reducer 实现相比需要使用更为复杂的解决方案。

最小值/最大值/计数示例，以及给定字段的最小值/最大值/计数均可视为聚合模式的良好应用。

提示：

之前讨论的 SingleMapperReducer 作业可视为聚合模式较好的示例。

取决于具体的用例，可自定义聚合模式，进而生成期望的输出结果。

3.3.2　过滤模式

过滤模式也称作转换模式，用于获取数据的子集（无论数据的大小，例如前 10 个列表结果，或者重复删除结果），如图 3.21 所示。

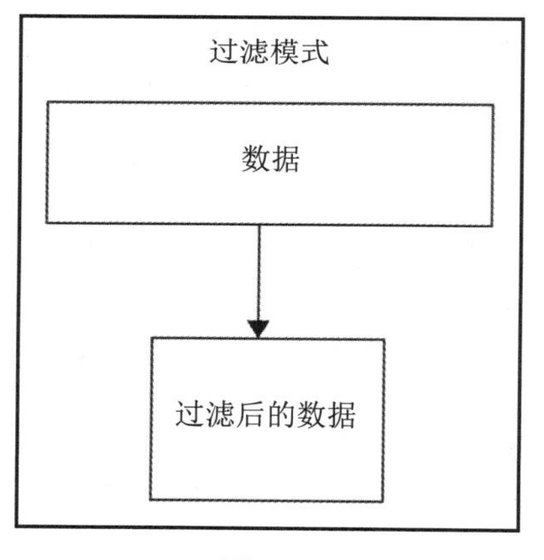

图 3.21

作为最基本的模式，过滤模式可作为其他模式的一些抽象模式。过滤模式简单地计算各条记录，并根据某些条件进行决策。也就是说，过滤掉某些不再关注的记录，同时保持有效的记录。考查一个计算函数 f，该函数接收一条记录，并返回 TRUE 或 FALSE 布尔值。如果该函数返回 TRUE，则保留记录；否则丢弃该记录。

提示：

之前讨论的 SingleMapper 作业可视为过滤模式一个较好的示例。

取决于具体的用例，还可自定义转换模式，进而生成期望的输出结果。

3.3.3　连接模式

数据随处可见。虽然数据本身具有一定的价值，但当开始分析数据集时，我们可以发现一些有趣的关系，这也是连接关系的用武之地。通过较小的参考集，可使用连接进一步丰富数据，也可以用来过滤或选择某种特定列表中的记录。

为了进一步理解此类模式及其实现方式，下面再次考查之前讨论的 MultipleMappersReducer 作业。

下列内容显示了部分代码，其中定义了两个 mapper 和一个 reducer。

```
public class MultipleMappersReducer
{
    public static void main(String[] args) throws Exception {
        Configuration conf = new Configuration();
        Job job = new Job(conf, "City Temperature Job");
        job.setMapperClass(TemperatureMapper.class);
        MultipleInputs.addInputPath(job, new Path(args[0]),
TextInputFormat.class, CityMapper.class);
        MultipleInputs.addInputPath(job, new Path(args[1]),
TextInputFormat.class, TemperatureMapper.class);

        job.setMapOutputKeyClass(Text.class);
        job.setMapOutputValueClass(Text.class);
        job.setReducerClass(TemperatureReducer.class);
        job.setOutputKeyClass(Text.class);
        job.setOutputValueClass(IntWritable.class);

        FileOutputFormat.setOutputPath(job, new Path(args[2]));

        System.exit(job.waitForCompletion(true) ? 0 : 1);
    }

    /*
    Id,City
    1,Boston
    2,New York
    */
    private static class CityMapper
            extends Mapper<Object, Text, Text, Text> {

        public void map(Object key, Text value, Context context)
```

```
                throws IOException, InterruptedException {
            String txt = value.toString();
            String[] tokens = txt.split(",");
            String id = tokens[0].trim();
            String name = tokens[1].trim();
            if (name.compareTo("City") != 0)
                context.write(new Text(id), new Text(name));
        }
    }

    /*
    Date,Id,Temperature
    2018-01-01,1,21
    2018-01-01,2,22
    */
    private static class TemperatureMapper
            extends Mapper<Object, Text, Text, Text> {

        public void map(Object key, Text value, Context context)
                throws IOException, InterruptedException {
            String txt = value.toString();
            String[] tokens = txt.split(",");
            String date = tokens[0];
            String id = tokens[1].trim();
            String temperature = tokens[2].trim();
            if (temperature.compareTo("Temperature") != 0)
                context.write(new Text(id), new Text(temperature));
        }
    }

    private static class TemperatureReducer
            extends Reducer<Text, Text, Text, IntWritable> {
        private IntWritable result = new IntWritable();
        private Text cityName = new Text("Unknown");
        public void reduce(Text key, Iterable<Text> values,
                           Context context) throws IOException,
InterruptedException {
            int sum = 0;
            int n = 0;

            cityName = new Text("city-"+key.toString());
```

```
            for (Text val : values) {
                String strVal = val.toString();
                if (strVal.length() <=3)
                {
                    sum += Integer.parseInt(strVal);
                    n +=1;
                } else {
                    cityName = new Text(strVal);
                }
            }
            if (n==0) n = 1;
            result.set(sum/n);
            context.write(cityName, result);
        }
    }
}
```

该作业的输出结果如下所示：

```
Boston 22
New York 23
Chicago 23
Philadelphia 23
San Francisco 22
city-6 22 //city ID 6 has no name in cities.csv only temperature
measurements
Las Vegas 0 // city of Las vegas has no temperature measurements in
temperature.csv
```

1．内连接

内连接要求左表和右表具有相同的列。如果左侧或右侧包含键的一个或多个副本，连接会迅速膨胀为某种笛卡儿连接，这将占用较长的计算时间。图 3.22 显示了内连接的状态。

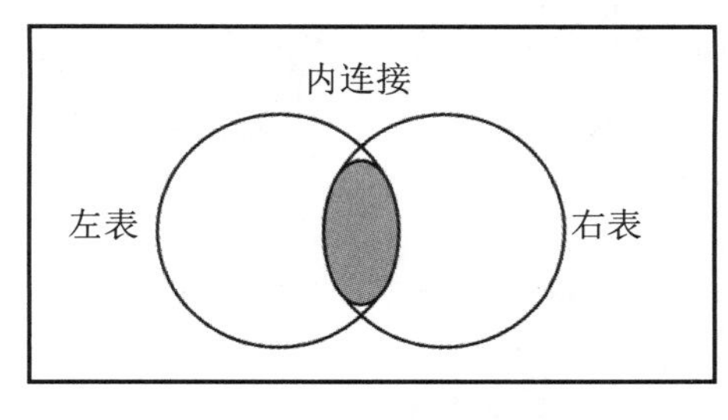

图 3.22

下面仅考查每所城市及其温度值。cityID 包含了下列代码所定义的两条记录：

```
private static class InnerJoinReducer
        extends Reducer<Text, Text, Text, IntWritable> {
    private IntWritable result = new IntWritable();
    private Text cityName = new Text("Unknown");
    public void reduce(Text key, Iterable<Text> values,
                       Context context) throws IOException,
InterruptedException {
        int sum = 0;
        int n = 0;
        for (Text val : values) {
            String strVal = val.toString();
            if (strVal.length() <=3)
            {
                sum += Integer.parseInt(strVal);
                n +=1;
            } else {
                cityName = new Text(strVal);
            }
        }
        if (n!=0 && cityName.toString().compareTo("Unknown") !=0) {
            result.set(sum / n);
            context.write(cityName, result);
        }
    }
}
```

输出结果如下所示（不包含 city-6 或 LasVegas）：

```
Boston 22
New York 23
Chicago 23
Philadelphia 23
San Francisco 22
```

2. 左反连接

左反连接可描述为：仅生成未出现于右表且来自左表中的行。当希望保留左表中的行，同时这些行未出现于右表中时，即可采用这种方案。该方案具有较好的性能——仅需整体考查一个表，而另一个表仅检查连接条件。图 3.23 显示了左反连接。

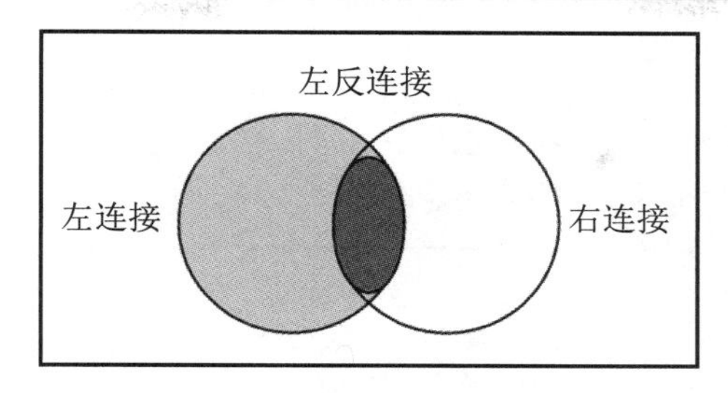

图 3.23

下面考查城市及其温度值。其中，cityID 仅包含名称且不包含温度记录，如下所示：

```
private static class LeftAntiJoinReducer
        extends Reducer<Text, Text, Text, IntWritable> {
    private IntWritable result = new IntWritable();
    private Text cityName = new Text("Unknown");
    public void reduce(Text key, Iterable<Text> values,
                       Context context) throws IOException,
InterruptedException {
        int sum = 0;
        int n = 0;

        for (Text val : values) {
            String strVal = val.toString();
            if (strVal.length() <=3)
            {
                sum += Integer.parseInt(strVal);
                n +=1;
            } else {
                cityName = new Text(strVal);
            }
        }
        if (n==0 ) {
            if (n==0) n=1;
            result.set(sum / n);
            context.write(cityName, result);
        }
    }
}
```

对应输出结果如下所示：

```
Las Vegas 0 // city of Las vegas has no temperature measurements in
temperature.csv
```

3. 左外连接

除了两个表公共部分（内连接）之外，左外连接还生成左表中的全部行。如果公共行较少，则最终结果相对庞大，因而会对性能产生影响。图 3.24 显示了左外连接。

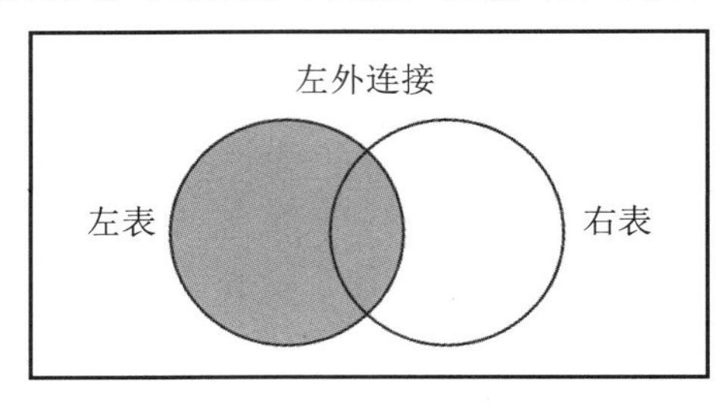

图 3.24

下面考查城市及其温度值。其中，cityID 包含两条记录，或者 cityID 仅位于 cities.csv 中，如下所示：

```
private static class LeftOuterJoinReducer
        extends Reducer<Text, Text, Text, IntWritable> {
    private IntWritable result = new IntWritable();
    private Text cityName = new Text("Unknown");
    public void reduce(Text key, Iterable<Text> values,
                       Context context) throws IOException,
InterruptedException {
        int sum = 0;
        int n = 0;

        for (Text val : values) {
            String strVal = val.toString();
            if (strVal.length() <=3)
            {
                sum += Integer.parseInt(strVal);
                n +=1;
            } else {
                cityName = new Text(strVal);
            }
        }
        if (cityName.toString().compareTo("Unknown") !=0)) {
            if (n==0) n = 1;

            result.set(sum / n);
```

```
                context.write(cityName, result);
            }
    }
}
```

对应输出结果如下所示：

```
Boston 22
New York 23
Chicago 23
Philadelphia 23
San Francisco 22
Las Vegas 0 // city of Las vegas has no temperature measurements in
temperature.csv
```

4．右外连接

右外连接生成右表中的全部行，以及两个表中的公共行（内部连接）。使用该连接将得到右表中的全部行，以及左、右表中的公共行。如果左表中不存在对应行，则填充 NULL。当采用右外连接时，其性能类似于之前讨论的左外连接。图 3.25 显示了右外连接。

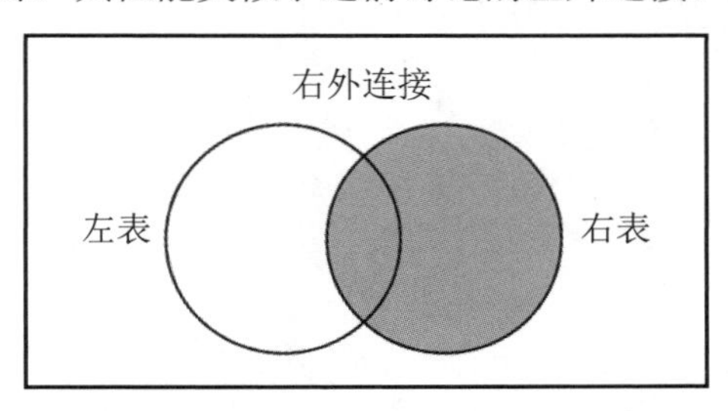

图 3.25

下面考查城市及其温度值。其中，cityID 包含两条记录，或者仅包含温度测量值，如下所示：

```
private static class RightOuterJoinReducer
        extends Reducer<Text, Text, Text, IntWritable> {
    private IntWritable result = new IntWritable();
    private Text cityName = new Text("Unknown");
    public void reduce(Text key, Iterable<Text> values,
                        Context context) throws IOException,
InterruptedException {
        int sum = 0;
        int n = 0;
        for (Text val : values) {
```

```
            String strVal = val.toString();
            if (strVal.length() <=3)
            {
                sum += Integer.parseInt(strVal);
                n +=1;
            } else {
                cityName = new Text(strVal);
            }
        }
      if (n !=0) {
          result.set(sum / n);
          context.write(cityName, result);
        }
    }
}
```

对应输出结果如下所示：

```
Boston 22
New York 23
Chicago 23
Philadelphia 23
San Francisco 22
city-6 22 //city ID 6 has no name in cities.csv only temperature
measurements
```

5. 全外连接

全外连接将生成连接子句的左、右两侧表中的全部行（包括匹配和不匹配的内容）。当需要保留两个表中的全部行时，可采用这一连接方式。如果表中包含了较少的公共行，最终结果将较为庞大，且性能也会受到一定程度的影响。图 3.26 显示了全外连接。

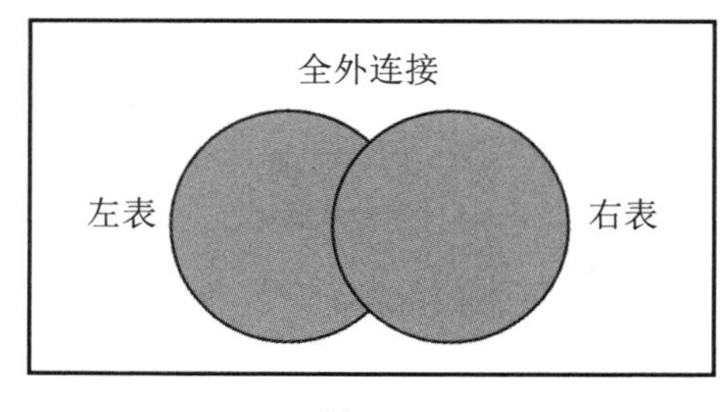

图 3.26

下面考查城市及其温度值。其中，cityID 包含两条记录，或者 cityID 仅存在于某一个表中，如下所示：

```
private static class FullOuterJoinReducer
        extends Reducer<Text, Text, Text, IntWritable> {
    private IntWritable result = new IntWritable();
    private Text cityName = new Text("Unknown");
    public void reduce(Text key, Iterable<Text> values,
                       Context context) throws IOException,
InterruptedException {
        int sum = 0;
        int n = 0;

        for (Text val : values) {
            String strVal = val.toString();
            if (strVal.length() <=3)
            {
                sum += Integer.parseInt(strVal);
                n +=1;
            } else {
                cityName = new Text(strVal);
            }
        }
        if (n==0) n = 1;
        result.set(sum/n);
        context.write(cityName, result);
    }
}
```

对应输出结果如下所示：

```
Boston 22
New York 23
Chicago 23
Philadelphia 23
San Francisco 22
city-6 22 //city ID 6 has no name in cities.csv only temperature
measurements
Las Vegas 0 // city of Las vegas has no temperature measurements in
temperature.csv
```

6．左半连接

左半连接仅生成左表中的行，当且仅当这些行位于右表中。这与之前讨论的左反连接刚好相反，且不包含右表中的数值。由于仅考查一个表，且另一个表仅执行条件连接检查，因而左半连接具有较好的性能。图 3.27 显示了左半连接。

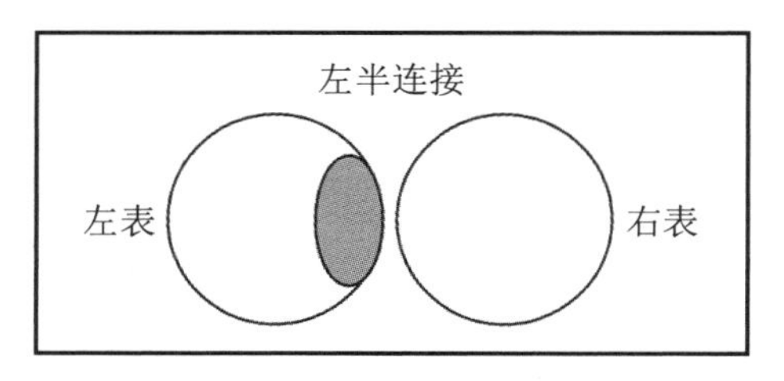

图 3.27

除了仅输出 cities.csv 中的表记录之外，左半连接的操作过程与左外连接较为类似。

7．交叉连接

交叉连接利用右表中的各行匹配左表中的每一行，并生成笛卡儿叉积。需要注意的是，由于交叉连接是性能最差的连接方式，因而仅在某些特殊用例中加以使用。图 3.28 显示了交叉连接。

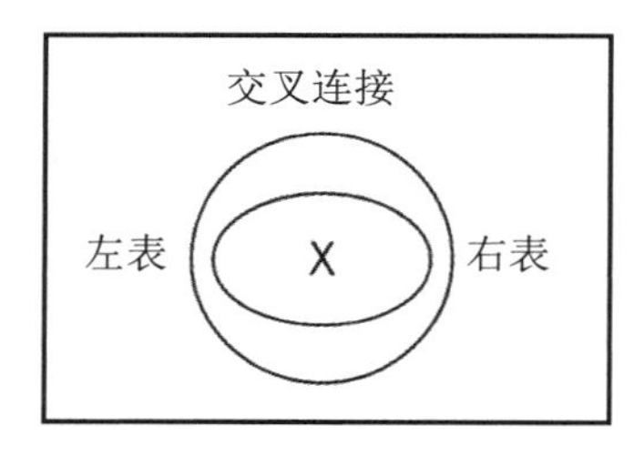

图 3.28

这将生成全部城市的温度值，其中包括 6×6 条记录（即 36 条输出记录）。由于输出结果相对庞大，因而交叉连接较少使用，且在大多数时候，该连接方式并不十分有用。

综上所述，我们可利用多种 mapper 方案实现不同的连接方式。

提示：

之前讨论的 MultipleMappersReducer 作业即是较好的连接模式示例。

根据具体的用例，还可自定义连接模式，进而生成期望的输出结果。

3.4　本 章 小 结

本章讨论了 MapReduce 框架、MapReduce 中的各种组件、MapReduce 范例中的各种模式，进而可用于设计和开发 MapReduce 代码，以满足特定的目标。

第 4 章将讨论 Python 语言，该语言可用于执行大数据分析。

第 4 章　Python-Hadoop 科学计算和大数据分析

本章将简要介绍 Python 语言，并利用 Hadoop 和 Python 数据包对大数据进行分析。其中涉及基本的 Python 安装、开启 Jupyter Notebook，并围绕相关示例展开讨论。

本章主要涉及下列主题：

- 安装操作。
  - 下载并安装 Python。
  - 下载并安装 Anaconda。
  - 安装 Jupyter Notebook。
- 数据分析。

4.1　安装操作

本节主要介绍各项安装步骤，并利用 Python 解释器设置 Jupyter Notebook，进而执行大数据分析。

4.1.1　安装 Python

读者可在浏览器中访问 http://www.python.org/download/，并进入下载页面。随后可看到，Python 支持 Windows、macOS 和 Linux 环境，并包含了不同的安装包，如下所示：

- Windows 版本，对应网址为 https://www.python.org/downloads/windows/。
- macOS 版本，对应网址为 https://www.python.org/downloads/mac-osx/。
- 源代码版本（Linux 和 Unix），对应网址为 https://www.python.org/downloads/source/。

当单击下载链接时，将显示如图 4.1 所示的页面。

当单击特定的版本时，例如 3.6.5，将显示如图 4.2 所示的页面。

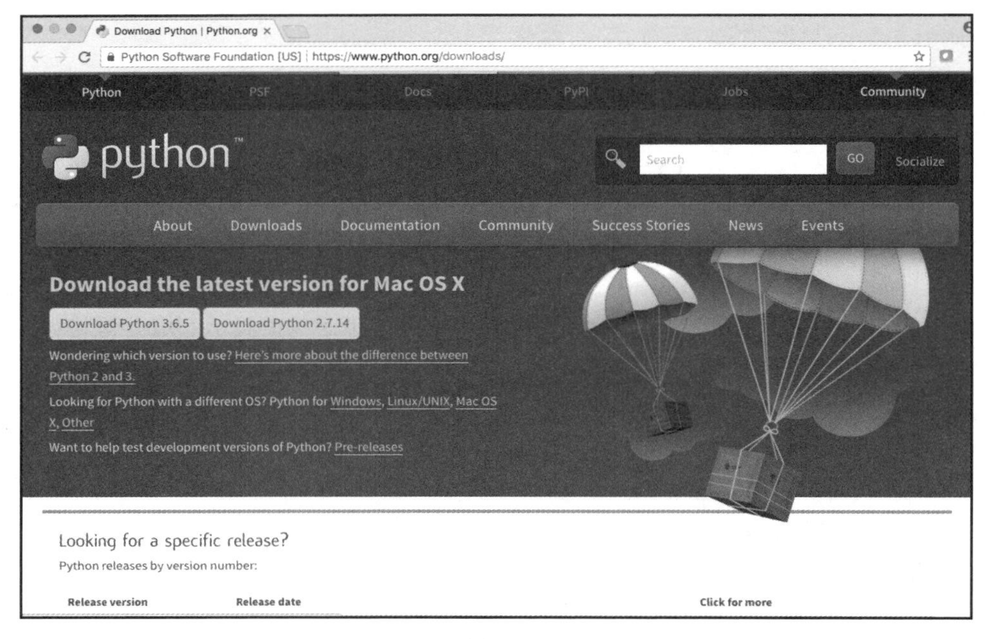
Download Python | Python.org
Python Software Foundation [US] | https://www.python.org/downloads/
Python
PSF
Docs
PyPI
Jobs
Community
python
Search
GO
Socialize
About
Downloads
Documentation
Community
Success Stories
News
Events
Download the latest version for Mac OS X
Download Python 3.6.5
Download Python 2.7.14
Wondering which version to use? Here's more about the difference between Python 2 and 3.
Looking for Python with a different OS? Python for Windows, Linux/UNIX, Mac OS X, Other
Want to help test development versions of Python? Pre-releases
Looking for a specific release?
Python releases by version number:
Release version
Release date
Click for more

图 4.1

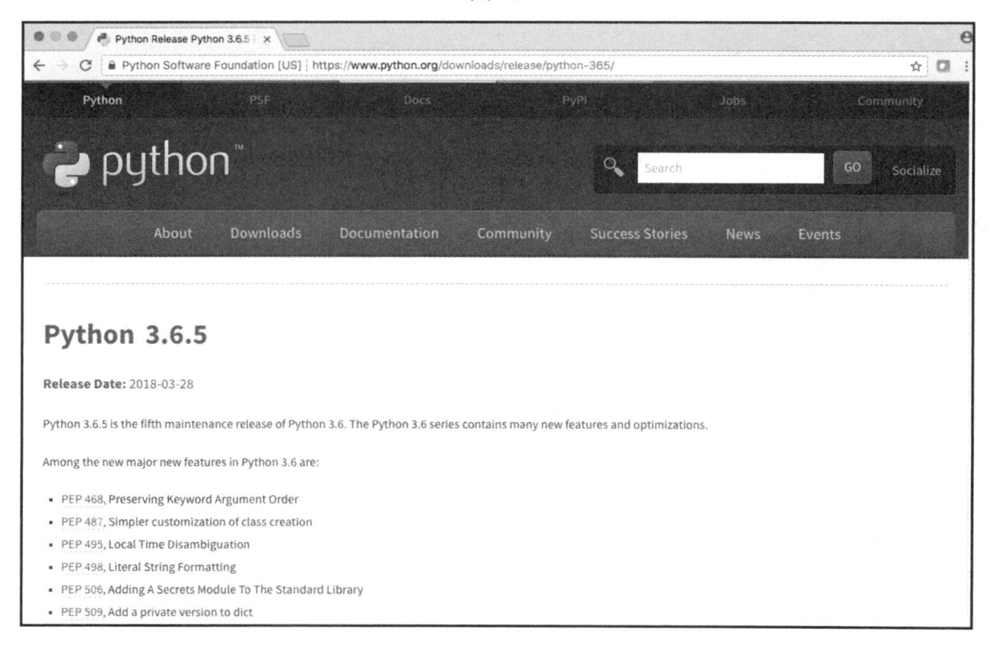
Python Release Python 3.6.5
Python Software Foundation [US] | https://www.python.org/downloads/release/python-365/
Python
PSF
Docs
PyPI
Jobs
Community
python
Search
GO
Socialize
About
Downloads
Documentation
Community
Success Stories
News
Events
Python 3.6.5
Release Date: 2018-03-28
Python 3.6.5 is the fifth maintenance release of Python 3.6. The Python 3.6 series contains many new features and optimizations.
Among the new major new features in Python 3.6 are:
PEP 468, Preserving Keyword Argument Order
PEP 487, Simpler customization of class creation
PEP 495, Local Time Disambiguation
PEP 498, Literal String Formatting
PEP 506, Adding A Secrets Module To The Standard Library
PEP 509, Add a private version to dict

图 4.2

读者可阅读相应的版本信息，并下载对应的 Python 版本，如图 4.3 所示。

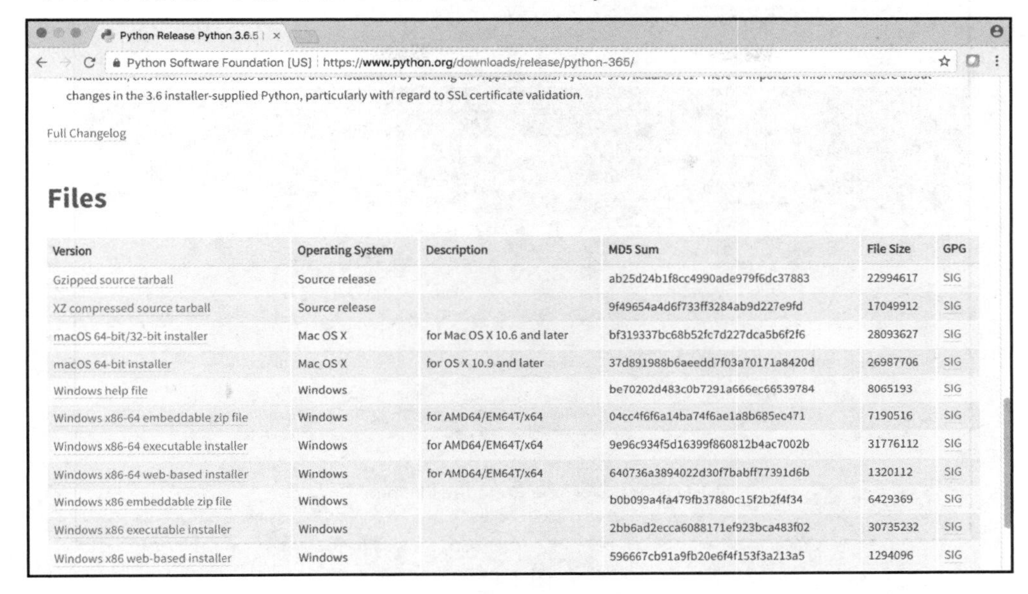

Version	Operating System	Description	MD5 Sum	File Size	GPG
Gzipped source tarball	Source release		ab25d24b1f8cc4990ade979f6dc37883	22994617	SIG
XZ compressed source tarball	Source release		9f49654a4d6f733ff3284ab9d227e9fd	17049912	SIG
macOS 64-bit/32-bit installer	Mac OS X	for Mac OS X 10.6 and later	bf319337bc68b52fc7d227dca5b6f2f6	28093627	SIG
macOS 64-bit installer	Mac OS X	for OS X 10.9 and later	37d891988b6aeedd7f03a70171a8420d	26987706	SIG
Windows help file	Windows		be70202d483c0b7291a666ec66539784	8065193	SIG
Windows x86-64 embeddable zip file	Windows	for AMD64/EM64T/x64	04cc4f6f6a14ba74f6ae1a8b685ec471	7190516	SIG
Windows x86-64 executable installer	Windows	for AMD64/EM64T/x64	9e96c934f5d16399f860812b4ac7002b	31776112	SIG
Windows x86-64 web-based installer	Windows	for AMD64/EM64T/x64	640736a3894022d30f7babff77391d6b	1320112	SIG
Windows x86 embeddable zip file	Windows		b0b099a4fa479fb37880c15f2b2f4f34	6429369	SIG
Windows x86 executable installer	Windows		2bb6ad2ecca6088171ef923bca483f02	30735232	SIG
Windows x86 web-based installer	Windows		596667cb91a9fb20e6f4f153f3a213a5	1294096	SIG

图 4.3

对于相应的操作系统，可单击正确的版本并下载安装程序。待下载完毕后，即可在计算机设备上安装 Python。

4.1.2 安装 Anaconda

标准的 Python 安装仍存在一些局限性，因而还需要安装 Jupyter、pip 以及其他数据包。Anaconda 是一个完整的安装程序，主要用于科学计算，并包含了 Python、标准库和许多有用的第三方库。

对此，访问 https://www.anaconda.com/download/，将会看到如图 4.4 所示的下载页面。

针对当前平台，可下载相应的 Anaconda 版本，并按照 https://docs.anaconda.com/anaconda/install/中的指示进行安装。

待安装完毕后，即可打开 Anaconda Navigator（在 Windows 环境下，Anaconda Navigator 位于“开始”菜单中；在 Mac 环境下，简单地搜索该程序即可）。

提示：

在 Linux 环境下，一般可采用命令行启动 Jupyter Notebook。

例如，在 Mac 环境下，Anaconda Navigator 如图 4.5 所示。

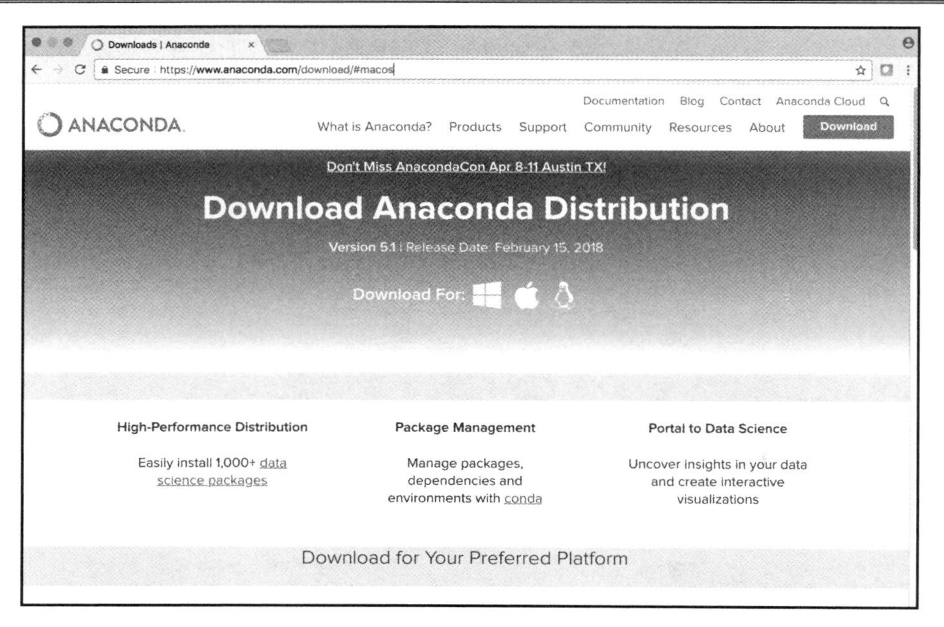
Downloads | Anaconda
Secure https://www.anaconda.com/download/#macos
Documentation Blog Contact Anaconda Cloud
ANACONDA.
What is Anaconda? Products Support Community Resources About Download
Don't Miss AnacondaCon Apr 8-11 Austin TX!
Download Anaconda Distribution
Version 5.1 | Release Date: February 15, 2018
Download For:
High-Performance Distribution
Easily install 1,000+ data science packages
Package Management
Manage packages, dependencies and environments with conda
Portal to Data Science
Uncover insights in your data and create interactive visualizations
Download for Your Preferred Platform

图 4.4

Anaconda Navigator
ANACONDA NAVIGATOR
Upgrade Now
Sign in to Anaconda Cloud
Home
Environments
Learning
Community
Applications on
base (root)
Channels
Refresh
jupyterlab
0.31.12
An extensible environment for interactive and reproducible computing, based on the Jupyter Notebook and Architecture.
Launch
notebook
5.4.1
Web-based, interactive computing notebook environment. Edit and run human-readable docs while describing the data analysis.
Launch
qtconsole
4.3.1
PyQt GUI that supports inline figures, proper multiline editing with syntax highlighting, graphical calltips, and more.
Launch
spyder
3.2.8
Scientific PYthon Development EnviRonment. Powerful Python IDE with advanced editing, interactive testing, debugging and introspection features
Launch
vscode
1.22.1
Streamlined code editor with support for development operations like debugging, task running and version control.
Launch
glueviz
0.12.4
Multidimensional data visualization across files. Explore relationships within and among related datasets.
Install
Documentation
Developer Blog
Feedback

图 4.5

在使用 Anaconda Navigator 时，可简单地单击 Jupyter Notebook 中的 Launch 按钮并启动 Jupyter，如图 4.6 所示。

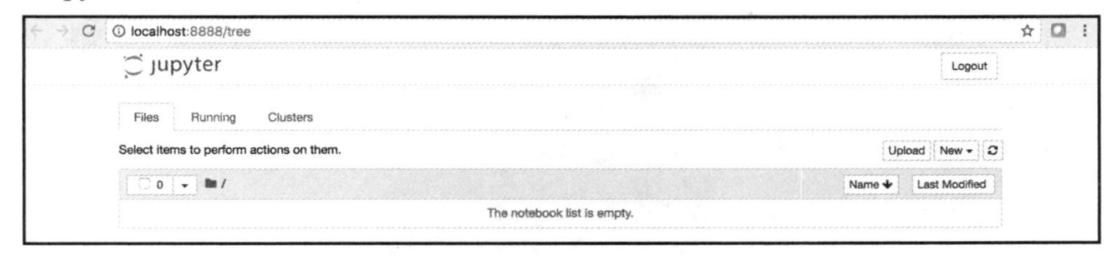

图 4.6

Conda 命令是目前最为简单、实用的工具，并可成功地对 Python 进行设置。Conda 支持协同存在的多种环境，因而可同时设置 Python 2.7 和 Python 3.6 环境。对于深度学习来说，还可将 TensorFlow 设置为独立环境，等等。

提示：

读者可访问 https://conda.io/docs/user-guide/install/index. html 下载 Conda。

Conda 下载页面如图 4.7 所示。

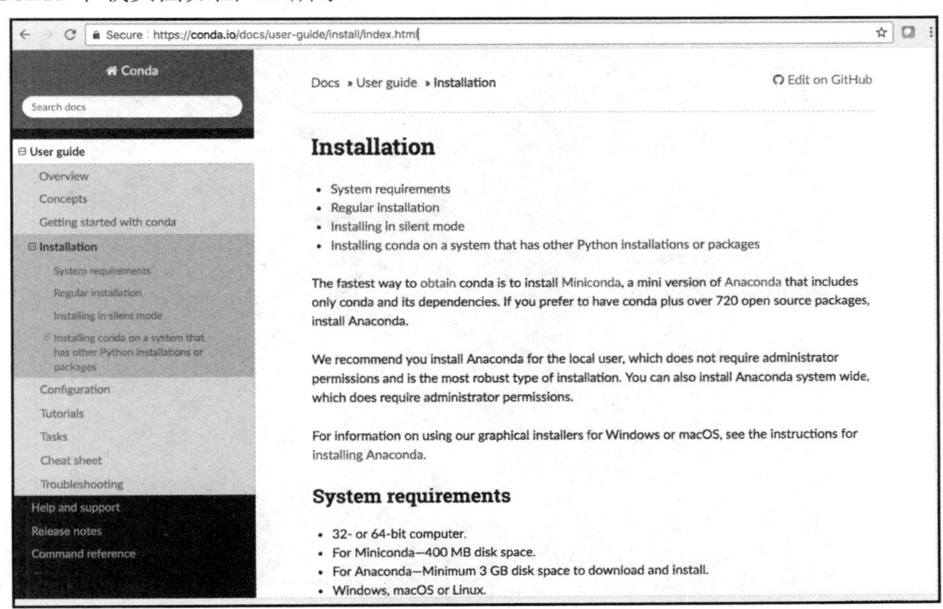

图 4.7

待 Conda 下载完毕后，可遵循对应的指令将 Conda 安装至本地机器上，如图 4.8 所示。

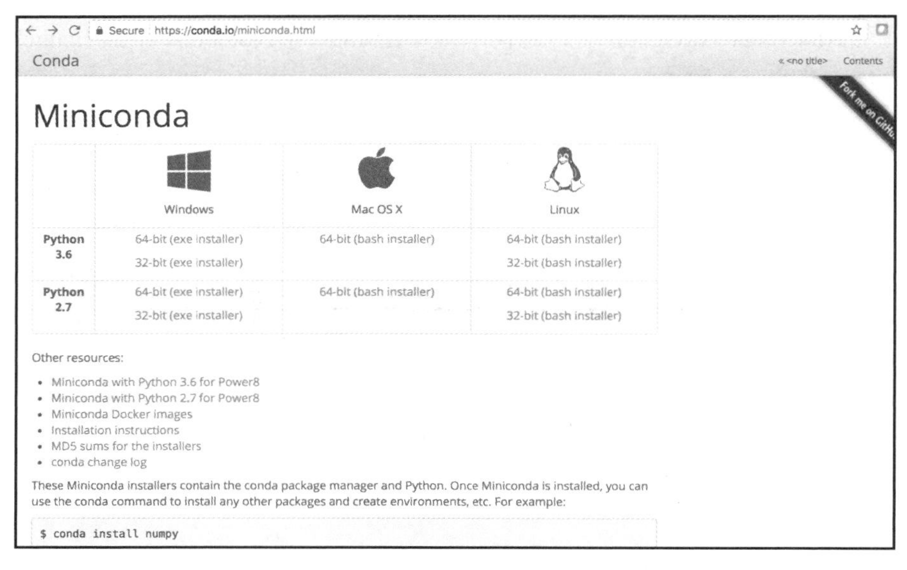

图 4.8

在命令行中输入 conda list 将显示所安装的数据包，进而可查看安装包的版本，如图 4.9 所示。

```
sridhars-MacBook-Pro-2:~ sridharalla$ conda list
# packages in environment at /anaconda2:
#
_ipyw_jlab_nb_ext_conf    0.1.0           py27h172cb35_0
affine                    2.2.0                     py_0    conda-forge
alabaster                 0.7.10                  py27_1    conda-forge
anaconda                  custom          py27h2cfa9e9_0
anaconda-client           1.6.14                    py_0    conda-forge
anaconda-navigator        1.8.1                   py27_0
anaconda-project          0.8.2                   py27_0    conda-forge
appnope                   0.1.0                   py27_0    conda-forge
appscript                 1.0.1                   py27_0    conda-forge
asn1crypto                0.24.0                  py27_0    conda-forge
astroid                   1.6.2                   py27_0    conda-forge
astropy                   2.0.5                   py27_1    conda-forge
attrs                     17.4.0                    py_0    conda-forge
babel                     2.5.3                   py27_0    conda-forge
backports                 1.0                     py27_1    conda-forge
backports.functools_lru_cache 1.5                   py27_0    conda-forge
backports.shutil_get_terminal_size 1.0.0                   py_3    conda-forge
backports_abc             0.5                     py27_0    conda-forge
beautifulsoup4            4.6.0                   py27_0    conda-forge
bitarray                  0.8.1                   py27_0    conda-forge
bkcharts                  0.2                     py27_0    conda-forge
blas                      1.1                   openblas    conda-forge
blaze                     0.11.3                  py27_0    conda-forge
bleach                    2.1.3                     py_0    conda-forge
blinker                   1.4                       py_0    conda-forge
bokeh                     0.12.14                 py27_0    conda-forge
```

图 4.9

基于 conda 命令的数据包的安装过程较为简单，可采用 conda install <packagename>

这一形式。

例如，当输入下列命令时

```
conda install scikit-learn
```

输出结果如图 4.10 所示。

```
sridhars-MacBook-Pro-2:~ sridharalla$ conda install scikit-learn
Fetching package metadata ............
Solving package specifications: .

Package plan for installation in environment /anaconda2:

The following packages will be UPDATED:

    conda: 4.3.34-py27_0 conda-forge --> 4.5.1-py27_0 conda-forge

Proceed ([y]/n)? y

conda-4.5.1-py 100% |###########################################################################################| Time: 0:00:00   1.73 MB/s
```

图 4.10

需要注意的是，conda install Jupyter 将安装 Jupyter Notebook，该过程需要使用到其他数据包，如图 4.11 所示。

```
    - jupyter

The following packages will be downloaded:

    package                    |            build
    ---------------------------|-----------------
    libxml2-2.9.8              |       hf84eae3_0         2.0 MB
    backports_abc-0.5          |   py27h7b3c97b_0          12 KB
    jsonschema-2.6.0           |   py27h7ed5aa4_0          61 KB
    markupsafe-1.0             |   py27h97b2822_1          24 KB
    notebook-5.4.1             |          py27_0         6.6 MB
    libpng-1.6.34              |       hb9fc6fc_0         334 KB
    bleach-2.1.3               |          py27_0          32 KB
    ca-certificates-2018.03.07 |                0         124 KB
    decorator-4.3.0            |          py27_0          15 KB
    xz-5.2.3                   |       h55aa19d_2         357 KB
    gstreamer-1.14.0           |       hb453b48_1         3.8 MB
    pyzmq-17.0.0               |   py27h14c3975_1         440 KB
    jupyter_console-5.2.0      |   py27hc6bee7e_1          34 KB
    icu-58.2                   |       h9c2bf20_1        22.5 MB
    ipython-5.6.0              |          py27_0        1020 KB
    pandocfilters-1.4.2        |   py27h428e1e5_1          12 KB
    tornado-5.0.1              |          py27_1         617 KB
    prompt_toolkit-1.0.15      |   py27h1b593e1_0         333 KB
    sip-4.19.8                 |   py27hf484d3e_0         291 KB
    ptyprocess-0.5.2           |   py27h4ccb14c_0          22 KB
    widgetsnbextension-3.2.0   |          py27_0         1.7 MB
    openssl-1.0.2o             |       h20670df_0         3.4 MB
    simplegeneric-0.8.1        |          py27_2           9 KB
    freetype-2.8               |       hab7d2ae_1         804 KB
    pickleshare-0.7.4          |   py27h09770e1_0          11 KB
    pygments-2.2.0             |   py27h4a8b6f5_0         1.3 MB
    libsodium-1.0.16           |       h1bed415_0         302 KB
    gst-plugins-base-1.14.0    |       hbbd80ab_1         6.3 MB
    zeromq-4.2.5               |       h439df22_0         567 KB
```

图 4.11

下面尝试使用另一个较为重要的命令，如下所示：

```
conda install pandas
```

对应输出结果如图 4.12 所示。

```
[root@4b726275a804 /]# conda install pandas
Solving environment: done

## Package Plan ##

  environment location: /root/miniconda2

  added / updated specs:
    - pandas


The following packages will be downloaded:

    package                    |            build
    ---------------------------|-----------------
    intel-openmp-2018.0.0      |                8         620 KB
    mkl_fft-1.0.1              |   py27h3010b51_0         137 KB
    pytz-2018.4                |           py27_0         211 KB
    mkl-2018.0.2               |                1       205.2 MB
    pandas-0.22.0              |   py27hf484d3e_0        10.5 MB
    mkl_random-1.0.1           |   py27h629b387_0         361 KB
    libgfortran-ng-7.2.0       |       hdf63c60_3         1.2 MB
    numpy-1.14.2               |   py27hdbf6ddf_1         4.1 MB
    ------------------------------------------------------------
                                           Total:       222.3 MB

The following NEW packages will be INSTALLED:

    intel-openmp:   2018.0.0-8
    libgfortran-ng: 7.2.0-hdf63c60_3
    mkl:            2018.0.2-1
    mkl_fft:        1.0.1-py27h3010b51_0
    mkl_random:     1.0.1-py27h629b387_0
    numpy:          1.14.2-py27hdbf6ddf_1
    pandas:         0.22.0-py27hf484d3e_0
    pytz:           2018.4-py27_0
```

图 4.12

其他较为重要的数据包还包括：

```
conda install scikit-learn
conda install matplotlib
conda install seaborn
```

除此之外，还需要安装其他数据包，进而访问 HDFS（Hadoop）以及打开文件，如下所示：

```
pip install hdfs
pip install pyarrow
```

通过运行下列命令，可生成 Jupyter Notebook 配置内容：

```
[root@4b726275a804 /]# jupyter notebook --generate-config
 Writing default config to: /root/.jupyter/jupyter_notebook_config.py
```

Jupyter 需要身份验证，即默认的令牌。然而，如果需要创建基于密码的验证，可运行下列命令创建相应的密码：

```
[root@4b726275a804 /]# jupyter notebook password
 Enter password:
 Verify password:
 [NotebookPasswordApp] Wrote hashed password to
/root/.jupyter/jupyter_notebook_config.json
```

输入下列命令可启动 Jupyter Notebook：

```
jupyter notebook --allow-root --no-browser --ip=* --port=8888
```

该命令运行时，将显示如图 4.13 所示的内容。

```
[root@4b726275a804 /]# jupyter notebook --allow-root --no-browser --ip=* --port=8888
[W 02:43:01.935 NotebookApp] WARNING: The notebook server is listening on all IP addresses and not using encryption. This is not recommended.
[I 02:43:01.955 NotebookApp] Serving notebooks from local directory: /
[I 02:43:01.955 NotebookApp] 0 active kernels
[I 02:43:01.955 NotebookApp] The Jupyter Notebook is running at:
[I 02:43:01.955 NotebookApp] http://[all ip addresses on your system]:8888/
[I 02:43:01.955 NotebookApp] Use Control-C to stop this server and shut down all kernels (twice to skip confirmation).
```

图 4.13

当访问 localhost:8888 时，浏览器将开启登录页面，并输入之前创建的密码，如图 4.14 所示。

图 4.14

在输入密码后，Jupyter Notebook 将显示现有的 Notebook。当前尚不存在相关 Notebook，下面将对此予以创建。对此，单击 New 按钮，并对即将创建的 Notebook 选择 Python 2，如图 4.15 所示。

图 4.16 显示了新创建的 Notebook，并可于其中输入某些测试代码。

至此，我们安装了 Python 和 Jupyter Notebook，下面将以此进行数据分析。在 4.2 节中，我们将深入考查不同的数据分析类型。

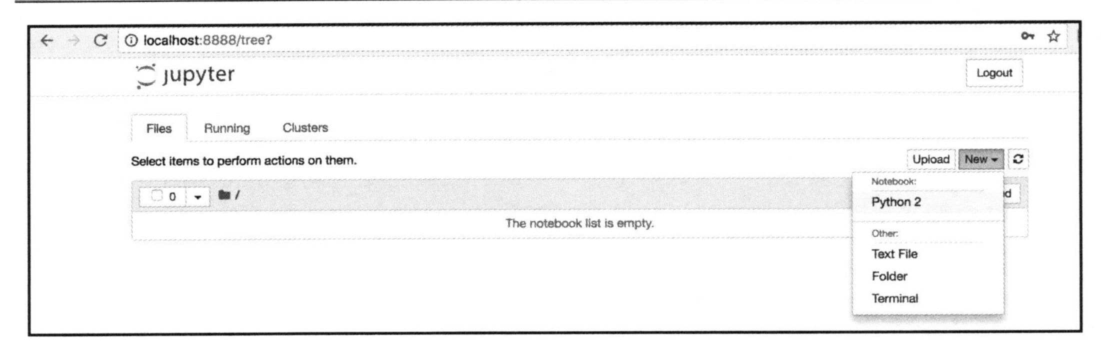

图 4.15

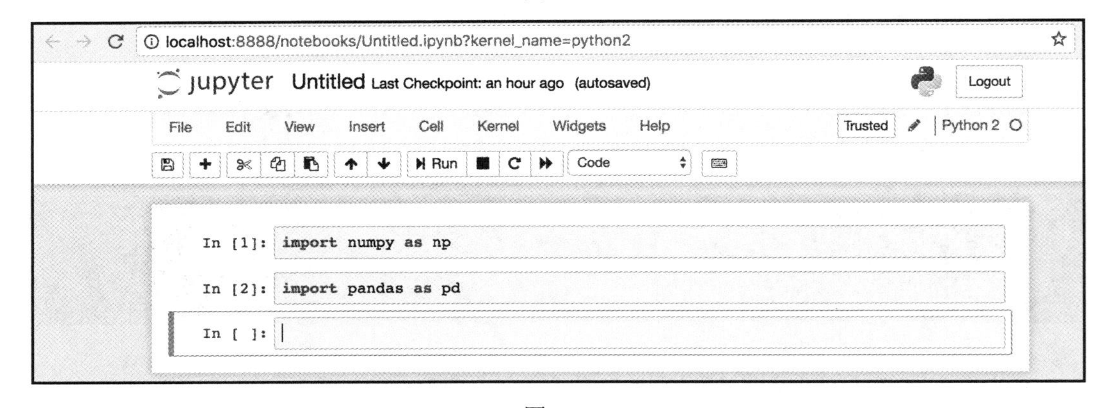

图 4.16

4.2　数据分析

在本书辅助网站中下载 OnlineRetail.csv 文件，随后利用 Pandas 加载文件。

基于 Pandas 的本地文件读取操作如下所示：

```
import pandas as pd
path = '/Users/sridharalla/Documents/OnlineRetail.csv'
df = pd.read_csv(path)
```

然而，考虑到我们将在 Hadoop 集群中分析数据，因而应使用 hdfs，而不是本地系统。下列代码显示了如何将 hdfs 文件加载至 pandas DataFrame 中：

```
import pandas as pd
from hdfs import InsecureClient
client_hdfs = InsecureClient('http://localhost:9870')
```

```
with client_hdfs.read('/user/normal/OnlineRetail.csv', encoding = 'utf-8')
as reader:
 df = pd.read_csv(reader,index_col=0)
```

输入下列命令：

```
df.head(3)
```

对应输出结果如图 4.17 所示。

In [11]: df.head(3)

Out[11]:

	InvoiceNo	StockCode	Description	Quantity	InvoiceDate	UnitPrice	CustomerID	Country
0	536365	85123A	WHITE HANGING HEART T-LIGHT HOLDER	6	12/1/10 8:26	2.55	17850.0	United Kingdom
1	536365	71053	WHITE METAL LANTERN	6	12/1/10 8:26	3.39	17850.0	United Kingdom
2	536365	84406B	CREAM CUPID HEARTS COAT HANGER	8	12/1/10 8:26	2.75	17850.0	United Kingdom

图 4.17

图 4.17 中显示了 DataFrame 中的前 3 项内容。

下面可对相关数据进行试验，输入下列命令：

```
len(df)
```

对应输出结果如下所示：

```
65499
```

这表示为 DataFrame 的长度（或尺寸）。具体来说，当前在整个文件中存在 65499 项条目。

执行下列操作：

```
df2 = df.loc[df.UnitPrice > 3.0]
df2.head(3)
```

此处定义了名为 df2 的新 DataFrame 并将其设置为：DataFrame 中全部数据项的单价均大于 3。

随后显示前 3 项内容，如图 4.18 所示。

	InvoiceNo	StockCode	Description	Quantity	InvoiceDate	UnitPrice	CustomerID	Country
1	536365	71053	WHITE METAL LANTERN	6	12/1/10 8:26	3.39	17850.0	United Kingdom
3	536365	84029G	KNITTED UNION FLAG HOT WATER BOTTLE	6	12/1/10 8:26	3.39	17850.0	United Kingdom
4	536365	84029E	RED WOOLLY HOTTIE WHITE HEART.	6	12/1/10 8:26	3.39	17850.0	United Kingdom

图 4.18

下列代码将选取单价大于 3 的数据索引，并将其描述内容设置于 Miscellaneous 中。随后将显示前 3 项内容，如下所示：

```
df.loc[df.UnitPrice > 3.0, ['Description']] = 'Miscellaneous'
df.head(3)
```

对应结果如图 4.19 所示。

```
In [45]: df.loc[df.UnitPrice > 3.0, ['Description']] = 'Miscellaneous'
         df.head(3)
Out[45]:
```

	InvoiceNo	StockCode	Description	Quantity	InvoiceDate	UnitPrice	CustomerID	Country
0	536365	85123A	WHITE HANGING HEART T-LIGHT HOLDER	1	12/1/10 8:26	2.55	17850.0	United Kingdom
1	536365	71053	Miscellaneous	1	12/1/10 8:26	3.39	17850.0	United Kingdom
2	536365	84406B	CREAM CUPID HEARTS COAT HANGER	1	12/1/10 8:26	2.75	17850.0	United Kingdom

图 4.19

不难发现，由于单价为 3.39 美元（大于 3），因而项目 2（对应的索引为 1）包含了 Miscellaneous 修改后的描述内容。

下列代码输出包含索引 2 的数据：

```
df.loc[2]
```

输出结果如图 4.20 所示。

```
In [55]: df.loc[2]
Out[55]: InvoiceNo                               536365
         StockCode                               84406B
         Description     CREAM CUPID HEARTS COAT HANGER
         Quantity                                     8
         InvoiceDate                       12/1/10 8:26
         UnitPrice                                 2.75
         CustomerID                               17850
         Country                         United Kingdom
         Name: 2, dtype: object
```

图 4.20

最后，还将创建 Quantity 列的示意图，如下所示：

```
df['Quantity'].plot()
```

对应输出结果如图 4.21 所示。

除此之外，还存在其他一些函数需要我们进一步地了解，例如.append()函数。

下面定义一个新的 df 对象，即 df3，我们将其设置为 df 的前 10 行并加上 df 的 200～209 行。换而言之，将 200～209 行添加至 df 的 0～9 行之后，如下所示：

```
df3 = df[0:10].append(df[200:210])
df3
```

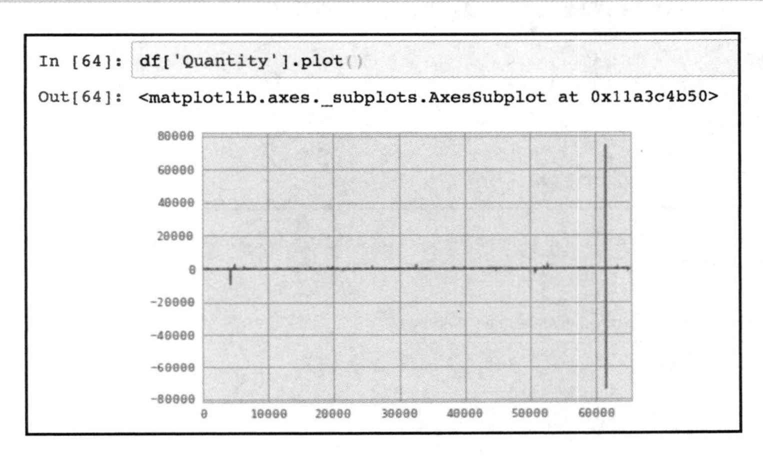

图 4.21

对应输出结果如图 4.22 所示。

```
In [35]: df3 = df[0:10].append(df[200:210])
         df3
Out[35]:
```

	InvoiceNo	StockCode	Description	Quantity	InvoiceDate	UnitPrice	CustomerID	Country
0	536365	85123A	WHITE HANGING HEART T-LIGHT HOLDER	6	12/1/10 8:26	2.55	17850.0	United Kingdom
1	536365	71053	Miscellaneous	6	12/1/10 8:26	3.39	17850.0	United States
2	536365	84406B	CREAM CUPID HEARTS COAT HANGER	8	12/1/10 8:26	2.75	17850.0	United Kingdom
3	536365	84029G	Miscellaneous	6	12/1/10 8:26	3.39	17850.0	United States
4	536365	84029E	Miscellaneous	6	12/1/10 8:26	3.39	17850.0	United States
5	536365	22752	Miscellaneous	2	12/1/10 8:26	7.65	17850.0	United States
6	536365	21730	Miscellaneous	6	12/1/10 8:26	4.25	17850.0	United States
7	536366	22633	HAND WARMER UNION JACK	6	12/1/10 8:28	1.85	17850.0	United Kingdom
8	536366	22632	HAND WARMER RED POLKA DOT	6	12/1/10 8:28	1.85	17850.0	United Kingdom
9	536367	84879	ASSORTED COLOUR BIRD ORNAMENT	32	12/1/10 8:34	1.69	13047.0	United Kingdom
200	536389	35004C	Miscellaneous	6	12/1/10 10:03	5.45	12431.0	United States
201	536389	35004G	Miscellaneous	4	12/1/10 10:03	6.35	12431.0	United States
202	536389	85014B	Miscellaneous	6	12/1/10 10:03	5.95	12431.0	United States
203	536389	85014A	Miscellaneous	3	12/1/10 10:03	5.95	12431.0	United States
204	536389	22193	Miscellaneous	2	12/1/10 10:03	8.50	12431.0	United States
205	536389	22726	Miscellaneous	4	12/1/10 10:03	3.75	12431.0	United States
206	536389	22727	Miscellaneous	4	12/1/10 10:03	3.75	12431.0	United States
207	536389	22192	Miscellaneous	2	12/1/10 10:03	8.50	12431.0	United States
208	536389	22191	Miscellaneous	2	12/1/10 10:03	8.50	12431.0	United States
209	536389	22195	LARGE HEART MEASURING SPOONS	24	12/1/10 10:03	1.65	12431.0	Australia

图 4.22

假设仅考查 StockCode、Quantity、InvoiceDate 和 UnitPrice 列，对此可定义一个新的 DataFrame 对象，且仅包含这几个数据列，如下所示：

```
df4 = pd.DataFrame(df, columns=['StockCode', 'Quantity',
'InvoiceDate','UnitPrice']
df4.head(3)
```

对应结果如图 4.23 所示。

```
In [40]: df4 = pd.DataFrame(df, columns=['StockCode', 'Quantity', 'InvoiceDate', 'UnitPrice'])
         df4.head(3)
Out[40]:
```

	StockCode	Quantity	InvoiceDate	UnitPrice
0	85123A	6	12/1/10 8:26	2.55
1	71053	6	12/1/10 8:26	3.39
2	84406B	8	12/1/10 8:26	2.75

图 4.23

Pandas 提供了不同的数据组合方式。特别地，我们可以执行合并、连接、附加操作。之前已经讨论了附加操作，下面将查看连接（concatenate）操作。

考查下列代码：

```
d1 = df[0:10]
d2 = df[10:20]

d3 = pd.concat([d1, d2])
d3
```

基本上，可将 d1 设置为 DataFrame 对象，其中包含了 df 中的前 10 个索引。随后，可将 d2 设置为 df 中后续的 10 个索引。最后，将 d3 设置为 d1 和 d2 的连接结果，即连接后的结果，如图 4.24 所示。

此外，还可指定相应的键，进而方便地区分 d1 和 d2。考查下列代码行：

```
d3 = pd.concat([d1, d2], keys=['d1', 'd2'])
```

对应结果如图 4.25 所示。

不难发现，这可方便地区分两个数据集。我们可以任意调用键，甚至像 x 和 y 这样简单的键也可以。如果定义了 3 个数据集 d1、d2、d3，则对应键表示为(x, y, z)，并以此区分这 3 个数据集。

```
In [166]: d1 = df[0:10]
          d2 = df[10:20]

          d3 = pd.concat([d1, d2])
          d3
Out[166]:
```

	InvoiceNo	StockCode	Description	Quantity	InvoiceDate	UnitPrice	CustomerID	Country
0	536365	85123A	WHITE HANGING HEART T-LIGHT HOLDER	6	12/1/10 8:26	2.55	17850.0	United Kingdom
1	536365	71053	WHITE METAL LANTERN	6	12/1/10 8:26	3.39	17850.0	United Kingdom
2	536365	84406B	CREAM CUPID HEARTS COAT HANGER	8	12/1/10 8:26	2.75	17850.0	United Kingdom
3	536365	84029G	KNITTED UNION FLAG HOT WATER BOTTLE	6	12/1/10 8:26	3.39	17850.0	United Kingdom
4	536365	84029E	RED WOOLLY HOTTIE WHITE HEART.	6	12/1/10 8:26	3.39	17850.0	United Kingdom
5	536365	22752	SET 7 BABUSHKA NESTING BOXES	2	12/1/10 8:26	7.65	17850.0	United Kingdom
6	536365	21730	GLASS STAR FROSTED T-LIGHT HOLDER	6	12/1/10 8:26	4.25	17850.0	United Kingdom
7	536366	22633	HAND WARMER UNION JACK	6	12/1/10 8:28	1.85	17850.0	United Kingdom
8	536366	22632	HAND WARMER RED POLKA DOT	6	12/1/10 8:28	1.85	17850.0	United Kingdom
9	536367	84879	ASSORTED COLOUR BIRD ORNAMENT	32	12/1/10 8:34	1.69	13047.0	United Kingdom
10	536367	22745	POPPY'S PLAYHOUSE BEDROOM	6	12/1/10 8:34	2.10	13047.0	United Kingdom
11	536367	22748	POPPY'S PLAYHOUSE KITCHEN	6	12/1/10 8:34	2.10	13047.0	United Kingdom
12	536367	22749	FELTCRAFT PRINCESS CHARLOTTE DOLL	8	12/1/10 8:34	3.75	13047.0	United Kingdom
13	536367	22310	IVORY KNITTED MUG COSY	6	12/1/10 8:34	1.65	13047.0	United Kingdom
14	536367	84969	BOX OF 6 ASSORTED COLOUR TEASPOONS	6	12/1/10 8:34	4.25	13047.0	United Kingdom
15	536367	22623	BOX OF VINTAGE JIGSAW BLOCKS	3	12/1/10 8:34	4.95	13047.0	United Kingdom
16	536367	22622	BOX OF VINTAGE ALPHABET BLOCKS	2	12/1/10 8:34	9.95	13047.0	United Kingdom
17	536367	21754	HOME BUILDING BLOCK WORD	3	12/1/10 8:34	5.95	13047.0	United Kingdom
18	536367	21755	LOVE BUILDING BLOCK WORD	3	12/1/10 8:34	5.95	13047.0	United Kingdom
19	536367	21777	RECIPE BOX WITH METAL HEART	4	12/1/10 8:34	7.95	13047.0	United Kingdom

图 4.24

```
In [179]: d1 = df[0:10]
          d2 = df[10:20]

          d3 = pd.concat([d1, d2], keys=['d1', 'd2'])
          d3
Out[179]:
```

		InvoiceNo	StockCode	Description	Quantity	InvoiceDate	UnitPrice	CustomerID	Country
d1	0	536365	85123A	WHITE HANGING HEART T-LIGHT HOLDER	6	12/1/10 8:26	2.55	17850.0	United Kingdom
	1	536365	71053	WHITE METAL LANTERN	6	12/1/10 8:26	3.39	17850.0	United Kingdom
	2	536365	84406B	CREAM CUPID HEARTS COAT HANGER	8	12/1/10 8:26	2.75	17850.0	United Kingdom
	3	536365	84029G	KNITTED UNION FLAG HOT WATER BOTTLE	6	12/1/10 8:26	3.39	17850.0	United Kingdom
	4	536365	84029E	RED WOOLLY HOTTIE WHITE HEART.	6	12/1/10 8:26	3.39	17850.0	United Kingdom
	5	536365	22752	SET 7 BABUSHKA NESTING BOXES	2	12/1/10 8:26	7.65	17850.0	United Kingdom
	6	536365	21730	GLASS STAR FROSTED T-LIGHT HOLDER	6	12/1/10 8:26	4.25	17850.0	United Kingdom
	7	536366	22633	HAND WARMER UNION JACK	6	12/1/10 8:28	1.85	17850.0	United Kingdom
	8	536366	22632	HAND WARMER RED POLKA DOT	6	12/1/10 8:28	1.85	17850.0	United Kingdom
	9	536367	84879	ASSORTED COLOUR BIRD ORNAMENT	32	12/1/10 8:34	1.69	13047.0	United Kingdom
d2	10	536367	22745	POPPY'S PLAYHOUSE BEDROOM	6	12/1/10 8:34	2.10	13047.0	United Kingdom
	11	536367	22748	POPPY'S PLAYHOUSE KITCHEN	6	12/1/10 8:34	2.10	13047.0	United Kingdom
	12	536367	22749	FELTCRAFT PRINCESS CHARLOTTE DOLL	8	12/1/10 8:34	3.75	13047.0	United Kingdom
	13	536367	22310	IVORY KNITTED MUG COSY	6	12/1/10 8:34	1.65	13047.0	United Kingdom
	14	536367	84969	BOX OF 6 ASSORTED COLOUR TEASPOONS	6	12/1/10 8:34	4.25	13047.0	United Kingdom
	15	536367	22623	BOX OF VINTAGE JIGSAW BLOCKS	3	12/1/10 8:34	4.95	13047.0	United Kingdom
	16	536367	22622	BOX OF VINTAGE ALPHABET BLOCKS	2	12/1/10 8:34	9.95	13047.0	United Kingdom
	17	536367	21754	HOME BUILDING BLOCK WORD	3	12/1/10 8:34	5.95	13047.0	United Kingdom
	18	536367	21755	LOVE BUILDING BLOCK WORD	3	12/1/10 8:34	5.95	13047.0	United Kingdom
	19	536367	21777	RECIPE BOX WITH METAL HEART	4	12/1/10 8:34	7.95	13047.0	United Kingdom

图 4.25

下面考查基于不同列的连接操作。默认状态下，concat()函数使用外连接，这意味着将组合所有的列。考查两个数据集 A 和 B，其中，集合 A 包含了隶属于 d1 中的所有列名；集合 B 包含了隶属于 d2 的所有列名。如果采用之前的代码连接 d1 和 d2，列将通过 A 和 B 的并集加以表示。

除此之外，还可指定使用内连接，即 A 和 B 的交集。考查下列代码行：

```
d4 = pd.DataFrame(df, columns=['InvoiceNo', 'StockCode',
'Description'])[0:10]
d5 = pd.DataFrame(df, columns=['StockCode', 'Description',
'Quantity'])[0:10]

pd.concat([d4, d5])
```

对应结果如图 4.26 所示。

```
In [183]: d4 = pd.DataFrame(df, columns=['InvoiceNo', 'StockCode', 'Description'])[0:10]
          d5 = pd.DataFrame(df, columns=['StockCode', 'Description', 'Quantity'])[0:10]

          pd.concat([d4, d5])
Out[183]:
```

	Description	InvoiceNo	Quantity	StockCode
0	WHITE HANGING HEART T-LIGHT HOLDER	536365	NaN	85123A
1	WHITE METAL LANTERN	536365	NaN	71053
2	CREAM CUPID HEARTS COAT HANGER	536365	NaN	84406B
3	KNITTED UNION FLAG HOT WATER BOTTLE	536365	NaN	84029G
4	RED WOOLLY HOTTIE WHITE HEART.	536365	NaN	84029E
5	SET 7 BABUSHKA NESTING BOXES	536365	NaN	22752
6	GLASS STAR FROSTED T-LIGHT HOLDER	536365	NaN	21730
7	HAND WARMER UNION JACK	536366	NaN	22633
8	HAND WARMER RED POLKA DOT	536366	NaN	22632
9	ASSORTED COLOUR BIRD ORNAMENT	536367	NaN	84879
0	WHITE HANGING HEART T-LIGHT HOLDER	NaN	6.0	85123A
1	WHITE METAL LANTERN	NaN	6.0	71053
2	CREAM CUPID HEARTS COAT HANGER	NaN	8.0	84406B
3	KNITTED UNION FLAG HOT WATER BOTTLE	NaN	6.0	84029G
4	RED WOOLLY HOTTIE WHITE HEART.	NaN	6.0	84029E
5	SET 7 BABUSHKA NESTING BOXES	NaN	2.0	22752
6	GLASS STAR FROSTED T-LIGHT HOLDER	NaN	6.0	21730
7	HAND WARMER UNION JACK	NaN	6.0	22633
8	HAND WARMER RED POLKA DOT	NaN	6.0	22632
9	ASSORTED COLOUR BIRD ORNAMENT	NaN	32.0	84879

图 4.26

通过观察可知，其中使用了全部列标记。

如前所述，默认状态下，concat()函数使用外连接。因此，pd.concat([d4, d5])等同于：

```
pd.concat([d4, d5], join='outer')
```

下面尝试使用内连接。保持其他内容保持不变，仅修改 concat()函数调用，如下所示：

```
pd.concat([d4, d5], join='inner')
```

对应输出结果如图 4.27 所示。

```
In [184]: d4 = pd.DataFrame(df, columns=['InvoiceNo', 'StockCode', 'Description'])[0:10]
          d5 = pd.DataFrame(df, columns=['StockCode', 'Description', 'Quantity'])[0:10]

          pd.concat([d4, d5], join = 'inner')
Out[184]:
```

	StockCode	Description
0	85123A	WHITE HANGING HEART T-LIGHT HOLDER
1	71053	WHITE METAL LANTERN
2	84406B	CREAM CUPID HEARTS COAT HANGER
3	84029G	KNITTED UNION FLAG HOT WATER BOTTLE
4	84029E	RED WOOLLY HOTTIE WHITE HEART.
5	22752	SET 7 BABUSHKA NESTING BOXES
6	21730	GLASS STAR FROSTED T-LIGHT HOLDER
7	22633	HAND WARMER UNION JACK
8	22632	HAND WARMER RED POLKA DOT
9	84879	ASSORTED COLOUR BIRD ORNAMENT
0	85123A	WHITE HANGING HEART T-LIGHT HOLDER
1	71053	WHITE METAL LANTERN
2	84406B	CREAM CUPID HEARTS COAT HANGER
3	84029G	KNITTED UNION FLAG HOT WATER BOTTLE
4	84029E	RED WOOLLY HOTTIE WHITE HEART.
5	22752	SET 7 BABUSHKA NESTING BOXES
6	21730	GLASS STAR FROSTED T-LIGHT HOLDER
7	22633	HAND WARMER UNION JACK
8	22632	HAND WARMER RED POLKA DOT
9	84879	ASSORTED COLOUR BIRD ORNAMENT

图 4.27

可以看到，这一次仅包含了 d4 和 d5 所共有的列标记。再次说明，我们可以添加键，进而可方便地区分两个表中的数据集。

合并过程则稍显复杂。此处，我们可以选择外部连接、内部连接、左连接和右连接；除此之外，还可选取所合并的列。

下面修改 d4 和 d5 的原始定义，如下所示：

```
d4 = pd.DataFrame(df, columns=['InvoiceNo', 'StockCode',
'Description'])[0:11]
d5 = pd.DataFrame(df, columns=['StockCode', 'Description',
'Quantity'])[10:20]
```

其中，d4 结尾处的中括号表示特定 DataFrame 的前 11 个元素；d5 结尾处的中括号表示，将 10～20 的元素置于 d5 中，而不是全部元素。

值得注意的是，其中包含了重叠的元素，且很快即会发挥其功效。

下面首先讨论 merge()函数，并执行 d4 和 d5 的左连接，如下所示：

```
pd.merge(d4, d5, how='left')
```

对应结果如图 4.28 所示。

```
In [222]: pd.merge(d4, d5, how='left')

Out[222]:
```

	InvoiceNo	StockCode	Description	Quantity
0	536365	85123A	WHITE HANGING HEART T-LIGHT HOLDER	NaN
1	536365	71053	WHITE METAL LANTERN	NaN
2	536365	84406B	CREAM CUPID HEARTS COAT HANGER	NaN
3	536365	84029G	KNITTED UNION FLAG HOT WATER BOTTLE	NaN
4	536365	84029E	RED WOOLLY HOTTIE WHITE HEART.	NaN
5	536365	22752	SET 7 BABUSHKA NESTING BOXES	NaN
6	536365	21730	GLASS STAR FROSTED T-LIGHT HOLDER	NaN
7	536366	22633	HAND WARMER UNION JACK	NaN
8	536366	22632	HAND WARMER RED POLKA DOT	NaN
9	536367	84879	ASSORTED COLOUR BIRD ORNAMENT	NaN
10	536367	22745	POPPY'S PLAYHOUSE BEDROOM	6.0

图 4.28

此处使用了 d4-d5 对中左 DataFrame 的所有列，并将 d5 的列添加至其中。可以看到，由于定义了 10～20 元素，因而不存在 0～10 索引对应的量值。然而，元素 11 均位于 d5 和 d4 中，因而可以看到 Quantity 下对应的数据值。

类似地，可针对右连接执行相同的操作，如下所示：

```
pd.merge(d4, d5, how='right')
```

对应结果如图 4.29 所示。

In [223]: pd.merge(d4, d5, how='right')

Out[223]:

	InvoiceNo	StockCode	Description	Quantity
0	536367	22745	POPPY'S PLAYHOUSE BEDROOM	6
1	NaN	22748	POPPY'S PLAYHOUSE KITCHEN	6
2	NaN	22749	FELTCRAFT PRINCESS CHARLOTTE DOLL	8
3	NaN	22310	IVORY KNITTED MUG COSY	6
4	NaN	84969	BOX OF 6 ASSORTED COLOUR TEASPOONS	6
5	NaN	22623	BOX OF VINTAGE JIGSAW BLOCKS	3
6	NaN	22622	BOX OF VINTAGE ALPHABET BLOCKS	2
7	NaN	21754	HOME BUILDING BLOCK WORD	3
8	NaN	21755	LOVE BUILDING BLOCK WORD	3
9	NaN	21777	RECIPE BOX WITH METAL HEART	4

图 4.29

当前使用了 d5 中的列标记，以及 d5 中的数据（跨越元素 10～20）。通过观察可知，索引 0 处的数据与 d4 实现了共享，因而结束于当前特定表中，其原因在于，元素 11（索引为 10）与 d5 的第一个元素重叠。

下面执行内部连接，如下所示：

```
pd.merge(d4, d5, how='inner')
```

对应结果如图 4.30 所示。

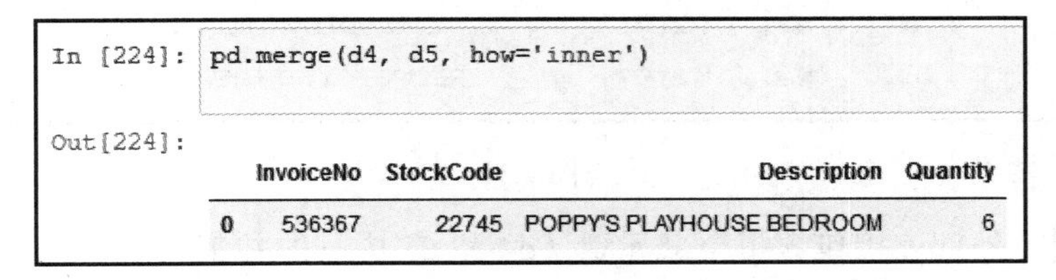

In [224]: pd.merge(d4, d5, how='inner')

Out[224]:

	InvoiceNo	StockCode	Description	Quantity
0	536367	22745	POPPY'S PLAYHOUSE BEDROOM	6

图 4.30

内部连接意味着它只包含两个 DataFrames 共有的元素。在当前示例中，对应元素表示为元素 11，其索引为 df 中的索引 10。由于该元素位于 d4 和 d5 中，因而包含了 InvoiceNo

和 Quantity 的数据（InvoiceNo 的数据位于 d4 中；Quantity 的数据位于 d5 中）。

下面执行外部连接，如下所示：

```
pd.merge(d4, d5, how='outer')
```

对应结果如图 4.31 所示。

```
In [225]: pd.merge(d4, d5, how='outer')
Out[225]:
```

	InvoiceNo	StockCode	Description	Quantity
0	536365	85123A	WHITE HANGING HEART T-LIGHT HOLDER	NaN
1	536365	71053	WHITE METAL LANTERN	NaN
2	536365	84406B	CREAM CUPID HEARTS COAT HANGER	NaN
3	536365	84029G	KNITTED UNION FLAG HOT WATER BOTTLE	NaN
4	536365	84029E	RED WOOLLY HOTTIE WHITE HEART.	NaN
5	536365	22752	SET 7 BABUSHKA NESTING BOXES	NaN
6	536365	21730	GLASS STAR FROSTED T-LIGHT HOLDER	NaN
7	536366	22633	HAND WARMER UNION JACK	NaN
8	536366	22632	HAND WARMER RED POLKA DOT	NaN
9	536367	84879	ASSORTED COLOUR BIRD ORNAMENT	NaN
10	536367	22745	POPPY'S PLAYHOUSE BEDROOM	6.0
11	NaN	22748	POPPY'S PLAYHOUSE KITCHEN	6.0
12	NaN	22749	FELTCRAFT PRINCESS CHARLOTTE DOLL	8.0
13	NaN	22310	IVORY KNITTED MUG COSY	6.0
14	NaN	84969	BOX OF 6 ASSORTED COLOUR TEASPOONS	6.0
15	NaN	22623	BOX OF VINTAGE JIGSAW BLOCKS	3.0
16	NaN	22622	BOX OF VINTAGE ALPHABET BLOCKS	2.0
17	NaN	21754	HOME BUILDING BLOCK WORD	3.0
18	NaN	21755	LOVE BUILDING BLOCK WORD	3.0
19	NaN	21777	RECIPE BOX WITH METAL HEART	4.0

图 4.31

不难发现，外部连接包含了全部元素（d4 和 d5 中列的并集）。

任何不存在的数据值均标记为 NaN。例如，d5 中不存在标记为 InvoiceNo 的列，因而此类数据值均显示为 NaN。

下面讨论列上的连接。对此，可在函数调用中引入新的参数“on=”。下列代码显示了 StockCode 列上的合并操作：

```
pd.merge(d4, d5, on='StockCode', how='left')
```

对应结果如图 4.32 所示。

```
In [227]: pd.merge(d4, d5, on='StockCode', how='left')

Out[227]:
```

	InvoiceNo	StockCode	Description_x	Description_y	Quantity
0	536365	85123A	WHITE HANGING HEART T-LIGHT HOLDER	NaN	NaN
1	536365	71053	WHITE METAL LANTERN	NaN	NaN
2	536365	84406B	CREAM CUPID HEARTS COAT HANGER	NaN	NaN
3	536365	84029G	KNITTED UNION FLAG HOT WATER BOTTLE	NaN	NaN
4	536365	84029E	RED WOOLLY HOTTIE WHITE HEART.	NaN	NaN
5	536365	22752	SET 7 BABUSHKA NESTING BOXES	NaN	NaN
6	536365	21730	GLASS STAR FROSTED T-LIGHT HOLDER	NaN	NaN
7	536366	22633	HAND WARMER UNION JACK	NaN	NaN
8	536366	22632	HAND WARMER RED POLKA DOT	NaN	NaN
9	536367	84879	ASSORTED COLOUR BIRD ORNAMENT	NaN	NaN
10	536367	22745	POPPY'S PLAYHOUSE BEDROOM	POPPY'S PLAYHOUSE BEDROOM	6.0

图 4.32

图 4.32 类似于使用左连接合并 d4 和 d5 时生成的表。但是，由于 Description 为 d4 和 d5 的共有列，因而二者均被添加进来，并通过_x 和_y 对其加以区分。

正如在最后一项中所看到的，该列由 d4 和 d5 共享，因此 Description_x 和 Description_y 彼此相同。

注意，仅可输入两个 DataFrame 共有的列名，因而可使用 StockCode 或 Description 执行合并操作。

当在 Description 列上进行合并时，对应代码如下所示：

```
pd.merge(d4, d5, on='Description', how='left')
```

对应结果如图 4.33 所示。

再次强调，通过加入_x 和_y 以表示 d4 和 d5，进而对共享列予以区分。

实际上，可传递一个列名列表，而不是单一的列名，如下所示：

```
pd.merge(d4, d5, on=['StockCode', 'Description'], how='left')
```

对应结果如图 4.34 所示。

```
In [228]: pd.merge(d4, d5, on='Description', how='left')
```

Out[228]:

	InvoiceNo	StockCode_x	Description	StockCode_y	Quantity
0	536365	85123A	WHITE HANGING HEART T-LIGHT HOLDER	NaN	NaN
1	536365	71053	WHITE METAL LANTERN	NaN	NaN
2	536365	84406B	CREAM CUPID HEARTS COAT HANGER	NaN	NaN
3	536365	84029G	KNITTED UNION FLAG HOT WATER BOTTLE	NaN	NaN
4	536365	84029E	RED WOOLLY HOTTIE WHITE HEART.	NaN	NaN
5	536365	22752	SET 7 BABUSHKA NESTING BOXES	NaN	NaN
6	536365	21730	GLASS STAR FROSTED T-LIGHT HOLDER	NaN	NaN
7	536366	22633	HAND WARMER UNION JACK	NaN	NaN
8	536366	22632	HAND WARMER RED POLKA DOT	NaN	NaN
9	536367	84879	ASSORTED COLOUR BIRD ORNAMENT	NaN	NaN
10	536367	22745	POPPY'S PLAYHOUSE BEDROOM	22745	6.0

图 4.33

```
In [229]: pd.merge(d4, d5, on=['StockCode', 'Description'], how='left')
```

Out[229]:

	InvoiceNo	StockCode	Description	Quantity
0	536365	85123A	WHITE HANGING HEART T-LIGHT HOLDER	NaN
1	536365	71053	WHITE METAL LANTERN	NaN
2	536365	84406B	CREAM CUPID HEARTS COAT HANGER	NaN
3	536365	84029G	KNITTED UNION FLAG HOT WATER BOTTLE	NaN
4	536365	84029E	RED WOOLLY HOTTIE WHITE HEART.	NaN
5	536365	22752	SET 7 BABUSHKA NESTING BOXES	NaN
6	536365	21730	GLASS STAR FROSTED T-LIGHT HOLDER	NaN
7	536366	22633	HAND WARMER UNION JACK	NaN
8	536366	22632	HAND WARMER RED POLKA DOT	NaN
9	536367	84879	ASSORTED COLOUR BIRD ORNAMENT	NaN
10	536367	22745	POPPY'S PLAYHOUSE BEDROOM	6.0

图 4.34

然而，在当前情况下，可以看到这将是同一个表，如下所示：

```
pd.merge(d4, d5, how='left')
```

在这种特殊情况下，所传递的列表包含了二者所共享的列名。如果共享了 3 个列，且我们只传递了两个列，则不会出现这种情况。

为了对此加以说明，考查下列代码：

```
d4 = pd.DataFrame(df, columns=['InvoiceNo', 'StockCode',
'Description','UnitPrice'])[0:11]
d5 = pd.DataFrame(df, columns=['StockCode', 'Description',
'Quantity','UnitPrice'])[10:20]
```

并再次执行下列代码：

```
pd.merge(d4, d5, on=['StockCode', 'Description'], how='left')
```

图 4.35 显示了当前表的状态。

```
In [232]: pd.merge(d4, d5, on=['StockCode', 'Description'], how='left')

Out[232]:
```

	InvoiceNo	StockCode	Description	UnitPrice_x	Quantity	UnitPrice_y
0	536365	85123A	WHITE HANGING HEART T-LIGHT HOLDER	2.55	NaN	NaN
1	536365	71053	WHITE METAL LANTERN	3.39	NaN	NaN
2	536365	84406B	CREAM CUPID HEARTS COAT HANGER	2.75	NaN	NaN
3	536365	84029G	KNITTED UNION FLAG HOT WATER BOTTLE	3.39	NaN	NaN
4	536365	84029E	RED WOOLLY HOTTIE WHITE HEART.	3.39	NaN	NaN
5	536365	22752	SET 7 BABUSHKA NESTING BOXES	7.65	NaN	NaN
6	536365	21730	GLASS STAR FROSTED T-LIGHT HOLDER	4.25	NaN	NaN
7	536366	22633	HAND WARMER UNION JACK	1.85	NaN	NaN
8	536366	22632	HAND WARMER RED POLKA DOT	1.85	NaN	NaN
9	536367	84879	ASSORTED COLOUR BIRD ORNAMENT	1.69	NaN	NaN
10	536367	22745	POPPY'S PLAYHOUSE BEDROOM	2.10	6.0	2.1

图 4.35

除此之外，还可指定希望所有列都予以显示，甚至是共享类。

考查下列代码：

```
pd.merge(d4, d5, left_index = True, right_index=True, how='outer')
```

可以指定任意连接类型，且仍然会显示所有列。但是，当前示例将使用外连接，如图 4.36 所示。

```
In [237]: pd.merge(d4, d5, left_index = True, right_index=True, how='outer')
```

Out[237]:

	InvoiceNo	StockCode_x	Description_x	UnitPrice_x	StockCode_y	Description_y	Quantity	UnitPrice_y
0	536365	85123A	WHITE HANGING HEART T-LIGHT HOLDER	2.55	NaN	NaN	NaN	NaN
1	536365	71053	WHITE METAL LANTERN	3.39	NaN	NaN	NaN	NaN
2	536365	84406B	CREAM CUPID HEARTS COAT HANGER	2.75	NaN	NaN	NaN	NaN
3	536365	84029G	KNITTED UNION FLAG HOT WATER BOTTLE	3.39	NaN	NaN	NaN	NaN
4	536365	84029E	RED WOOLLY HOTTIE WHITE HEART.	3.39	NaN	NaN	NaN	NaN
5	536365	22752	SET 7 BABUSHKA NESTING BOXES	7.65	NaN	NaN	NaN	NaN
6	536365	21730	GLASS STAR FROSTED T-LIGHT HOLDER	4.25	NaN	NaN	NaN	NaN
7	536366	22633	HAND WARMER UNION JACK	1.85	NaN	NaN	NaN	NaN
8	536366	22632	HAND WARMER RED POLKA DOT	1.85	NaN	NaN	NaN	NaN
9	536367	84879	ASSORTED COLOUR BIRD ORNAMENT	1.69	NaN	NaN	NaN	NaN
10	536367	22745	POPPY'S PLAYHOUSE BEDROOM	2.10	22745	POPPY'S PLAYHOUSE BEDROOM	6.0	2.10
11	NaN	NaN	NaN	NaN	22748	POPPY'S PLAYHOUSE KITCHEN	6.0	2.10
12	NaN	NaN	NaN	NaN	22749	FELTCRAFT PRINCESS CHARLOTTE DOLL	8.0	3.75
13	NaN	NaN	NaN	NaN	22310	IVORY KNITTED MUG COSY	6.0	1.65
14	NaN	NaN	NaN	NaN	84969	BOX OF 6 ASSORTED COLOUR TEASPOONS	6.0	4.25
15	NaN	NaN	NaN	NaN	22623	BOX OF VINTAGE JIGSAW BLOCKS	3.0	4.95
16	NaN	NaN	NaN	NaN	22622	BOX OF VINTAGE ALPHABET BLOCKS	2.0	9.95
17	NaN	NaN	NaN	NaN	21754	HOME BUILDING BLOCK WORD	3.0	5.95
18	NaN	NaN	NaN	NaN	21755	LOVE BUILDING BLOCK WORD	3.0	5.95
19	NaN	NaN	NaN	NaN	21777	RECIPE BOX WITH METAL HEART	4.0	7.95

图 4.36

接下来讨论 join()函数。需要注意的是，如果共享某个列名，则不可连接两个 DataFrame。因此，下列代码将不予支持：

```
d4 = pd.DataFrame(df, columns=['StockCode', 'Description',
'UnitPrice'])[0:11]
d5 = pd.DataFrame(df, columns=[ 'Description', 'Quantity',
'InvoiceNo'])[10:20]
d4.join(d5)
```

考查下列代码：

```
d4 = pd.DataFrame(df, columns=['StockCode', 'UnitPrice'])[0:11]
d5 = pd.DataFrame(df, columns=[ 'Description', 'Quantity'])[10:20]
d4.join(d5)
```

对应表如图 4.37 所示。

此处使用了 d4 表，并添加了列以及 d5 中的对应数据。由于 d5 未包含描述数据，或者索引 0～9 的量值，因而全部显示为 NaN。由于 d5 和 d4 均包含索引 10 的数据，则该元素的所有数据都显示在相应的列中。

```
In [242]: d4 = pd.DataFrame(df, columns=['StockCode', 'UnitPrice'])[0:11]
          d5 = pd.DataFrame(df, columns=[ 'Description', 'Quantity'])[10:20]
          d4.join(d5)
Out[242]:
```

	StockCode	UnitPrice	Description	Quantity
0	85123A	2.55	NaN	NaN
1	71053	3.39	NaN	NaN
2	84406B	2.75	NaN	NaN
3	84029G	3.39	NaN	NaN
4	84029E	3.39	NaN	NaN
5	22752	7.65	NaN	NaN
6	21730	4.25	NaN	NaN
7	22633	1.85	NaN	NaN
8	22632	1.85	NaN	NaN
9	84879	1.69	NaN	NaN
10	22745	2.10	POPPY'S PLAYHOUSE BEDROOM	6.0

图 4.37

此外，也可采用相反方向进行连接，如下所示：

```
d4 = pd.DataFrame(df, columns=['StockCode', 'UnitPrice'])[0:11]
d5 = pd.DataFrame(df, columns=[ 'Description', 'Quantity'])[10:20]
d5.join(d4)
```

对应结果如图 4.38 所示。

```
In [243]: d4 = pd.DataFrame(df, columns=['StockCode', 'UnitPrice'])[0:11]
          d5 = pd.DataFrame(df, columns=[ 'Description', 'Quantity'])[10:20]
          d5.join(d4)
Out[243]:
```

	Description	Quantity	StockCode	UnitPrice
10	POPPY'S PLAYHOUSE BEDROOM	6	22745	2.1
11	POPPY'S PLAYHOUSE KITCHEN	6	NaN	NaN
12	FELTCRAFT PRINCESS CHARLOTTE DOLL	8	NaN	NaN
13	IVORY KNITTED MUG COSY	6	NaN	NaN
14	BOX OF 6 ASSORTED COLOUR TEASPOONS	6	NaN	NaN
15	BOX OF VINTAGE JIGSAW BLOCKS	3	NaN	NaN
16	BOX OF VINTAGE ALPHABET BLOCKS	2	NaN	NaN
17	HOME BUILDING BLOCK WORD	3	NaN	NaN
18	LOVE BUILDING BLOCK WORD	3	NaN	NaN
19	RECIPE BOX WITH METAL HEART	4	NaN	NaN

图 4.38

除了 d4 中的列和对应的数据被添加到 d5 表中，逻辑上并无太大变化。

下面利用 combine_first()组合数据，考查下列代码：

```
d6 = pd.DataFrame.copy(df)[0:5]
d7 = pd.DataFrame.copy(df)[2:8]

d6.loc[3, ['Quantity']] = 110
d6.loc[4, ['Quantity']] = 110

d7.loc[3, ['Quantity']] = 210
d7.loc[4, ['Quantity']] = 210
pd.concat([d6, d7], keys=['d6', 'd7'])
```

在 pd.DataFrame 之后添加.copy 可确保生成原 df 的副本，而不是在原始 df 上进行编辑。通过这种方式，d6 将索引 3 和 4 的量值修改为 110 将不会对 d7 造成影响，反之亦然。需要注意的是，如果传递一个要选择的列表，那么，该方式将不会起到任何作用。因此，下列代码无法正常工作：

```
pd.DataFrame(df, columns=['Quantity', 'UnitPrice'])
```

运行上述代码后，对应表如图 4.39 所示。

```
In [59]: d6 = pd.DataFrame.copy(df)[0:5]
         d7 = pd.DataFrame.copy(df)[2:8]


         d6.loc[3, ['Quantity']] = 110
         d6.loc[4, ['Quantity']] = 110

         d7.loc[3, ['Quantity']] = 210
         d7.loc[4, ['Quantity']] = 210
         pd.concat([d6, d7], keys=['d6', 'd7'])
```

Out[59]:

		InvoiceNo	StockCode	Description	Quantity	InvoiceDate	UnitPrice	CustomerID	Country
d6	0	536365	85123A	WHITE HANGING HEART T-LIGHT HOLDER	6	12/1/10 8:26	2.55	17850.0	United Kingdom
	1	536365	71053	WHITE METAL LANTERN	6	12/1/10 8:26	3.39	17850.0	United Kingdom
	2	536365	84406B	CREAM CUPID HEARTS COAT HANGER	8	12/1/10 8:26	2.75	17850.0	United Kingdom
	3	536365	84029G	KNITTED UNION FLAG HOT WATER BOTTLE	110	12/1/10 8:26	3.39	17850.0	United Kingdom
	4	536365	84029E	RED WOOLLY HOTTIE WHITE HEART.	110	12/1/10 8:26	3.39	17850.0	United Kingdom
d7	2	536365	84406B	CREAM CUPID HEARTS COAT HANGER	8	12/1/10 8:26	2.75	17850.0	United Kingdom
	3	536365	84029G	KNITTED UNION FLAG HOT WATER BOTTLE	210	12/1/10 8:26	3.39	17850.0	United Kingdom
	4	536365	84029E	RED WOOLLY HOTTIE WHITE HEART.	210	12/1/10 8:26	3.39	17850.0	United Kingdom
	5	536365	22752	SET 7 BABUSHKA NESTING BOXES	2	12/1/10 8:26	7.65	17850.0	United Kingdom
	6	536365	21730	GLASS STAR FROSTED T-LIGHT HOLDER	6	12/1/10 8:26	4.25	17850.0	United Kingdom
	7	536366	22633	HAND WARMER UNION JACK	6	12/1/10 8:28	1.85	17850.0	United Kingdom

图 4.39

注意，d6 和 d7 包含了共有元素，即索引为 2～4 的元素。

考查下列代码：

```
d6.combine_first(d7)
```

对应结果如图 4.40 所示。

```
In [61]: d6.combine_first(d7)
Out[61]:
```

	InvoiceNo	StockCode	Description	Quantity	InvoiceDate	UnitPrice	CustomerID	Country
0	536365	85123A	WHITE HANGING HEART T-LIGHT HOLDER	6.0	12/1/10 8:26	2.55	17850.0	United Kingdom
1	536365	71053	WHITE METAL LANTERN	6.0	12/1/10 8:26	3.39	17850.0	United Kingdom
2	536365	84406B	CREAM CUPID HEARTS COAT HANGER	8.0	12/1/10 8:26	2.75	17850.0	United Kingdom
3	536365	84029G	KNITTED UNION FLAG HOT WATER BOTTLE	110.0	12/1/10 8:26	3.39	17850.0	United Kingdom
4	536365	84029E	RED WOOLLY HOTTIE WHITE HEART.	110.0	12/1/10 8:26	3.39	17850.0	United Kingdom
5	536365	22752	SET 7 BABUSHKA NESTING BOXES	2.0	12/1/10 8:26	7.65	17850.0	United Kingdom
6	536365	21730	GLASS STAR FROSTED T-LIGHT HOLDER	6.0	12/1/10 8:26	4.25	17850.0	United Kingdom
7	536366	22633	HAND WARMER UNION JACK	6.0	12/1/10 8:28	1.85	17850.0	United Kingdom

图 4.40

这样做的结果是将 d7 数据与 d6 数据相结合，但优先考虑 d6。我们在 d6 中将索引 3 和 4 的量值设置为 110。可以看到，d6 的数据被保存在两个数据集具有公共索引的地方。考查下列代码：

```
d7.combine_first(d6)
```

对应结果如图 4.41 所示。

```
In [63]: d7.combine_first(d6)
Out[63]:
```

	InvoiceNo	StockCode	Description	Quantity	InvoiceDate	UnitPrice	CustomerID	Country
0	536365	85123A	WHITE HANGING HEART T-LIGHT HOLDER	6.0	12/1/10 8:26	2.55	17850.0	United Kingdom
1	536365	71053	WHITE METAL LANTERN	6.0	12/1/10 8:26	3.39	17850.0	United Kingdom
2	536365	84406B	CREAM CUPID HEARTS COAT HANGER	8.0	12/1/10 8:26	2.75	17850.0	United Kingdom
3	536365	84029G	KNITTED UNION FLAG HOT WATER BOTTLE	210.0	12/1/10 8:26	3.39	17850.0	United Kingdom
4	536365	84029E	RED WOOLLY HOTTIE WHITE HEART.	210.0	12/1/10 8:26	3.39	17850.0	United Kingdom
5	536365	22752	SET 7 BABUSHKA NESTING BOXES	2.0	12/1/10 8:26	7.65	17850.0	United Kingdom
6	536365	21730	GLASS STAR FROSTED T-LIGHT HOLDER	6.0	12/1/10 8:26	4.25	17850.0	United Kingdom
7	536366	22633	HAND WARMER UNION JACK	6.0	12/1/10 8:28	1.85	17850.0	United Kingdom

图 4.41

图 4.41 中可以看到，两个元素具有共同的索引（在索引 3 和 4 处），d7 的数据被保存。

通过 value_counts()，还可在选择类别中获得每项数值的出现次数。考查下列代码：

```
pd.value_counts(df['Country'])
```

对应结果如图 4.42 所示。

```
In [8]:  pd.value_counts(df['Country'])

Out[8]:  United Kingdom     61186
         Germany              982
         France               967
         EIRE                 504
         Spain                355
         Portugal             212
         Netherlands          186
         Switzerland          175
         Norway               147
         Australia            142
         Belgium              142
         Italy                112
         Cyprus                99
         Japan                 69
         Sweden                41
         Lithuania             35
         Poland                33
         Iceland               31
         Denmark               20
         Channel Islands       17
         Finland               17
         Israel                16
         Austria                9
         Bahrain                2
         Name: Country, dtype: int64
```

图 4.42

在合并过程中需要考虑的一件事是：可能会遇到重复的数据值。对此，可使用.drop_duplicates()。

考查下列代码：

```
d1 = pd.DataFrame(df, columns = ['InvoiceNo', 'StockCode',
'Description'])[0:100]
d2 = pd.DataFrame(df, columns = ['Description', 'InvoiceDate',
'Quantity'])[0:100]

pd.merge(d1, d2)
```

对应结果如图 4.43 所示。

查看如图 4.44 所示的下方内容。

```
In [48]: d1 = pd.DataFrame(df, columns = ['InvoiceNo', 'StockCode', 'Description'])[0:100]
         d2 = pd.DataFrame(df, columns = ['Description', 'InvoiceDate', 'Quantity'])[0:100]

         pd.merge(d1, d2)

Out[48]:
```

	InvoiceNo	StockCode	Description	InvoiceDate	Quantity
0	536365	85123A	WHITE HANGING HEART T-LIGHT HOLDER	12/1/10 8:26	6
1	536365	85123A	WHITE HANGING HEART T-LIGHT HOLDER	12/1/10 9:02	6
2	536365	85123A	WHITE HANGING HEART T-LIGHT HOLDER	12/1/10 9:32	6
3	536373	85123A	WHITE HANGING HEART T-LIGHT HOLDER	12/1/10 8:26	6
4	536373	85123A	WHITE HANGING HEART T-LIGHT HOLDER	12/1/10 9:02	6
5	536373	85123A	WHITE HANGING HEART T-LIGHT HOLDER	12/1/10 9:32	6
6	536375	85123A	WHITE HANGING HEART T-LIGHT HOLDER	12/1/10 8:26	6
7	536375	85123A	WHITE HANGING HEART T-LIGHT HOLDER	12/1/10 9:02	6
8	536375	85123A	WHITE HANGING HEART T-LIGHT HOLDER	12/1/10 9:32	6
9	536365	71053	WHITE METAL LANTERN	12/1/10 8:26	6
10	536365	71053	WHITE METAL LANTERN	12/1/10 9:02	6
11	536365	71053	WHITE METAL LANTERN	12/1/10 9:32	6
12	536373	71053	WHITE METAL LANTERN	12/1/10 8:26	6
13	536373	71053	WHITE METAL LANTERN	12/1/10 9:02	6
14	536373	71053	WHITE METAL LANTERN	12/1/10 9:32	6
15	536375	71053	WHITE METAL LANTERN	12/1/10 8:26	6
16	536375	71053	WHITE METAL LANTERN	12/1/10 9:02	6
17	536375	71053	WHITE METAL LANTERN	12/1/10 9:32	6
18	536365	84406B	CREAM CUPID HEARTS COAT HANGER	12/1/10 8:26	8
19	536365	84406B	CREAM CUPID HEARTS COAT HANGER	12/1/10 9:02	8
20	536365	84406B	CREAM CUPID HEARTS COAT HANGER	12/1/10 9:32	8
21	536373	84406B	CREAM CUPID HEARTS COAT HANGER	12/1/10 8:26	8

图 4.43

158	536378	22386	JUMBO BAG PINK POLKADOT	12/1/10 9:37	10
159	536378	85099C	JUMBO BAG BAROQUE BLACK WHITE	12/1/10 9:37	10
160	536378	21033	JUMBO BAG CHARLIE AND LOLA TOYS	12/1/10 9:37	10
161	536378	20723	STRAWBERRY CHARLOTTE BAG	12/1/10 9:37	10
162	536378	84997B	RED 3 PIECE RETROSPOT CUTLERY SET	12/1/10 9:37	12
163	536378	84997C	BLUE 3 PIECE POLKADOT CUTLERY SET	12/1/10 9:37	6
164	536378	21094	SET/6 RED SPOTTY PAPER PLATES	12/1/10 9:37	12
165	536378	20725	LUNCH BAG RED RETROSPOT	12/1/10 9:37	10
166	536378	21559	STRAWBERRY LUNCH BOX WITH CUTLERY	12/1/10 9:37	6
167	536378	22352	LUNCH BOX WITH CUTLERY RETROSPOT	12/1/10 9:37	6
168	536378	21212	PACK OF 72 RETROSPOT CAKE CASES	12/1/10 9:37	120
169	536378	21975	PACK OF 60 DINOSAUR CAKE CASES	12/1/10 9:37	24
170	536378	21977	PACK OF 60 PINK PAISLEY CAKE CASES	12/1/10 9:37	24
171	536378	84991	60 TEATIME FAIRY CAKE CASES	12/1/10 9:37	24

172 rows × 5 columns

图 4.44

不难发现，其中存在多项的重复数据。对此，可采用 drop_duplicates()移除这些数据。另外，还可进一步指定可用的列数据，进而确定哪些数据项可被移除。例如，可利用 StockCode 移除全部复制项，此处假设每项均包含唯一的股票代码。除此之外，还可针对每项设置唯一的描述，并以这种方式删除对应的数据项。考查下列代码：

```
d1 = pd.DataFrame(df, columns = ['InvoiceNo', 'StockCode',
'Description'])[0:100]
d2 = pd.DataFrame(df, columns = ['Description', 'InvoiceDate',
'Quantity'])[0:100]

pd.merge(d1, d2).drop_duplicates(['StockCode'])
```

对应结果如图 4.45 所示。

```
In [62]: d1 = pd.DataFrame(df, columns = ['InvoiceNo', 'StockCode', 'Description'])[0:100]
         d2 = pd.DataFrame(df, columns = ['Description', 'InvoiceDate', 'Quantity'])[0:100]

         pd.merge(d1, d2).drop_duplicates(['StockCode'])
```

Out[62]:

	InvoiceNo	StockCode	Description	InvoiceDate	Quantity
0	536365	85123A	WHITE HANGING HEART T-LIGHT HOLDER	12/1/10 8:26	6
9	536365	71053	WHITE METAL LANTERN	12/1/10 8:26	6
18	536365	84406B	CREAM CUPID HEARTS COAT HANGER	12/1/10 8:26	8
27	536365	84029G	KNITTED UNION FLAG HOT WATER BOTTLE	12/1/10 8:26	6
36	536365	84029E	RED WOOLLY HOTTIE WHITE HEART.	12/1/10 8:26	6
45	536365	22752	SET 7 BABUSHKA NESTING BOXES	12/1/10 8:26	2
54	536365	21730	GLASS STAR FROSTED T-LIGHT HOLDER	12/1/10 8:26	6
63	536366	22633	HAND WARMER UNION JACK	12/1/10 8:28	6
72	536366	22632	HAND WARMER RED POLKA DOT	12/1/10 8:28	6
81	536367	84879	ASSORTED COLOUR BIRD ORNAMENT	12/1/10 8:34	32
82	536367	22745	POPPY'S PLAYHOUSE BEDROOM	12/1/10 8:34	6
83	536367	22748	POPPY'S PLAYHOUSE KITCHEN	12/1/10 8:34	6
84	536367	22749	FELTCRAFT PRINCESS CHARLOTTE DOLL	12/1/10 8:34	8
85	536367	22310	IVORY KNITTED MUG COSY	12/1/10 8:34	6
86	536367	84969	BOX OF 6 ASSORTED COLOUR TEASPOONS	12/1/10 8:34	6
87	536367	22623	BOX OF VINTAGE JIGSAW BLOCKS	12/1/10 8:34	3
88	536367	22622	BOX OF VINTAGE ALPHABET BLOCKS	12/1/10 8:34	2

图 4.45

查看如图 4.46 所示的下方内容。

从图 4.46 中可以看到，多个复制项已被移除。此外，还可传入 Description 和 StockCode，或者 Description，这将生成相同的结果。

165	536378	20725	LUNCH BAG RED RETROSPOT	12/1/10 9:37	10
166	536378	21559	STRAWBERRY LUNCH BOX WITH CUTLERY	12/1/10 9:37	6
167	536378	22352	LUNCH BOX WITH CUTLERY RETROSPOT	12/1/10 9:37	6
168	536378	21212	PACK OF 72 RETROSPOT CAKE CASES	12/1/10 9:37	120
169	536378	21975	PACK OF 60 DINOSAUR CAKE CASES	12/1/10 9:37	24
170	536378	21977	PACK OF 60 PINK PAISLEY CAKE CASES	12/1/10 9:37	24
171	536378	84991	60 TEATIME FAIRY CAKE CASES	12/1/10 9:37	24

73 rows × 5 columns

图 4.46

需要注意的是，当前，索引遍布于各处。对此，可使用 reset_index()解决此类问题。考查下列代码：

```
d1 = pd.DataFrame(df, columns = ['InvoiceNo', 'StockCode',
'Description'])[0:100]
d2 = pd.DataFrame(df, columns = ['Description', 'InvoiceDate',
'Quantity'])[0:100]

d3 = pd.merge(d1, d2).drop_duplicates(['StockCode'])
d3.reset_index()
```

对应结果如图 4.47 所示。

```
In [66]: d3.reset_index()
Out[66]:
```

	index	InvoiceNo	StockCode	Description	InvoiceDate	Quantity
0	0	536365	85123A	WHITE HANGING HEART T-LIGHT HOLDER	12/1/10 8:26	6
1	9	536365	71053	WHITE METAL LANTERN	12/1/10 8:26	6
2	18	536365	84406B	CREAM CUPID HEARTS COAT HANGER	12/1/10 8:26	8
3	27	536365	84029G	KNITTED UNION FLAG HOT WATER BOTTLE	12/1/10 8:26	6
4	36	536365	84029E	RED WOOLLY HOTTIE WHITE HEART.	12/1/10 8:26	6
5	45	536365	22752	SET 7 BABUSHKA NESTING BOXES	12/1/10 8:26	2
6	54	536365	21730	GLASS STAR FROSTED T-LIGHT HOLDER	12/1/10 8:26	6
7	63	536366	22633	HAND WARMER UNION JACK	12/1/10 8:28	6
8	72	536366	22632	HAND WARMER RED POLKA DOT	12/1/10 8:28	6
9	81	536367	84879	ASSORTED COLOUR BIRD ORNAMENT	12/1/10 8:34	32
10	82	536367	22745	POPPY'S PLAYHOUSE BEDROOM	12/1/10 8:34	6
11	83	536367	22748	POPPY'S PLAYHOUSE KITCHEN	12/1/10 8:34	6
12	84	536367	22749	FELTCRAFT PRINCESS CHARLOTTE DOLL	12/1/10 8:34	8
13	85	536367	22310	IVORY KNITTED MUG COSY	12/1/10 8:34	6
14	86	536367	84969	BOX OF 6 ASSORTED COLOUR TEASPOONS	12/1/10 8:34	6

图 4.47

最终结果与期望结果保持一致。虽然这里重置了索引，但也将旧索引作为列添加了进来。对此，存在一种简单的修复方法，即引入新的参数。考查下列代码：

```
d3.reset_index(drop=True)
```

对应结果如图 4.48 所示。

```
In [67]: d3.reset_index(drop=True)
Out[67]:
```

	InvoiceNo	StockCode	Description	InvoiceDate	Quantity
0	536365	85123A	WHITE HANGING HEART T-LIGHT HOLDER	12/1/10 8:26	6
1	536365	71053	WHITE METAL LANTERN	12/1/10 8:26	6
2	536365	84406B	CREAM CUPID HEARTS COAT HANGER	12/1/10 8:26	8
3	536365	84029G	KNITTED UNION FLAG HOT WATER BOTTLE	12/1/10 8:26	6
4	536365	84029E	RED WOOLLY HOTTIE WHITE HEART.	12/1/10 8:26	6
5	536365	22752	SET 7 BABUSHKA NESTING BOXES	12/1/10 8:26	2
6	536365	21730	GLASS STAR FROSTED T-LIGHT HOLDER	12/1/10 8:26	6
7	536366	22633	HAND WARMER UNION JACK	12/1/10 8:28	6
8	536366	22632	HAND WARMER RED POLKA DOT	12/1/10 8:28	6
9	536367	84879	ASSORTED COLOUR BIRD ORNAMENT	12/1/10 8:34	32
10	536367	22745	POPPY'S PLAYHOUSE BEDROOM	12/1/10 8:34	6
11	536367	22748	POPPY'S PLAYHOUSE KITCHEN	12/1/10 8:34	6
12	536367	22749	FELTCRAFT PRINCESS CHARLOTTE DOLL	12/1/10 8:34	8
13	536367	22310	IVORY KNITTED MUG COSY	12/1/10 8:34	6
14	536367	84969	BOX OF 6 ASSORTED COLOUR TEASPOONS	12/1/10 8:34	6
15	536367	22623	BOX OF VINTAGE JIGSAW BLOCKS	12/1/10 8:34	3
16	536367	22622	BOX OF VINTAGE ALPHABET BLOCKS	12/1/10 8:34	2
17	536367	21754	HOME BUILDING BLOCK WORD	12/1/10 8:34	3
18	536367	21755	LOVE BUILDING BLOCK WORD	12/1/10 8:34	3

图 4.48

可以看出，情况已有所改善。默认状态下，drop=False。因此，如果不希望将旧索引作为新列添加至数据中，则可设置为 drop=True。

之前曾讨论了.plot()函数，该函数有助于实现 DataFrame 的可视化效果，特别是 DataFrame 较大时。

下列代码显示了单列示例：

```
d8 = pd.DataFrame(df, columns=['Quantity'])[0:100]
d8.plot()
```

这里仅选择了前 100 个元素，以实现更好的图像显示效果以及示例解释效果，如图 4.49

所示。

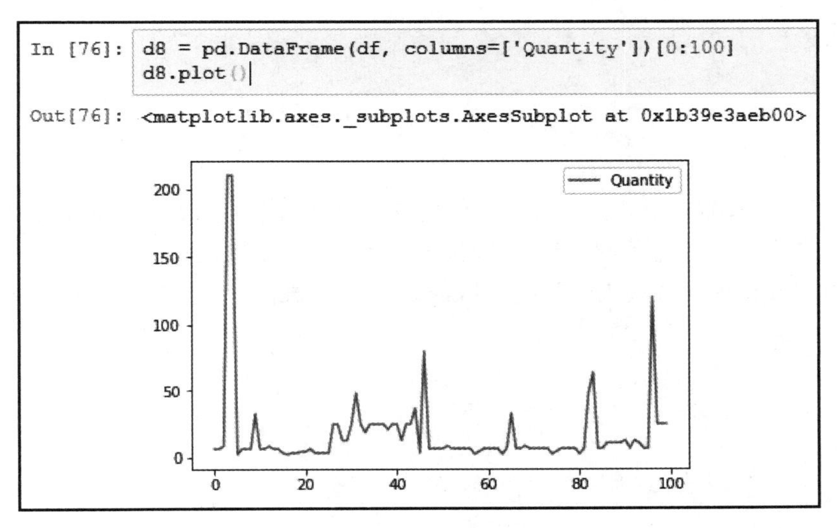

图 4.49

下列代码表示为多列显示效果：

```
d8 = pd.DataFrame(df, columns=['Quantity', 'UnitPrice'])[0:100]
d8.plot()
```

对应结果如图 4.50 所示。

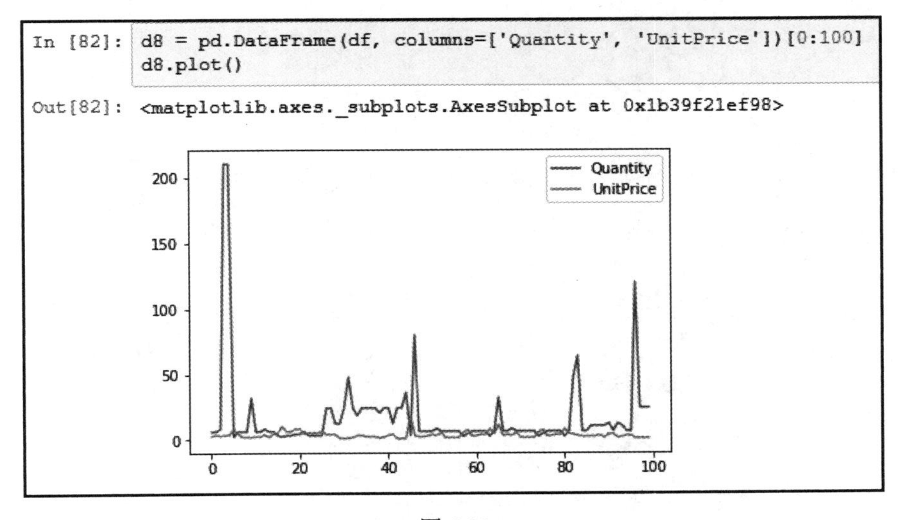

图 4.50

需要注意的是，上述代码并不会显示定性数据列，例如 Description，且仅显示 Quantity 和 UnitPrice 这一类数据。

4.3　本章小结

本章讨论了 Python 语言，以及如何使用该语言并借助于 Jupyter Notebook 进行数据分析。除此之外，本章还考查了基于 Python 语言的多种不同操作方案。

第 5 章将介绍另一种较为流行的分析语言，即 R 语言，以及如何利用 R 语言执行数据分析。

第 5 章　R-Hadoop 统计数据计算

本章讨论 R 语言及其应用方式，进而通过 Hadoop 针对大数据执行统计计算。除此之外，我们还将考查一些替代方案，例如工作站上的开源 R 系统、并行化的商业级产品，例如 Revolution R Enterprise；而在这两者间还存在许多其他选择方案，其中还会涉及数据的可伸缩性、性能、功能和易用性等问题。因此，最终的选择结果取决于数据的大小、预算、技能水平以及管理行为。

本章将探讨开源 R 的一些替代方法及其优点。另外，还将结合开源和商业技术，以实现更大规模、可靠以及易于开发的备选方案。

本章主要涉及以下主题：

- R 语言和 Hadoop 间的集成。
- R 语言与 Hadoop 间的整合方法。
- 利用 R 语言进行数据分析。

5.1　概　　述

对于 Hadoop 新手来说，本章的主要目的是帮助 R 用户理解并选择相应的评估解决方案。与大多数开源项目一样，首先要考虑的一般是资金问题。好消息是，其中存在多种免费的替代方案；同时，开源项目中提供了许多附加功能。

当利用开源栈实现 R 与 Hadoop 间的集成时，通常包含以下 4 种可选方案：

- 在工作站上安装 R，并连接 Hadoop 中的数据。
- 在共享服务器上安装 R，并连接至 Hadoop 上。
- 使用 Revolution R Open。
- 利用 RMR2 在 MapReduce 中执行 R。

下面逐一讨论每种可选方案。

5.1.1　在工作站上安装 R 并连接 Hadoop 中的数据

该方案最大的优势在于简单性和成本——该方案完全免费。利用开源组织发布的 Revolution，包括 rhdfs 和 rhbase，R 用户可以直接从 Hadoop 中的 hdfs 文件系统和 hbase

数据库子系统中摄取数据。这两个连接器都是由 Revolution 创建和维护的，同时也是 RHadoop 包中的一部分内容，并可视作一类首选方案。

除此之外，还存在一些附加选项。例如，RHive 包直接从 R 中执行 Hive SQL（类似于 SQL 查询语言），并提供了 Hive 元数据检索功能，例如数据库名、表名、列名等。特别值得一提的是，在 rhive 包中，数据操作将下推至 Hadoop 中，从而避免了数据移动和并行操作，从而大大提高了速度，这也可视为 rhive 包的一个优点。类似的下推操作还可通过 rhbase 予以实现。尽管如此，操作环境以及复杂的分析问题往往会反映出功能方面的欠缺。

除了有限的下推功能之外，R 最擅长处理从 hdfs、hbase 或 hive 中采样的少量数据；通过这种方式，R 用户可以快速使用 Hadoop。

5.1.2　在共享服务器上安装 R 并连接至 Hadoop

考虑到内存的局限性（例如在笔记本电脑上执行 R 语言任务），共享服务器是一种很自然的解决方案。借助于今天的技术，投入少量资金即可在用户间实现服务器共享。在 Windows 或 Linux 环境下（配置了 256 或 512MB 的内存空间），R 可以用于分析多达数百 GB 的文件，尽管速度上可能会存在欠缺。

类似于上述第 1 种方案，共享服务器上的 R 也可采用 rhbase 和 rhive 包提供的下推技术，实现并行机制并避免对数据进行移动。然而，类似于工作站，rhive 和 rhbase 下推技术仍存在一定的局限性。

大量的 RAM 可避免内存耗尽问题，且对计算性能不会产生负面影响（可能会需要一些技术上的支持）。基于此，共享服务器可看作是工作站 R 系统的一个很好的补充，但并非是一类替代方案。

5.1.3　利用 Revolution R Open

利用 Revolution R Open（RRO）替代 CARN 还可以进一步提升性能。类似于 R，RRO 采用 R 编写且完全开源，并可供用户免费下载。RRO 通过 Intel MATH 核心库改善了数学计算过程，同时 100%兼容于 CRAN 中的算法和其他存储库，例如 BioConductor。另外，R 脚本不需要做任何修改，而且 MKL 库针对脚本提供了不同等级的加速能力，其中使用了大量的数学和线性代数原语（primitives）。如果在该语言中执行数学运算，借助于 RPO，计算的平均性能能够成倍增长。同样，RRO 也可使用 rhdfs 这一类连接器，并通过 rhbase 和 rhive 执行 Hadoop 的连接、下推操作。

5.1.4　利用 RMR2 在 MapReduce 内执行 R

一旦发现问题集过大，或者用户的耐心在工作站或服务器上一点点地被消耗，rhbase 和 rhive 下推操作的限制正在阻碍事态的进展，则可考虑在 Hadoop 中运行 R。

开源 RHadoop 项目包含 rhdfs、rhbase、plyrmr，以及名为 rmr2 的数据包，以使 r 用户可通过 R 函数构建 Hadoop MapReduce 操作。当采用 mapper 时，R 函数可应用于 hdfs 文件、hbase 表以及其他数据集构成的所有数据块。对应结果将发送至一个 reducer 和 R 函数中，用以执行聚合或分析操作。所有工作都是在 Hadoop 中被引导的，但确实在 R 中被构建的。可以确定的是，针对每个 hdfs 文件片段使用 R 函数是一种较好的加速计算方式。但在大多数情况下，避免移动数据才是真正提高性能的因素。对此，rmr2 针对 Hadoop 节点上的数据运用了 R 函数，而不是将数据移至 R 所处的位置处。

虽然 rmr2 向数据科学家或统计人员提供了强大的功能，但用户的想法或许会很快产生变化——针对大型数据集，在 R 中针对算法执行计算过程。当通过这一方式使用 rmr2 时，对于 R 程序员来说，这将使得开发过程趋于复杂化，其原因在于，程序员需要编写算法的全部逻辑，或者适应现有的 CRAN 算法。随后，程序员还需验证算法的准确性，进而体现期望的数学结果，并针对少量场景编写代码，例如所丢失的数据。

rmr2 需要亲自编写代码对并行机制进行管理。对于数据转换操作、聚合等内容来说，该过程并不复杂；但如果打算在大型数据集上训练预测模型或构建分类器，那么，这一过程可能会相当枯燥。与其他方案相比，rmr2 在这一方面体现得尤为明显，但该方案却是真实可行的。大多数 R 程序员会发现，与基于 Java 的 Hadoop 的 mapper 和 reducer 相比，rmr2 则要容易得多。尽管如此，rmr2 仍具备以下特征：

- 完全开源。
- 有助于并行计算进而处理大型数据集。
- 可避免数据移动。
- 应用广泛。
- 完全免费。

rmr2 并非是这一领域内的唯一选择，rhive 包也可提供类似的功能。读者可访问 https://www.rhipe.com/downloadconfirmation/了解详细信息，并访问 GitHub 进行下载。

针对基于 Hadoop 的 R 系统，开源方案的范围正在不断扩大。例如，Apache Apark 社区通过 SparkR 改进了 R 集成方案。SparkR 提供了在 R 中访问 Spark 的能力，这与 Hadoop 的 MapReduce 十分类似。

可以预见，SparkR 团队还会加入对 Spark MLlib 机器学习算法的支持，并可在 R 系统中直接予以执行，但具体发布日期尚不明确。

对于平台供应商来说，R 语言已经变得越发重要。一些合作伙伴（例如 Cloudera、Hortonworks、MapR）和数据库供应商已经敏锐地意识到，R 语言在数据科学社区已呈迅速增长之势；对于构建于 Hadoop 之上的各类存储库，R 已成为获取商业价值的一种重要的手段。

在接下来的内容中，我们将对开源方案进行扩展，例如基于 Hadoop 的 Revolution R Enterprise，进而构建简洁、高性能、可移植、可伸缩的替代方案。

R 系统是一款令人称奇的数据科学编程工具，可在模型上运行统计数据分析，并可将分析结果转换为彩色图形。毫无疑问，对于统计人员、数据科学家、数据分析师和数据架构师来说，R 是一个首选编程工具。

在操作过程中，全部对象均会被加载至单机设备的主内存中，这也可视为 R 编程语言的一个缺点。以拍字节（petabyte）计算的大型数据集无法被载入 RAM 内存中，对此，与 R 集成的 Hadoop 是一种较为理想的方案。为了适应 R 编程语言在内存、单机方面的局限性，数据科学家不得不将他们的数据分析过程限制在大数据集的数据样本上。R 编程语言的这种局限性是处理大数据时的一大障碍。鉴于 R 语言的可伸缩性不强，核心 R 引擎只能处理有限数量的数据。

相反，分布式处理框架（例如 Hadoop）对于大型数据集（拍字节级别）上的复杂操作和任务具有一定的可伸缩性，但却缺乏强大的统计分析能力。Hadoop 作为一种流行的大数据处理框架，将其与 R 语言集成则是下一个逻辑操作步骤。在 Hadoop 上应用 R 语言提供了一种高度可伸缩性的数据分析平台，并根据数据集的大小予以实现。将 Hadoop 与 R 集成可以让数据科学家在大型数据集中以并行方式运行 R，因为在 R 语言中，没有一个数据科学库可以在大于其内存的数据集中运行。在商业硬件集群（用于垂直扩展）提供的成本-价值回报方面，基于 R 语言和 Hadoop 的大数据分析具有一定的竞争力。

5.2　R 语言和 Hadoop 间的集成方法

与 Hadoop 协同工作的数据分析师或数据科学家需要使用到 R 数据包或 R 脚本，进而用于数据处理。当结合 Hadoop 使用此类 R 脚本或 R 数据包时，需要利用 Java 编程语言（或实现了 Hadoop MapReduce 的其他编程语言）重写这些 R 脚本。这将是一个较为繁重的过程，可能会导致错误的出现。当利用 R 编程语言集成 Hadoop 时，需要使用到基于 R 语言编写的软件，且数据存储于 Hadoop 分布式存储中。当采用 R 语言执行大型计算时，存在多种解决方案，但所有方案均要求数据在分配至计算节点之前载入内存中。对于大型数据集来说，这并非是一种理想的方法。下列内容列出了常用的 Hadoop-R 集成方法，进而针对大型数据集发挥 R 语言分析的最佳功效。

5.2.1　RHadoop——在工作站上安装 R 并将数据连接至 Hadoop 中

对于 R 编程语言与 Hadoop 间的集成，较为常见的开源分析方案是 RHadoop，并可从 HBase 子系统和 HDFS 文件系统中直接获取数据。考虑到简洁性和成本优势，RHadoop 可视作是一种首选方案。RHadoop 是一个由 5 个不同数据包构成的集合，可使 Hadoop 用户利用 R 编程语言管理和分析数据。RHadoop 数据包兼容于开源 Hadoop，以及常见的 Hadoop 版本，例如 Cloudera、Hortonworks 和 MapR，如下所示。

- rhbase：rhbase 包使用 Thrift 服务器为 R 中的 HBase 提供数据库管理功能。rhbase 包需要安装在运行 R 客户端的节点上。当采用 rhbase 时，数据工程师和数据科学家可从 R 中读取、写入和修改 HBase 表中的数据。
- rhdfs：rhdfs 向 R 程序员提供了与 Hadoop 分布式文件系统间的连接能力，进而可读取、写入和修改存储于 Hadoop HDFS 中的数据。
- plyrmr：plyrmr 支持由 Hadoop 管理的大型数据集上的数据操控。具体来说，plyrmr（针对 MapReduce 的 plyr）针对 reshape2 和 plry 这一类常见的数据包提供了相应的操作。plyrmr 依赖于 Hadoop MapReduce 执行相关操作，但对大多数 MapReduce 细节内容进行了抽象。
- ravro：ravro 包可使用户读取和写入本地和 HDFS 文件系统中的 Avro 文件。
- rmr2（在 Hadoop MapReduce 中执行 R）：当使用 rmr2 包时，R 程序员可在存储于 Hadoop 集群上的数据执行统计分析。对于 rmr2 来说，R 与 Hadoop 间的集成可能稍显复杂，但是，许多 R 程序员发现，与使用基于 Java 的 Hadoop mapper 和 reducer 相比，采用 rmr2 则更加简单。rmr2 的使用过程可能稍显枯燥，但却避免了数据的移动，并有助于执行并行计算以处理大型数据集。

5.2.2　RHIPE——在 Hadoop MapReduce 中执行 R 语言

R 和 Hadoop 集成编程环境（RHIPE）是一个 R 语言库，以使用户可在 R 编程语言中运行 Hadoop MapReduce 作业。R 程序员需要编写 R Map 和 Reduce 函数；同时，RHIPE 库将对其进行转换，并调用相应的 Hadoop Map 和 Hadoop Reduce 任务。其间，RHIPE 使用了协议缓冲编码方案转换 Map 和 Reduce 输入。与其他并行 R 数据包相比，RHIPE 的优点在于：可较好地与 Hadoop 集成，并在机器集群间利用 HDFS 提供数据分布方案，进而实现容错机制并对处理器应用进行优化。

5.2.3　R 和 Hadoop 流

Hadoop 流 API 可使用户利用任何可执行的脚本运行 Hadoop MapReduce 作业，并作为 mapper 或 reducer 读取标准输入中的数据，或者是将数据写入标准输出中。因此，Hadoop 流式 API 可在 Map 或 Reduce 阶段与 R 语言编程脚本结合使用。这里，R 与 Hadoop 间的集成方法并不涉及任何客户端集成，其原因在于，流式作业通过 Hadoop 命令行启用。提交后的 MapReduce 作业通过 UNIX 标准流和序列化进行转换，无论程序员提供的输入脚本语言是什么，均可确保 Java 输入进入 Hadoop。

关于 R 语言与 Hadoop 间的集成，读者还可尝试提出自己的解决方案。

5.2.4　RHIVE——在工作站上安装 R 并连接至 Hadoop 数据

如果希望在 R 接口中执行 Have 查询，RHIVE 包则是一个首选方案，其中包含了元数据的检索功能，例如 Apache Hive 中的数据库名称、列名和表名。通过 R 函数扩展 HiveSQL，对于存储于 Hadoop 中的数据，RHIVE 提供了 R 编程语言中丰富的统计库和算法。针对 Apache Hive 编目后的、存储于 Hadoop 集群中的数据，RHIVE 函数支持用户使用 R 统计学习模型。针对 Hadoop-R 集成方案，RHIVE 的优点在于：鉴于数据操作下推至 Hadoop 中，因而可实现并行操作，并可避免数据的移动行为。

5.2.5　ORCH——基于 Hadoop 的 Oracle 连接器

ORCH 可用于非 Oracle Hadoop 集群或其他 Oracle 大数据装置上。mapper 和 reducer 采用 R 语言编写，而 MapReduce 则通过高级接口在 R 环境中执行。对于 R 语言的 Hadoop 集成，当采用 ORCH 时，R 程序员不必学习新的编程语言（如 Java）就可以深入了解 Hadoop 环境的细节，如 Hadoop 集群硬件或软件。ORCH 连接器还允许用户通过相同的函数调用在本地测试 MapReduce 程序的功能，这比将它们部署到 Hadoop 集群早得多。

对于 R-Hadoop 大数据分析来说，开源方案的数量仍处于持续增长中，但对于简单的 Hadoop MapReduce 来说，R-Hadoop 流已被证明是一类最佳处理方案。R 和 Hadoop 组合方案可视为大数据处理过程中不可或缺的工具，并可生成快速的预测分析结果，同时兼顾所需的性能、可伸缩性和灵活性。

R 语言的优点是它提供了用于统计和数据可视化的详尽的数据科学库列表，这一点已从许多 Hadoop 用户口中得到证实。但是，R 语言中的数据科学库其本质上是非并行的，这也使得数据检索较为耗时——这体现了 R 编程语言内在的局限性。如果抛开这一点，R 和 Hadoop 的整合方案对于大数据分析来说则极具优势。

5.3　数据分析

R 语言支持各种数据分析操作。在 Python 的 pandas 中所做的一切，我们也可以在 R

中加以实现。

考查下列代码：

```
df = read.csv(file=file.choose(), header=T, fill=T, sep=",",
stringsAsFactors=F)
```

其中，file.choose()表示将生成一个新窗口，并选择要打开的数据文件；header=T 意味着读取数据头；fill=T 表示对于未定义或丢失的数据值填写 NaN；最后，sep=","是指当前已知晓如何区分.csv 文件中不同的数据值。在当前示例中，将采用逗号对数值进行分隔；stringsAsFactors 表示将所有的字符串值视为字符串，而非因子。这也使得我们在后续操作过程中替换数据值。

对应结果如图 5.1 所示。

图 5.1

按下 Enter 键后，将显示如图 5.2 所示的窗口。

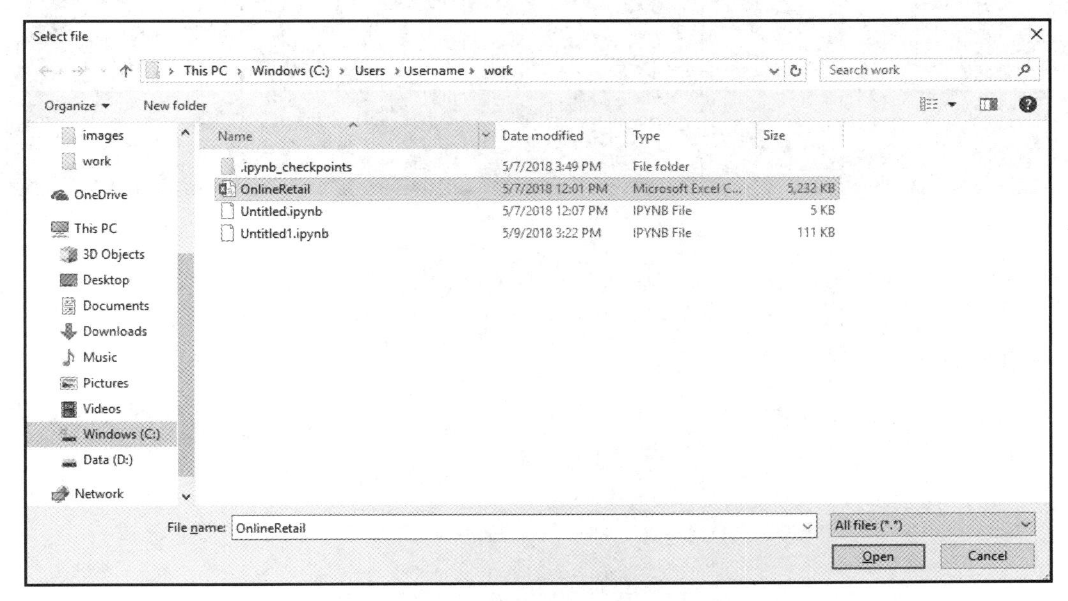

图 5.2

无论哪一种操作系统，都将会看到一个打开的窗口，并允许选择一个文件。接下来，将看到如图 5.3 所示的内容。

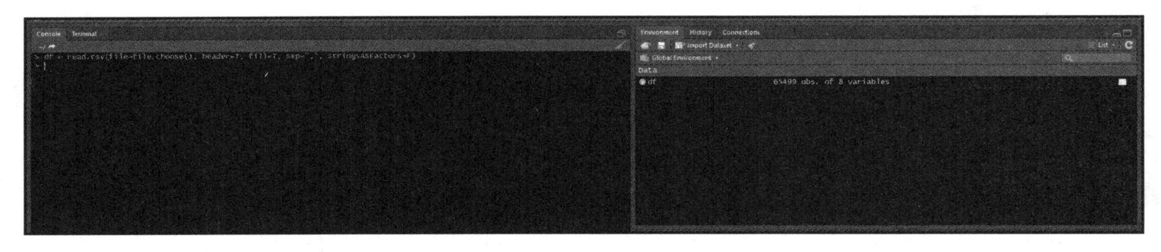

图 5.3

其中，右侧窗口中显示了一个名为 df 的新字段。单击该字段，将显示如图 5.4 所示的内容。

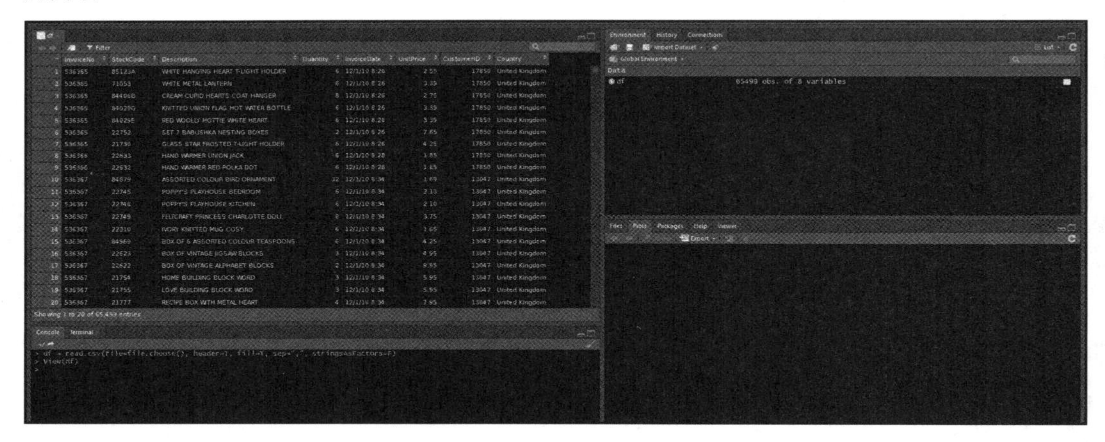

图 5.4

对于已创建的数据帧，下面开始执行分析工作。

此处，相关信息包括行-列数量、数据帧的长度以及列名。考查下列代码行及其相应的输出结果：

```
> is.data.frame(df)
[1] TRUE

> ncol(df)
[1] 8

> length(df)
[1] 8
```

```
> nrow(df)
[1] 27080

> names(df)
[1] "InvoiceNo" "StockCode" "Description" "Quantity" "InvoiceDate"
    "UnitPrice" "CustomerID" "Country"

> colnames(df)
[1] "InvoiceNo" "StockCode" "Description" "Quantity" "InvoiceDate"
    "UnitPrice" "CustomerID" "Country"
```

接下来考查如何生成数据子集。查看下列代码：

```
d1 = df[1:3]
```

对应结果如图 5.5 所示。

df　d1　Filter

	InvoiceNo	StockCode	Description
1	536365	85123A	WHITE HANGING HEART T-LIGHT HOLDER
2	536365	71053	WHITE METAL LANTERN
3	536365	84406B	CREAM CUPID HEARTS COAT HANGER
4	536365	84029G	KNITTED UNION FLAG HOT WATER BOTTLE
5	536365	84029E	RED WOOLLY HOTTIE WHITE HEART.
6	536365	22752	SET 7 BABUSHKA NESTING BOXES
7	536365	21730	GLASS STAR FROSTED T-LIGHT HOLDER
8	536366	22633	HAND WARMER UNION JACK
9	536366	22632	HAND WARMER RED POLKA DOT
10	536367	84879	ASSORTED COLOUR BIRD ORNAMENT
11	536367	22745	POPPYS PLAYHOUSE BEDROOM ,6,12/1/10 8:34,2....
12	536367	22749	FELTCRAFT PRINCESS CHARLOTTE DOLL
13	536367	22310	IVORY KNITTED MUG COSY
14	536367	84969	BOX OF 6 ASSORTED COLOUR TEASPOONS
15	536367	22623	BOX OF VINTAGE JIGSAW BLOCKS
16	536367	22622	BOX OF VINTAGE ALPHABET BLOCKS
17	536367	21754	HOME BUILDING BLOCK WORD
18	536367	21755	LOVE BUILDING BLOCK WORD
19	536367	21777	RECIPE BOX WITH METAL HEART
20	536367	48187	DOORMAT NEW ENGLAND

Showing 1 to 20 of 27,080 entries

图 5.5

基本上讲，我们选择了 1、2、3 列作为 d1 的数据集。除了列之外，还可以选取相应的行。下列代码定义了 d1：

```
d1 = df[1:10, c(1:3)]
```

对应结果如图 5.6 所示。

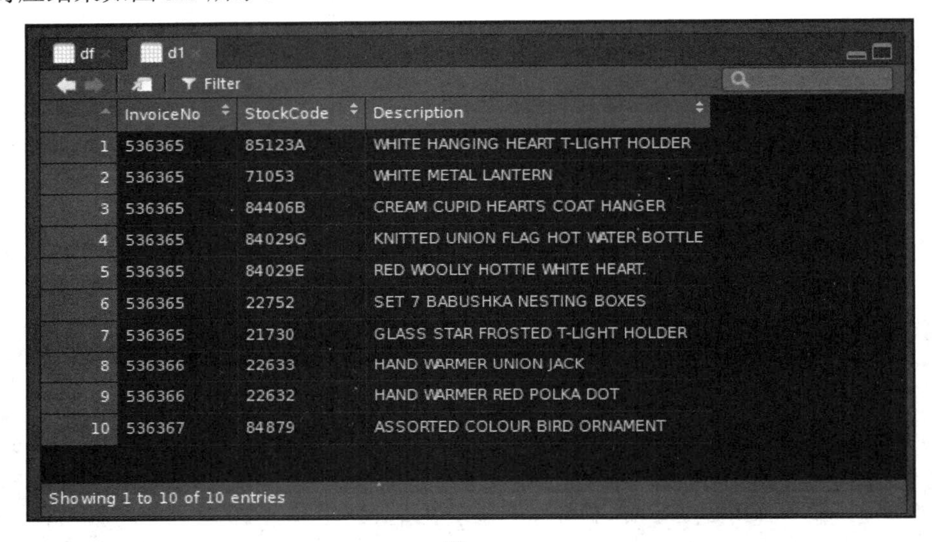

	InvoiceNo	StockCode	Description
1	536365	85123A	WHITE HANGING HEART T-LIGHT HOLDER
2	536365	71053	WHITE METAL LANTERN
3	536365	84406B	CREAM CUPID HEARTS COAT HANGER
4	536365	84029G	KNITTED UNION FLAG HOT WATER BOTTLE
5	536365	84029E	RED WOOLLY HOTTIE WHITE HEART.
6	536365	22752	SET 7 BABUSHKA NESTING BOXES
7	536365	21730	GLASS STAR FROSTED T-LIGHT HOLDER
8	536366	22633	HAND WARMER UNION JACK
9	536366	22632	HAND WARMER RED POLKA DOT
10	536367	84879	ASSORTED COLOUR BIRD ORNAMENT

Showing 1 to 10 of 10 entries

图 5.6

除此之外，还可以访问数据帧的某一列，如下所示：

```
v1 = df[[3]]
```

这将把数据的全部列赋予 v1。接下来访问 v1 的前 5 列元素，如下所示：

```
v1[1:5]
```

对应结果如图 5.7 所示。

```
> v1[1:5]
[1] "WHITE HANGING HEART T-LIGHT HOLDER"  "WHITE METAL LANTERN"                "CREAM CUPID HEARTS COAT HANGER"
[4] "KNITTED UNION FLAG HOT WATER BOTTLE" "RED WOOLLY HOTTIE WHITE HEART."
```

图 5.7

另外，还可执行下列操作：

```
v2 = df$Description
v2[1:5]
```

对应结果如图 5.8 所示。

```
> v2 = df$Description
> v2[1:5]
[1] "WHITE HANGING HEART T-LIGHT HOLDER"  "WHITE METAL LANTERN"             "CREAM CUPID HEARTS COAT HANGER"
[4] "KNITTED UNION FLAG HOT WATER BOTTLE" "RED WOOLLY HOTTIE WHITE HEART."
> |
```

图 5.8

我们甚至可对单一行进行访问（假设已知晓特定值）。考查下列股票代码：

```
d1[d1$StockCode == "85123A", ]
```

对应结果如图 5.9 所示。

```
> d1[d1$StockCode == "85123A", ]
  InvoiceNo StockCode                        Description
1    536365    85123A WHITE HANGING HEART T-LIGHT HOLDER
```

图 5.9

下列代码展示了如何访问特定行：

```
d1 = df[1:10, c(1:8)]
d1[2, c(1:8)]
```

对应结果如图 5.10 所示。

```
> d1 = df[1:10, c(1:8)]
> d1[2, c(1:8)]
  InvoiceNo StockCode         Description Quantity  InvoiceDate UnitPrice CustomerID        Country
2    536365     71053 WHITE METAL LANTERN        6 12/1/10 8:26      3.39      17850 United Kingdom
```

图 5.10

类似于 Python 中的.head()函数，R 中定义了一个 head()函数，如下所示：

```
head(df)
```

对应结果如图 5.11 所示。

```
> head(df)
  InvoiceNo StockCode                         Description Quantity  InvoiceDate UnitPrice CustomerID        Country
1    536365    85123A  WHITE HANGING HEART T-LIGHT HOLDER        6 12/1/10 8:26      2.55      17850 United Kingdom
2    536365     71053                 WHITE METAL LANTERN        6 12/1/10 8:26      3.39      17850 United Kingdom
3    536365    84406B      CREAM CUPID HEARTS COAT HANGER        8 12/1/10 8:26      2.75      17850 United Kingdom
4    536365    84029G KNITTED UNION FLAG HOT WATER BOTTLE        6 12/1/10 8:26      3.39      17850 United Kingdom
5    536365    84029E      RED WOOLLY HOTTIE WHITE HEART.        6 12/1/10 8:26      3.39      17850 United Kingdom
6    536365     22752        SET 7 BABUSHKA NESTING BOXES        2 12/1/10 8:26      7.65      17850 United Kingdom
> |
```

图 5.11

我们还可添加一个参数，进而选择希望显示的行数量，例如前 10 行数据。对应代码如下所示：

```
head(df, 10)
```

对应结果如图 5.12 所示。

```
> head(df, 10)
   InvoiceNo StockCode                        Description Quantity  InvoiceDate UnitPrice CustomerID        Country
1     536365    85123A  WHITE HANGING HEART T-LIGHT HOLDER        6 12/1/10 8:26      2.55      17850 United Kingdom
2     536365     71053                 WHITE METAL LANTERN        6 12/1/10 8:26      3.39      17850 United Kingdom
3     536365    84406B      CREAM CUPID HEARTS COAT HANGER        8 12/1/10 8:26      2.75      17850 United Kingdom
4     536365    84029G KNITTED UNION FLAG HOT WATER BOTTLE        6 12/1/10 8:26      3.39      17850 United Kingdom
5     536365    84029E      RED WOOLLY HOTTIE WHITE HEART.        6 12/1/10 8:26      3.39      17850 United Kingdom
6     536365     22752        SET 7 BABUSHKA NESTING BOXES        2 12/1/10 8:26      7.65      17850 United Kingdom
7     536365     21730   GLASS STAR FROSTED T-LIGHT HOLDER        6 12/1/10 8:26      4.25      17850 United Kingdom
8     536366     22633              HAND WARMER UNION JACK        6 12/1/10 8:28      1.85      17850 United Kingdom
9     536366     22632           HAND WARMER RED POLKA DOT        6 12/1/10 8:28      1.85      17850 United Kingdom
10    536367     84879       ASSORTED COLOUR BIRD ORNAMENT       32 12/1/10 8:34      1.69      13047 United Kingdom
>
```

图 5.12

相应地，还可将第 2 个参数设置为负数，如下所示：

```
head(d1, -2)
```

对应结果如图 5.13 所示。

```
> head(d1, -2)
  InvoiceNo StockCode                        Description Quantity  InvoiceDate UnitPrice CustomerID        Country
1    536365    85123A  WHITE HANGING HEART T-LIGHT HOLDER        6 12/1/10 8:26      2.55      17850 United Kingdom
2    536365     71053                 WHITE METAL LANTERN        6 12/1/10 8:26      3.39      17850 United Kingdom
3    536365    84406B      CREAM CUPID HEARTS COAT HANGER        8 12/1/10 8:26      2.75      17850 United Kingdom
4    536365    84029G KNITTED UNION FLAG HOT WATER BOTTLE        6 12/1/10 8:26      3.39      17850 United Kingdom
5    536365    84029E      RED WOOLLY HOTTIE WHITE HEART.        6 12/1/10 8:26      3.39      17850 United Kingdom
6    536365     22752        SET 7 BABUSHKA NESTING BOXES        2 12/1/10 8:26      7.65      17850 United Kingdom
7    536365     21730   GLASS STAR FROSTED T-LIGHT HOLDER        6 12/1/10 8:26      4.25      17850 United Kingdom
8    536366     22633              HAND WARMER UNION JACK        6 12/1/10 8:28      1.85      17850 United Kingdom
```

图 5.13

类似地，还可使用 tail()函数显示最后 *n* 行，如下所示：

```
tail(d1, 4)
```

对应结果如图 5.14 所示。

```
> d1 = df[0:10, c(1:8)]
> tail(d1, 4)
   InvoiceNo StockCode                       Description Quantity  InvoiceDate UnitPrice CustomerID        Country
7     536365     21730 GLASS STAR FROSTED T-LIGHT HOLDER        6 12/1/10 8:26      4.25      17850 United Kingdom
8     536366     22633            HAND WARMER UNION JACK        6 12/1/10 8:28      1.85      17850 United Kingdom
9     536366     22632         HAND WARMER RED POLKA DOT        6 12/1/10 8:28      1.85      17850 United Kingdom
10    536367     84879     ASSORTED COLOUR BIRD ORNAMENT       32 12/1/10 8:34      1.69      13047 United Kingdom
> |
```

图 5.14

与 head()函数类似，tail()函数中也可将第 2 个参数设置为负数，如下所示：

```
tail(d1, -2)
```

这将显示 *nrow*(d1) + *n* 行。其中，*n* 表示为传递至 tail()函数中的参数。对应结果如图 5.15 所示。

```
> tail(d1, -2)
   InvoiceNo StockCode                        Description Quantity  InvoiceDate UnitPrice CustomerID        Country
3     536365    84406B        CREAM CUPID HEARTS COAT HANGER        8 12/1/10 8:26      2.75      17850 United Kingdom
4     536365    84029G KNITTED UNION FLAG HOT WATER BOTTLE        6 12/1/10 8:26      3.39      17850 United Kingdom
5     536365    84029E      RED WOOLLY HOTTIE WHITE HEART.        6 12/1/10 8:26      3.39      17850 United Kingdom
6     536365     22752        SET 7 BABUSHKA NESTING BOXES        2 12/1/10 8:26      7.65      17850 United Kingdom
7     536365     21730   GLASS STAR FROSTED T-LIGHT HOLDER        6 12/1/10 8:26      4.25      17850 United Kingdom
8     536366     22633              HAND WARMER UNION JACK        6 12/1/10 8:28      1.85      17850 United Kingdom
9     536366     22632           HAND WARMER RED POLKA DOT        6 12/1/10 8:28      1.85      17850 United Kingdom
10    536367     84879       ASSORTED COLOUR BIRD ORNAMENT       32 12/1/10 8:34      1.69      13047 United Kingdom
>
```

图 5.15

我们可以对某一列进行一些基本的统计分析，但首先需要对数据进行转换。相应地，可执行 min()、max()、mean()等操作，如下所示：

```
min(as.numeric(df$UnitPrice))
[1] 0
min(df$UnitPrice)
[1] 0
```

其中，as.numeric()表示字符串数据将被转换为数字。当前示例并未涉及字符串值；否则，min(df$UnitPrice)将返回 0，如下所示：

```
max(df$UnitPrice)
[1] 16888.02

mean(df$UnitPrice)
[1] 5.857586

median(df$UnitPrice)
[1] 2.51

quantile(df$UnitPrice)
```

对应结果如图 5.16 所示。

```
> quantile(df$UnitPrice)
      0%      25%      50%      75%     100%
    0.00     1.25     2.51     4.24 16888.02
>
```

图 5.16

另外，还可添加另一个参数，进而自定义百分比值，如下所示：

```
quantile(df$UnitPrice, c(0, .1, .5, .9)
```

对应结果如图 5.17 所示。

sd(df$UnitPrice)的输出结果如图 5.18 所示。

```
> quantile(df$UnitPrice, c(0, .1, .5, .9))
  0%  10%  50%  90%
0.00 0.83 2.51 7.95
>
```

图 5.17

```
> sd(df$UnitPrice)
[1] 145.796
>
```

图 5.18

这表示为 df$UnitPrice 的标准偏差。此外，还可计算方差，如下所示：

```
var(df$UnitPrice)
```

对应结果如图 5.19 所示。

range(df$UnitPrice)的输出结果如图 5.20 所示。

```
> var(df$UnitPrice)
[1] 21256.46
>
```

图 5.19

> range(df$UnitPrice)
[1]   0.00 16888.02
>

图 5.20

除此之外，还可得到由 5 个数值构成的概括内容，即最小值、第 1 个分位数、中位数（即 50%标记）、第 3 个分位数（75%标记）以及最大值，如下所示：

```
fivenum(df$UnitPrice)
```

对应结果如图 5.21 所示。

```
> fivenum(df$UnitPrice)
[1]     0.00     1.25     2.51     4.24 16888.02
>
```

图 5.21

相应地，还可对所选列进行绘制，如下所示：

```
plot(df$UnitPrice)
```

对应结果如图 5.22 所示。

鉴于存在不同类型的绘制结果，因而可引入另一个参数并指定绘制类型。考查下列代码：

```
plot(df$UnitPrice, type="p")
```

对应结果如图 5.23 所示。

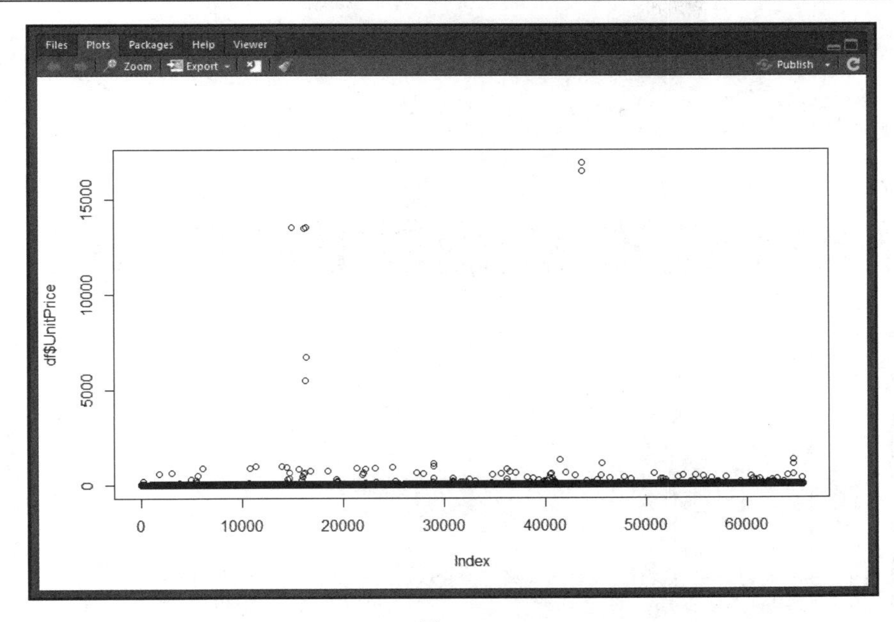

图 5.22

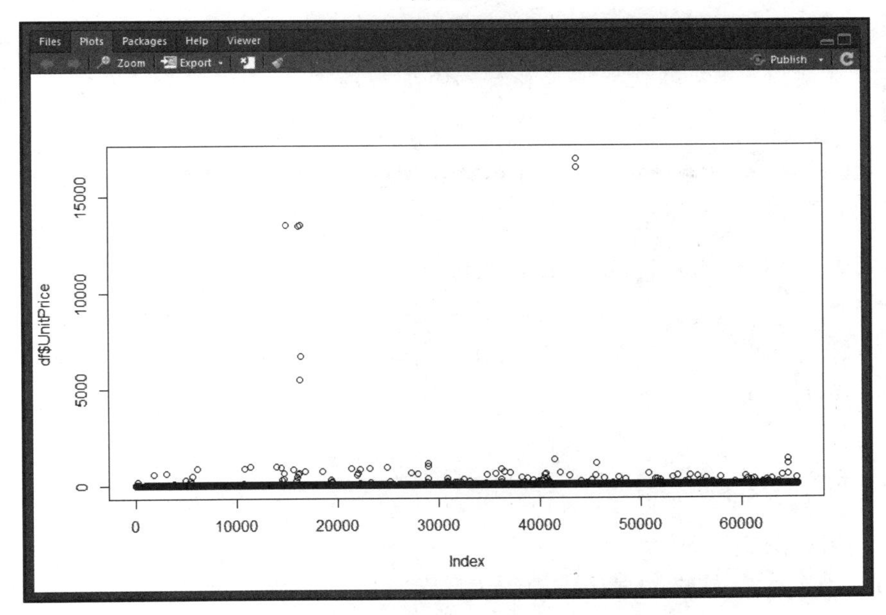

图 5.23

不难发现，图 5.23 与图 5.22 相同。另外，还可使用较小的范围，以使显示结果看起来不那么拥挤，如下所示：

```
d1 = df[0:30, c(1:8)]
plot(d1$UnitPrice)
```

对应结果如图 5.24 所示。

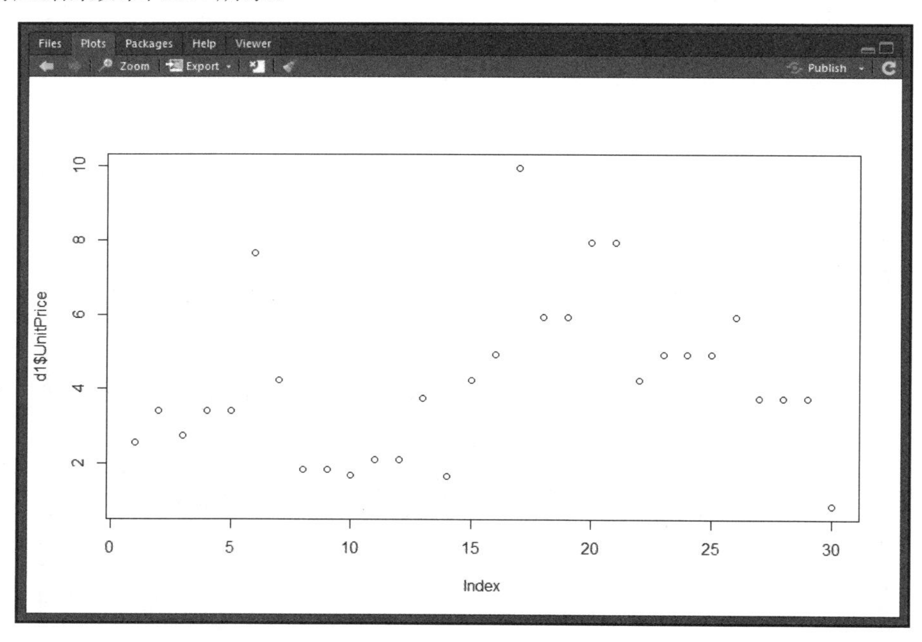

图 5.24

出于简单考虑，下面重新定义 d1，其仅包含 UnitPrice 列，如下所示：

```
d1 = d1$UnitPrice
plot(d1, type="p")
```

对应结果与图 5.24 相同。

考查下列代码：

```
plot(d1, type="l")
```

对应结果如图 5.25 所示。

下列代码显示了 d1 的效果图：

```
plot(d1, type="b")
```

对应结果如图 5.26 所示。

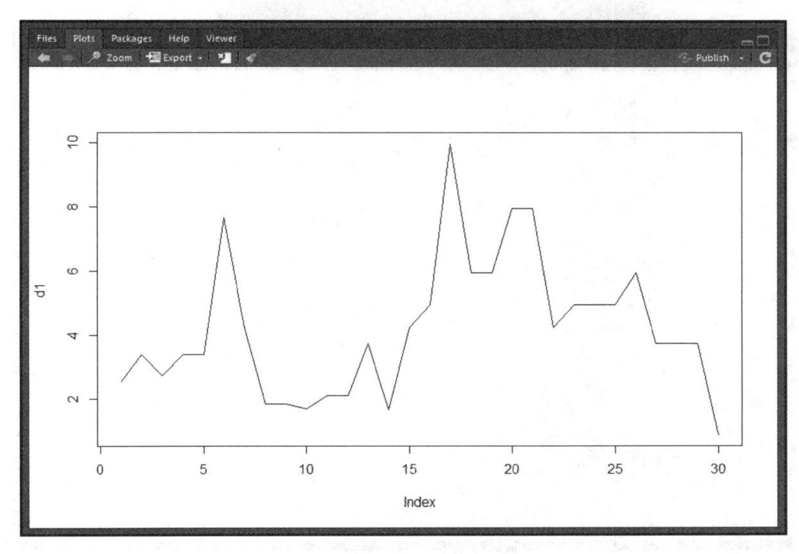

图 5.25

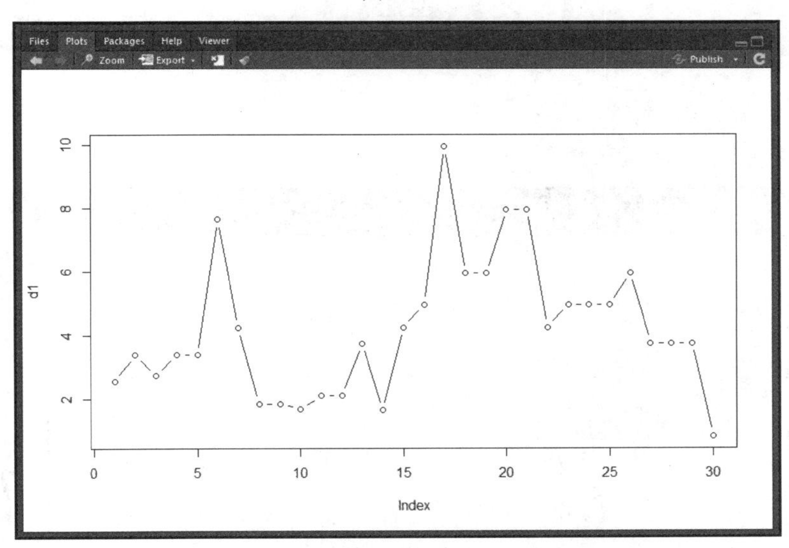

图 5.26

图 5.26 中显示了 d1 的点-线示意图，但在图中上方处，二者却并未重叠。对此，查看下列代码：

```
plot(d1, type="c")
```

对应结果如图 5.27 所示。

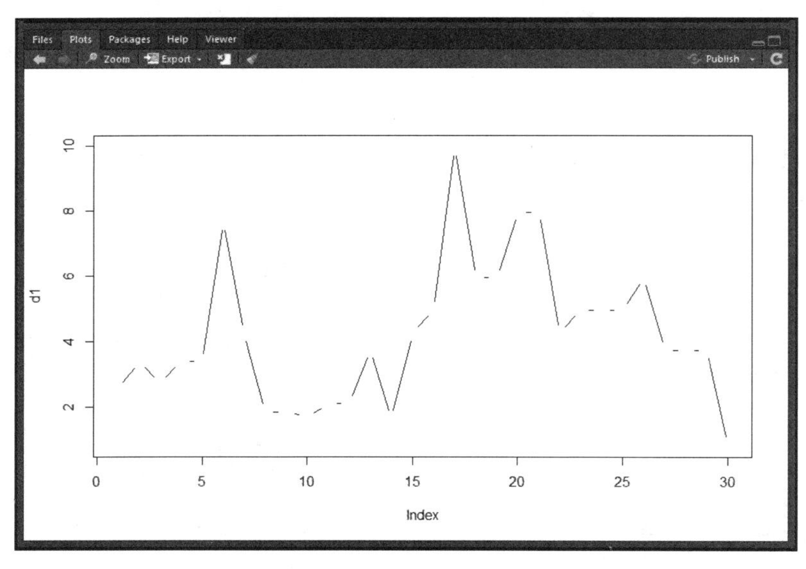

图 5.27

图 5.27 表示为之前 type="b"的直线图。考查下列代码：

```
plot(d1, type="o")
```

对应结果如图 5.28 所示。

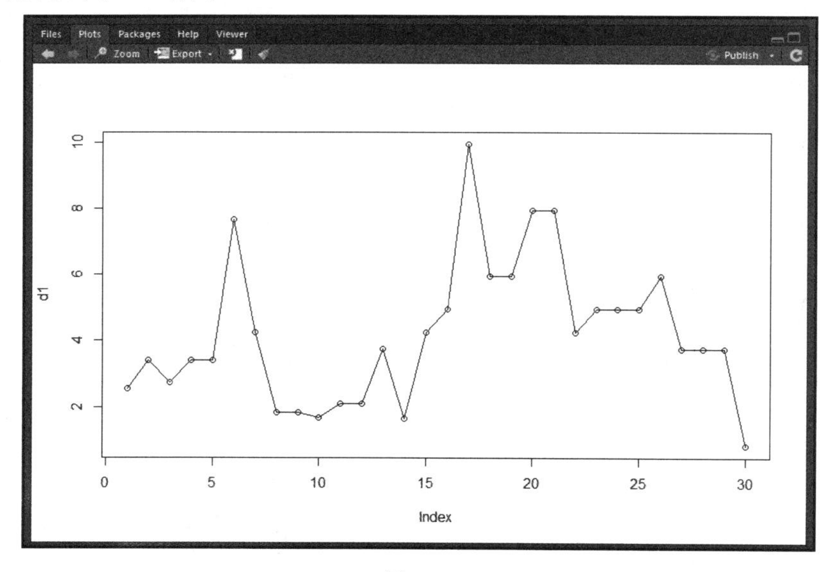

图 5.28

图 5.28 表示为 d1 的直线-点图，且在上方彼此重叠；而对于下列代码：

```
plot(d1, type="h")
```

对应结果如图 5.29 所示。

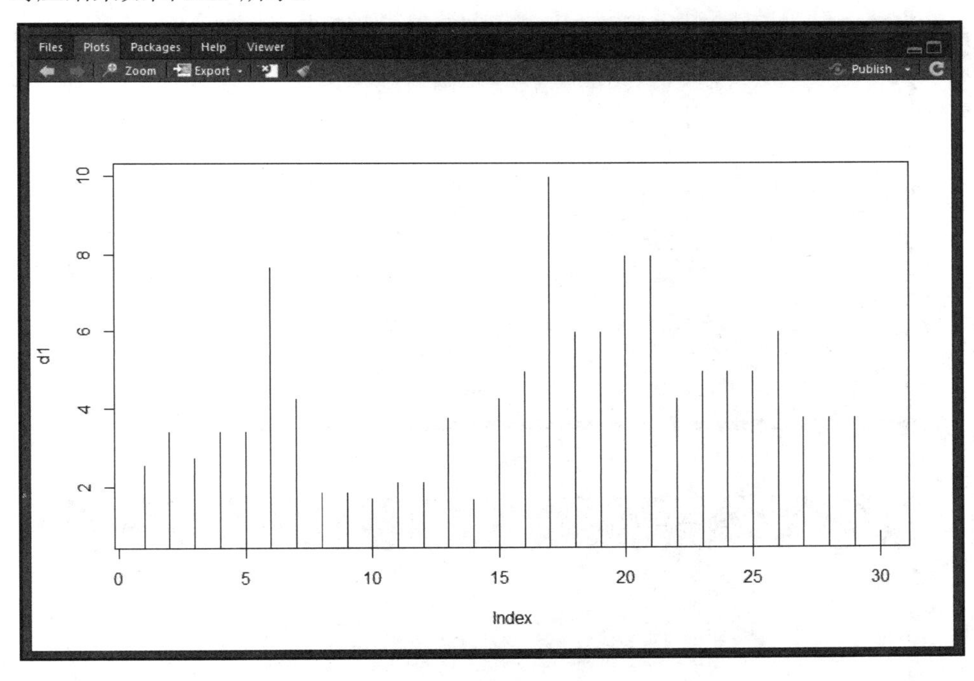

图 5.29

下列代码显示了 d1 的直方图：

```
plot(d1, type="s")
```

对应结果如图 5.30 所示。

下列代码显示了 d1 的阶状图：

```
plot(d1, type="S")
```

对应结果如图 5.31 所示。

图 5.30 和图 5.31 之间的差别在于：第 1 幅图像定义为 type="s"，因而首先在水平方向上进行绘制，随后是垂直方向。第 2 幅图则定义为 type="S"，且首先在垂直方向上进行绘制，随后是水平方向。这一差别在图中可以清楚地看到。

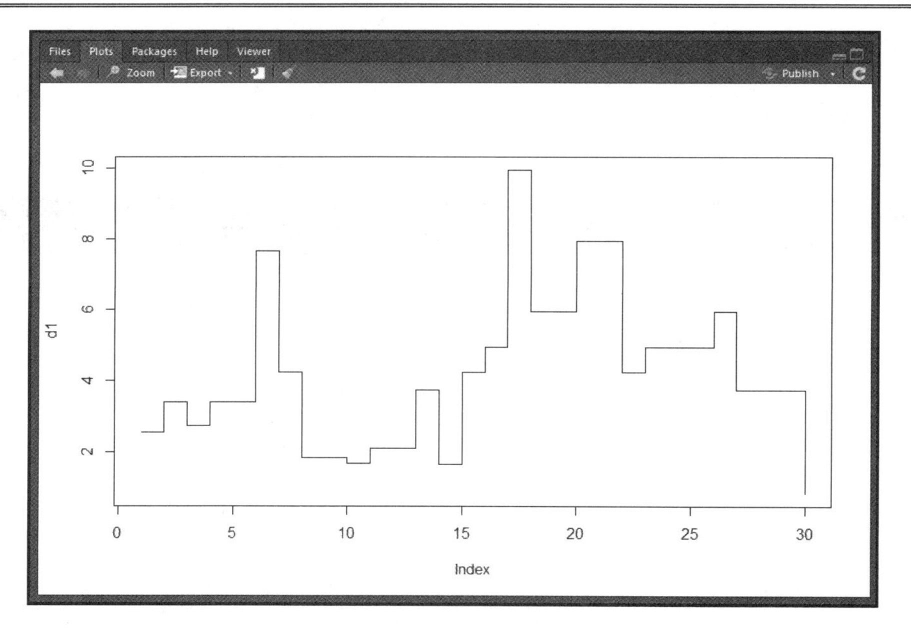

图 5.30

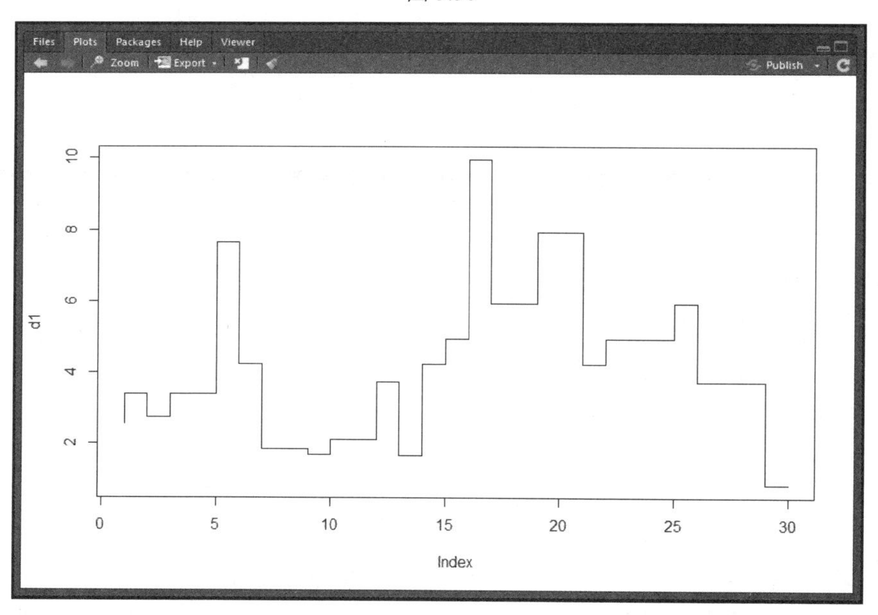

图 5.31

除此之外，还存在一些其他的可用参数，如下所示：

```
#Note: these are parameters, not individual lines of code.

#The title of the graph
main="Title"

#Subtitle for the graph
sub="title"

#Label for the x-axis
xlab="X Axis"

#Label for the y-axis
ylab="Y Axis"

#The aspect ratio between y and x.
asp=1
```

考查下列示例：

```
plot(d1, type="h", main="Graph of Unit Prices vs Index", sub ="First 30
Rows", xlab = "Row Index", ylab="Prices", asp=1.4)
```

对应结果如图 5.32 所示。

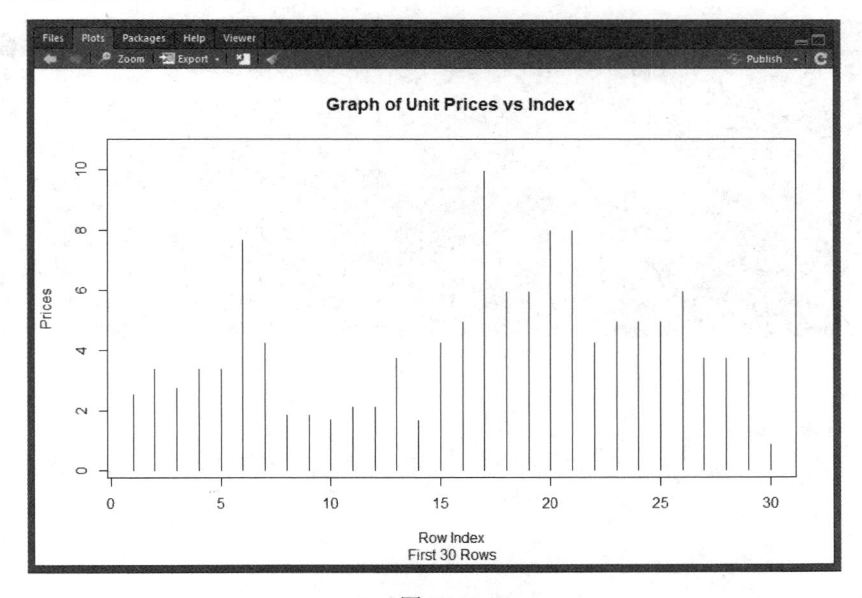

图 5.32

另外，使用 rbind()函数还可将两个不同的数据帧进行整合。

考查下列代码：

```
d2 = df[0:10, c(1:8)]
d3 = df[21:30, c(1:8)]
d4 = rbind(d2, d3)
```

d2 的结果如图 5.33 所示。

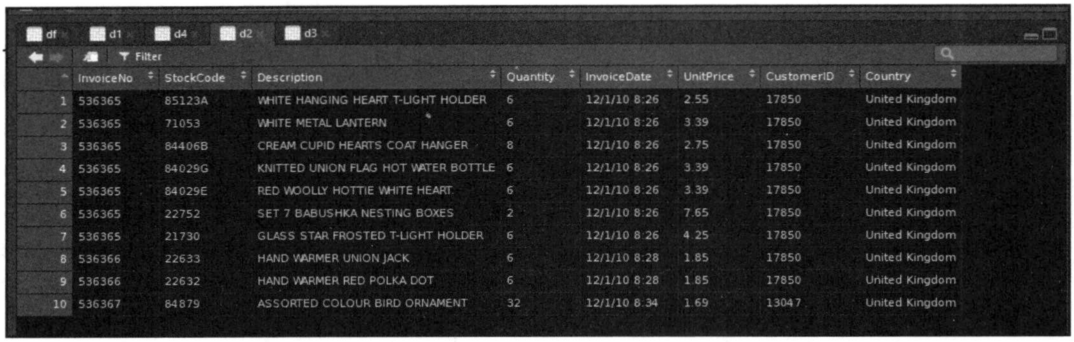

	InvoiceNo	StockCode	Description	Quantity	InvoiceDate	UnitPrice	CustomerID	Country
1	536365	85123A	WHITE HANGING HEART T-LIGHT HOLDER	6	12/1/10 8:26	2.55	17850	United Kingdom
2	536365	71053	WHITE METAL LANTERN	6	12/1/10 8:26	3.39	17850	United Kingdom
3	536365	84406B	CREAM CUPID HEARTS COAT HANGER	8	12/1/10 8:26	2.75	17850	United Kingdom
4	536365	84029G	KNITTED UNION FLAG HOT WATER BOTTLE	6	12/1/10 8:26	3.39	17850	United Kingdom
5	536365	84029E	RED WOOLLY HOTTIE WHITE HEART.	6	12/1/10 8:26	3.39	17850	United Kingdom
6	536365	22752	SET 7 BABUSHKA NESTING BOXES	2	12/1/10 8:26	7.65	17850	United Kingdom
7	536365	21730	GLASS STAR FROSTED T-LIGHT HOLDER	6	12/1/10 8:26	4.25	17850	United Kingdom
8	536366	22633	HAND WARMER UNION JACK	6	12/1/10 8:28	1.85	17850	United Kingdom
9	536366	22632	HAND WARMER RED POLKA DOT	6	12/1/10 8:28	1.85	17850	United Kingdom
10	536367	84879	ASSORTED COLOUR BIRD ORNAMENT	32	12/1/10 8:34	1.69	13047	United Kingdom

图 5.33

d3 的结果如图 5.34 所示。

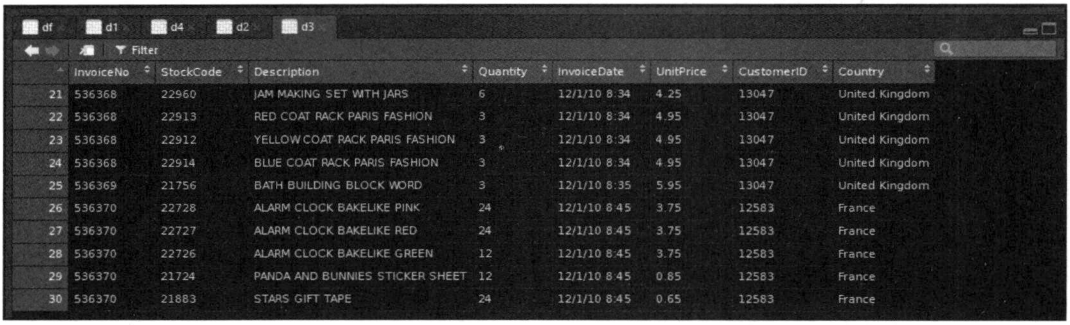

	InvoiceNo	StockCode	Description	Quantity	InvoiceDate	UnitPrice	CustomerID	Country
21	536368	22960	JAM MAKING SET WITH JARS	6	12/1/10 8:34	4.25	13047	United Kingdom
22	536368	22913	RED COAT RACK PARIS FASHION	3	12/1/10 8:34	4.95	13047	United Kingdom
23	536368	22912	YELLOW COAT RACK PARIS FASHION	3	12/1/10 8:34	4.95	13047	United Kingdom
24	536368	22914	BLUE COAT RACK PARIS FASHION	3	12/1/10 8:34	4.95	13047	United Kingdom
25	536369	21756	BATH BUILDING BLOCK WORD	3	12/1/10 8:35	5.95	13047	United Kingdom
26	536370	22728	ALARM CLOCK BAKELIKE PINK	24	12/1/10 8:45	3.75	12583	France
27	536370	22727	ALARM CLOCK BAKELIKE RED	24	12/1/10 8:45	3.75	12583	France
28	536370	22726	ALARM CLOCK BAKELIKE GREEN	12	12/1/10 8:45	3.75	12583	France
29	536370	21724	PANDA AND BUNNIES STICKER SHEET	12	12/1/10 8:45	0.85	12583	France
30	536370	21883	STARS GIFT TAPE	24	12/1/10 8:45	0.65	12583	France

图 5.34

d4 的结果如图 5.35 所示。

需要注意的一点是，传递至 rbind()中的全部数据帧须包含相同的列，而对顺序则无要求。

此外，还可对两个数据帧进行合并。

考查下列代码：

```
d2 = df[0:11, c("InvoiceNo", "StockCode", "Description")]
d3 = df[11:20, c("StockCode", "Description", "Quantity")]
d4 = merge(d2, d3)
```

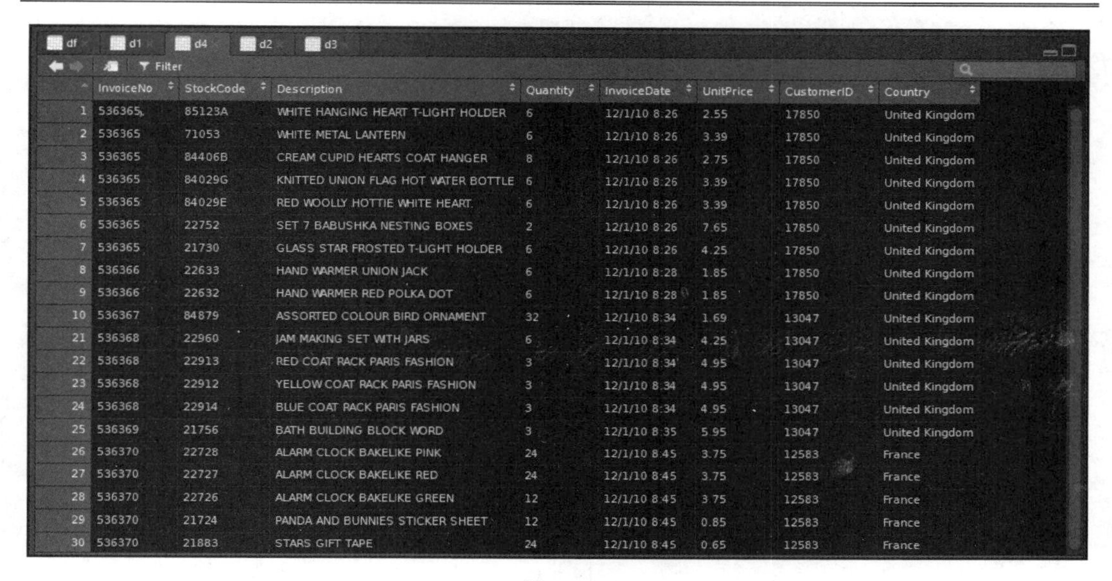

	InvoiceNo	StockCode	Description	Quantity	InvoiceDate	UnitPrice	CustomerID	Country
1	536365	85123A	WHITE HANGING HEART T-LIGHT HOLDER	6	12/1/10 8:26	2.55	17850	United Kingdom
2	536365	71053	WHITE METAL LANTERN	6	12/1/10 8:26	3.39	17850	United Kingdom
3	536365	84406B	CREAM CUPID HEARTS COAT HANGER	8	12/1/10 8:26	2.75	17850	United Kingdom
4	536365	84029G	KNITTED UNION FLAG HOT WATER BOTTLE	6	12/1/10 8:26	3.39	17850	United Kingdom
5	536365	84029E	RED WOOLLY HOTTIE WHITE HEART.	6	12/1/10 8:26	3.39	17850	United Kingdom
6	536365	22752	SET 7 BABUSHKA NESTING BOXES	2	12/1/10 8:26	7.65	17850	United Kingdom
7	536365	21730	GLASS STAR FROSTED T-LIGHT HOLDER	6	12/1/10 8:26	4.25	17850	United Kingdom
8	536366	22633	HAND WARMER UNION JACK	6	12/1/10 8:28	1.85	17850	United Kingdom
9	536366	22632	HAND WARMER RED POLKA DOT	6	12/1/10 8:28	1.85	17850	United Kingdom
10	536367	84879	ASSORTED COLOUR BIRD ORNAMENT	32	12/1/10 8:34	1.69	13047	United Kingdom
21	536368	22960	JAM MAKING SET WITH JARS	6	12/1/10 8:34	4.25	13047	United Kingdom
22	536368	22913	RED COAT RACK PARIS FASHION	3	12/1/10 8:34	4.95	13047	United Kingdom
23	536368	22912	YELLOW COAT RACK PARIS FASHION	3	12/1/10 8:34	4.95	13047	United Kingdom
24	536368	22914	BLUE COAT RACK PARIS FASHION	3	12/1/10 8:34	4.95	13047	United Kingdom
25	536369	21756	BATH BUILDING BLOCK WORD	3	12/1/10 8:35	5.95	13047	United Kingdom
26	536370	22728	ALARM CLOCK BAKELIKE PINK	24	12/1/10 8:45	3.75	12583	France
27	536370	22727	ALARM CLOCK BAKELIKE RED	24	12/1/10 8:45	3.75	12583	France
28	536370	22726	ALARM CLOCK BAKELIKE GREEN	12	12/1/10 8:45	3.75	12583	France
29	536370	21724	PANDA AND BUNNIES STICKER SHEET	12	12/1/10 8:45	0.85	12583	France
30	536370	21883	STARS GIFT TAPE	24	12/1/10 8:45	0.65	12583	France

图 5.35

d2 的结果如图 5.36 所示。

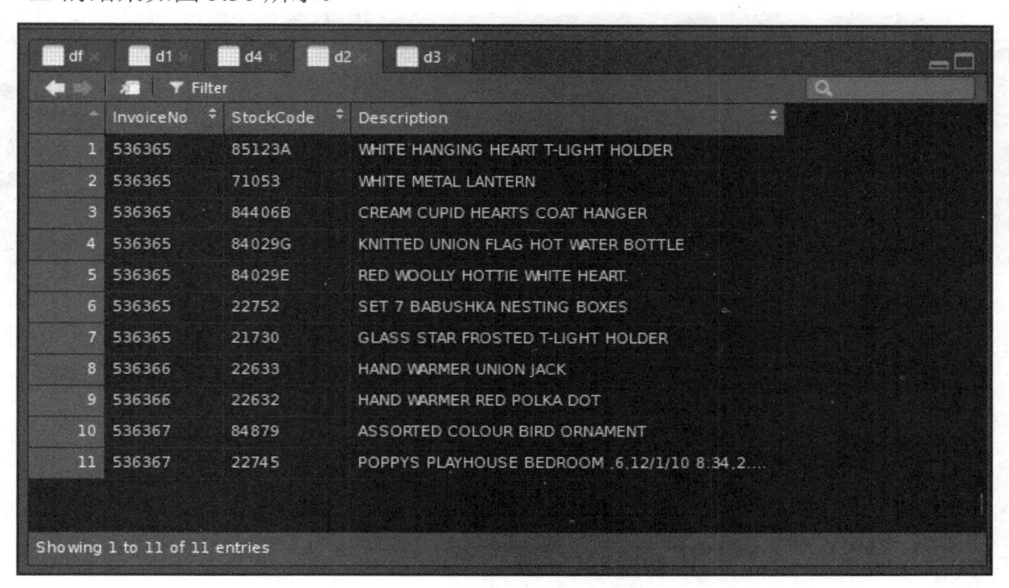

	InvoiceNo	StockCode	Description
1	536365	85123A	WHITE HANGING HEART T-LIGHT HOLDER
2	536365	71053	WHITE METAL LANTERN
3	536365	84406B	CREAM CUPID HEARTS COAT HANGER
4	536365	84029G	KNITTED UNION FLAG HOT WATER BOTTLE
5	536365	84029E	RED WOOLLY HOTTIE WHITE HEART.
6	536365	22752	SET 7 BABUSHKA NESTING BOXES
7	536365	21730	GLASS STAR FROSTED T-LIGHT HOLDER
8	536366	22633	HAND WARMER UNION JACK
9	536366	22632	HAND WARMER RED POLKA DOT
10	536367	84879	ASSORTED COLOUR BIRD ORNAMENT
11	536367	22745	POPPYS PLAYHOUSE BEDROOM ,6,12/1/10 8:34,2....

Showing 1 to 11 of 11 entries

图 5.36

d3 的结果如图 5.37 所示。

df　d1　d4　d2　d3

Filter

	StockCode	Description	Quantity
11	22745	POPPYS PLAYHOUSE BEDROOM ,6,12/1/10 8:34,2....	6
12	22749	FELTCRAFT PRINCESS CHARLOTTE DOLL	8
13	22310	IVORY KNITTED MUG COSY	6
14	84969	BOX OF 6 ASSORTED COLOUR TEASPOONS	6
15	22623	BOX OF VINTAGE JIGSAW BLOCKS	3
16	22622	BOX OF VINTAGE ALPHABET BLOCKS	2
17	21754	HOME BUILDING BLOCK WORD	3
18	21755	LOVE BUILDING BLOCK WORD	3
19	21777	RECIPE BOX WITH METAL HEART	4
20	48187	DOORMAT NEW ENGLAND	4

Showing 1 to 10 of 10 entries

图 5.37

d4 的结果如图 5.38 所示。

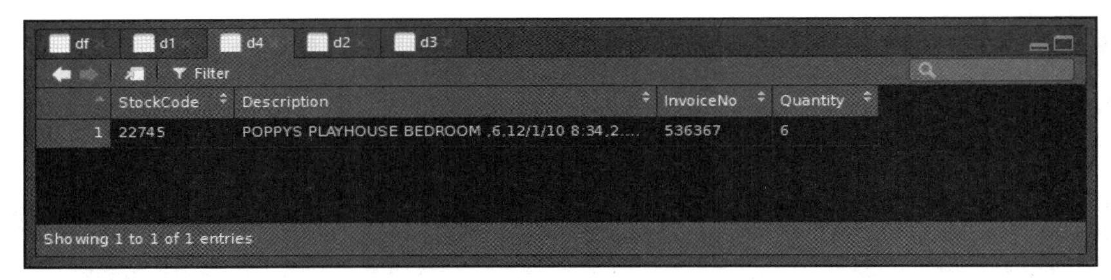

df　d1　d4　d2　d3

Filter

	StockCode	Description	InvoiceNo	Quantity
1	22745	POPPYS PLAYHOUSE BEDROOM ,6,12/1/10 8:34,2....	536367	6

Showing 1 to 1 of 1 entries

图 5.38

因此，在默认状态下，merge()采用了内连接。

接下来考查外连接，对应代码如下所示：

```
d4 = merge(d2, d3, all=T)
```

对应结果如图 5.39 所示。

下列代码显示了左外连接：

```
d4 = merge(d2, d3, all.x=T)
```

对应结果如图 5.40 所示。

	StockCode	Description	InvoiceNo	Quantity
1	21730	GLASS STAR FROSTED T-LIGHT HOLDER	536365	NA
2	21754	HOME BUILDING BLOCK WORD	NA	3
3	21755	LOVE BUILDING BLOCK WORD	NA	3
4	21777	RECIPE BOX WITH METAL HEART	NA	4
5	22310	IVORY KNITTED MUG COSY	NA	6
6	22622	BOX OF VINTAGE ALPHABET BLOCKS	NA	2
7	22623	BOX OF VINTAGE JIGSAW BLOCKS	NA	3
8	22632	HAND WARMER RED POLKA DOT	536366	NA
9	22633	HAND WARMER UNION JACK	536366	NA
10	22745	POPPYS PLAYHOUSE BEDROOM ,6,12/1/10 8:34,2....	536367	6
11	22749	FELTCRAFT PRINCESS CHARLOTTE DOLL	NA	8
12	22752	SET 7 BABUSHKA NESTING BOXES	536365	NA
13	48187	DOORMAT NEW ENGLAND	NA	4
14	71053	WHITE METAL LANTERN	536365	NA
15	84029E	RED WOOLLY HOTTIE WHITE HEART.	536365	NA
16	84029G	KNITTED UNION FLAG HOT WATER BOTTLE	536365	NA
17	84406B	CREAM CUPID HEARTS COAT HANGER	536365	NA
18	84879	ASSORTED COLOUR BIRD ORNAMENT	536367	NA
19	84969	BOX OF 6 ASSORTED COLOUR TEASPOONS	NA	6
20	85123A	WHITE HANGING HEART T-LIGHT HOLDER	536365	NA

Showing 1 to 20 of 20 entries

图 5.39

	StockCode	Description	InvoiceNo	Quantity
1	21730	GLASS STAR FROSTED T-LIGHT HOLDER	536365	NA
2	22632	HAND WARMER RED POLKA DOT	536366	NA
3	22633	HAND WARMER UNION JACK	536366	NA
4	22745	POPPYS PLAYHOUSE BEDROOM ,6,12/1/10 8:34,2....	536367	6
5	22752	SET 7 BABUSHKA NESTING BOXES	536365	NA
6	71053	WHITE METAL LANTERN	536365	NA
7	84029E	RED WOOLLY HOTTIE WHITE HEART.	536365	NA
8	84029G	KNITTED UNION FLAG HOT WATER BOTTLE	536365	NA
9	84406B	CREAM CUPID HEARTS COAT HANGER	536365	NA
10	84879	ASSORTED COLOUR BIRD ORNAMENT	536367	NA
11	85123A	WHITE HANGING HEART T-LIGHT HOLDER	536365	NA

Showing 1 to 11 of 11 entries

图 5.40

下列代码显示了右外连接：

```
d4 = merge(d2, d3, all.y=T)
```

对应结果如图 5.41 所示。

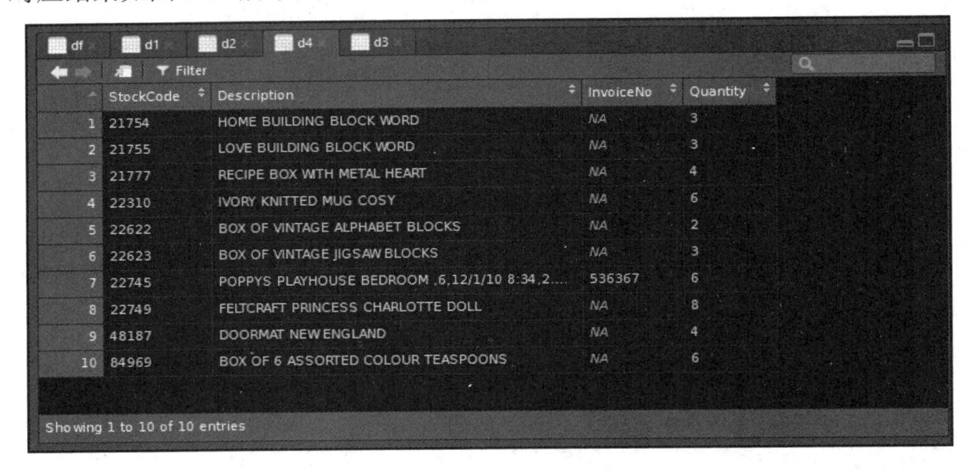

	StockCode	Description	InvoiceNo	Quantity
1	21754	HOME BUILDING BLOCK WORD	NA	3
2	21755	LOVE BUILDING BLOCK WORD	NA	3
3	21777	RECIPE BOX WITH METAL HEART	NA	4
4	22310	IVORY KNITTED MUG COSY	NA	6
5	22622	BOX OF VINTAGE ALPHABET BLOCKS	NA	2
6	22623	BOX OF VINTAGE JIGSAW BLOCKS	NA	3
7	22745	POPPYS PLAYHOUSE BEDROOM ,6,12/1/10 8:34,2....	536367	6
8	22749	FELTCRAFT PRINCESS CHARLOTTE DOLL	NA	8
9	48187	DOORMAT NEW ENGLAND	NA	4
10	84969	BOX OF 6 ASSORTED COLOUR TEASPOONS	NA	6

Showing 1 to 10 of 10 entries

图 5.41

最后是交叉连接，如下列代码所示：

```
d4 = merge(d2, d3, by=NULL)
```

对应结果如图 5.42 所示。

	InvoiceNo	StockCode.x	Description.x	StockCode.y	Description.y	Quantity
1	536365	85123A	WHITE HANGING HEART T-LIGHT HOLDER	22745	POPPYS PLAYHOUSE BEDROOM ,6,12/1/10 8:34,2....	6
2	536365	71053	WHITE METAL LANTERN	22745	POPPYS PLAYHOUSE BEDROOM ,6,12/1/10 8:34,2....	6
3	536365	84406B	CREAM CUPID HEARTS COAT HANGER	22745	POPPYS PLAYHOUSE BEDROOM ,6,12/1/10 8:34,2....	6
4	536365	84029G	KNITTED UNION FLAG HOT WATER BOTTLE	22745	POPPYS PLAYHOUSE BEDROOM ,6,12/1/10 8:34,2....	6
5	536365	84029E	RED WOOLLY HOTTIE WHITE HEART.	22745	POPPYS PLAYHOUSE BEDROOM ,6,12/1/10 8:34,2....	6
6	536365	22752	SET 7 BABUSHKA NESTING BOXES	22745	POPPYS PLAYHOUSE BEDROOM ,6,12/1/10 8:34,2....	6
7	536365	21730	GLASS STAR FROSTED T-LIGHT HOLDER	22745	POPPYS PLAYHOUSE BEDROOM ,6,12/1/10 8:34,2....	6
8	536366	22633	HAND WARMER UNION JACK	22745	POPPYS PLAYHOUSE BEDROOM ,6,12/1/10 8:34,2....	6
9	536366	22632	HAND WARMER RED POLKA DOT	22745	POPPYS PLAYHOUSE BEDROOM ,6,12/1/10 8:34,2....	6
10	536367	84879	ASSORTED COLOUR BIRD ORNAMENT	22745	POPPYS PLAYHOUSE BEDROOM ,6,12/1/10 8:34,2....	6
11	536367	22745	POPPYS PLAYHOUSE BEDROOM ,6,12/1/10 8:34,2....	22745	POPPYS PLAYHOUSE BEDROOM ,6,12/1/10 8:34,2....	6
12	536365	85123A	WHITE HANGING HEART T-LIGHT HOLDER	22749	FELTCRAFT PRINCESS CHARLOTTE DOLL	8
13	536365	71053	WHITE METAL LANTERN	22749	FELTCRAFT PRINCESS CHARLOTTE DOLL	8
14	536365	84406B	CREAM CUPID HEARTS COAT HANGER	22749	FELTCRAFT PRINCESS CHARLOTTE DOLL	8
15	536365	84029G	KNITTED UNION FLAG HOT WATER BOTTLE	22749	FELTCRAFT PRINCESS CHARLOTTE DOLL	8
16	536365	84029E	RED WOOLLY HOTTIE WHITE HEART.	22749	FELTCRAFT PRINCESS CHARLOTTE DOLL	8
17	536365	22752	SET 7 BABUSHKA NESTING BOXES	22749	FELTCRAFT PRINCESS CHARLOTTE DOLL	8
18	536365	21730	GLASS STAR FROSTED T-LIGHT HOLDER	22749	FELTCRAFT PRINCESS CHARLOTTE DOLL	8
19	536366	22633	HAND WARMER UNION JACK	22749	FELTCRAFT PRINCESS CHARLOTTE DOLL	8

Showing 1 to 20 of 110 entries

图 5.42

与 pandas 类似，可以使用“by=”在两个数据项间指定.x 和.y，而非_x 和_y。考查下列代码：

```
d4 = merge(d2, d3, by="StockCode", all=T)
```

这表示为 StockCode 列上的外连接，对应结果如图 5.43 所示。

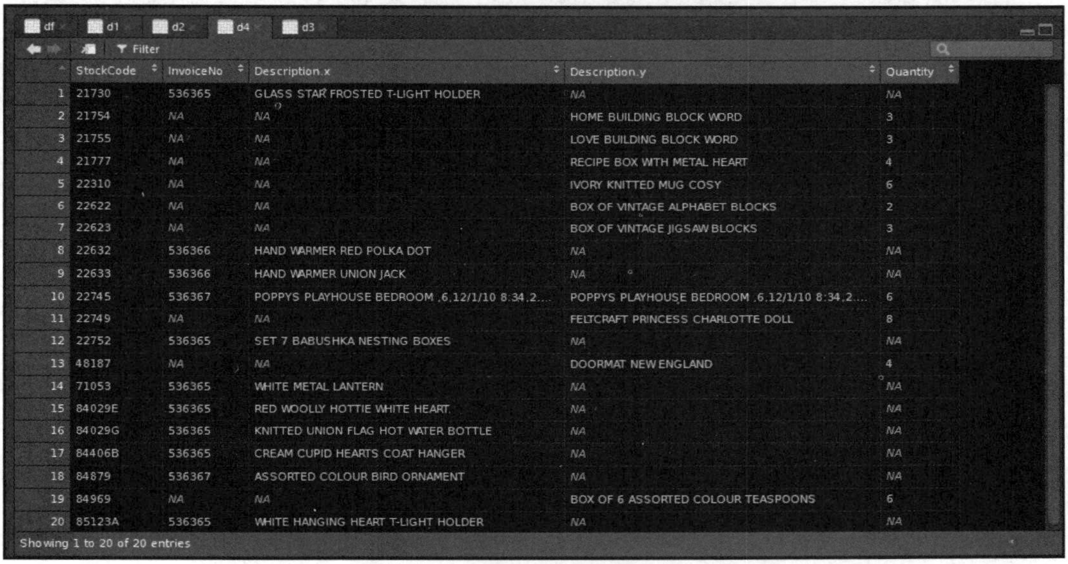

	StockCode	InvoiceNo	Description.x	Description.y	Quantity
1	21730	536365	GLASS STAR FROSTED T-LIGHT HOLDER	NA	NA
2	21754	NA	NA	HOME BUILDING BLOCK WORD	3
3	21755	NA	NA	LOVE BUILDING BLOCK WORD	3
4	21777	NA	NA	RECIPE BOX WITH METAL HEART	4
5	22310	NA	NA	IVORY KNITTED MUG COSY	6
6	22622	NA	NA	BOX OF VINTAGE ALPHABET BLOCKS	2
7	22623	NA	NA	BOX OF VINTAGE JIGSAW BLOCKS	3
8	22632	536366	HAND WARMER RED POLKA DOT	NA	NA
9	22633	536366	HAND WARMER UNION JACK	NA	NA
10	22745	536367	POPPYS PLAYHOUSE BEDROOM ,6,12/1/10 8:34,2....	POPPYS PLAYHOUSE BEDROOM ,6,12/1/10 8:34,2....	6
11	22749	NA	NA	FELTCRAFT PRINCESS CHARLOTTE DOLL	8
12	22752	536365	SET 7 BABUSHKA NESTING BOXES	NA	NA
13	48187	NA	NA	DOORMAT NEW ENGLAND	4
14	71053	536365	WHITE METAL LANTERN	NA	NA
15	84029E	536365	RED WOOLLY HOTTIE WHITE HEART.	NA	NA
16	84029G	536365	KNITTED UNION FLAG HOT WATER BOTTLE	NA	NA
17	84406B	536365	CREAM CUPID HEARTS COAT HANGER	NA	NA
18	84879	536367	ASSORTED COLOUR BIRD ORNAMENT	NA	NA
19	84969	NA	NA	BOX OF 6 ASSORTED COLOUR TEASPOONS	6
20	85123A	536365	WHITE HANGING HEART T-LIGHT HOLDER	NA	NA

Showing 1 to 20 of 20 entries

图 5.43

我们可以把所有的命令都记录下来，以备不时之需。执行下列代码可保存命令日志：

```
savehistory(file="logname.Rhistory")
```

执行下列代码可加载历史记录：

```
loadhistory(file="logname.Rhistory")
```

如果希望查看历史记录，则可简单地执行下列命令：

```
history()
```

对应结果如图 5.44 所示。

下列代码用于检查数据，以判断是否存在空数据：

```
colSums(is.na(df))
```

```
Environment  History  Connections
To Console  To Source
lines(d4$UnitPrice, type="p")
lines(d4$UnitPrice)
lines(d4$UnitPrice)
lines(d4$UnitPrice)
plot(d4$UnitPrice)
lines(d4$UnitPrice)
plot()
plot(d4$StockCode, d4$UnitPrice)
View(d4)
df = read.table(file=file.choose(), header=T, fill=T, sep=",")
View(df)
df = read.table(file=file.choose(), header=T, fill=T, sep=",")
history()
df = read.table(file=file.choose(), header=T, fill=T, sep=",")
d1 = df[0:10, c(1:8)]
savehistory(file=file.choose())
savehistory(file="log.Rhistory")
savehistory(file="logname.Rhistory")
```

图 5.44

对应结果如图 5.45 所示。

```
>     colSums(is.na(df))
  InvoiceNo   StockCode Description    Quantity InvoiceDate   UnitPrice  CustomerID     Country
          0           0           0           0           0           0           0           0
```

图 5.45

回忆一下，当合并两个数据帧时，某些数据值包含了 NaN，如下所示：

```
d2 = df[0:11, c("InvoiceNo", "StockCode", "Description")]
d3 = df[11:20, c("StockCode", "Description", "Quantity")]
```

接下来，对其进行外部合并，如下所示：

```
d4 = merge(d2, d3, all=T)
```

并尝试执行下列代码：

```
colSums(is.na(d4))
```

对应结果如图 5.46 所示。

```
> colSums(is.na(d4))
  StockCode Description   InvoiceNo    Quantity
          0           0           9          10
```

图 5.46

除此之外，还可对数据值进行替换。

假设希望将数据项（价格大于 3）的描述内容修改为"Miscellaneous"，考查下列代码：

```
d1 = df[0:30, c(1:8)]
```

对应结果如图 5.47 所示。

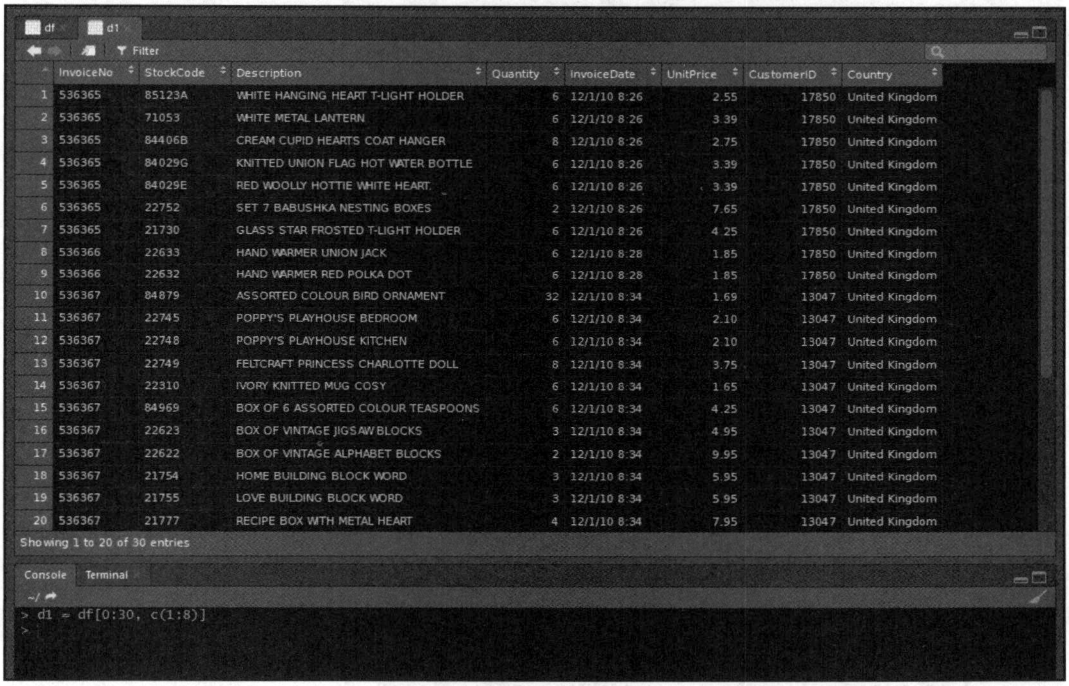

	InvoiceNo	StockCode	Description	Quantity	InvoiceDate	UnitPrice	CustomerID	Country
1	536365	85123A	WHITE HANGING HEART T-LIGHT HOLDER	6	12/1/10 8:26	2.55	17850	United Kingdom
2	536365	71053	WHITE METAL LANTERN	6	12/1/10 8:26	3.39	17850	United Kingdom
3	536365	84406B	CREAM CUPID HEARTS COAT HANGER	8	12/1/10 8:26	2.75	17850	United Kingdom
4	536365	84029G	KNITTED UNION FLAG HOT WATER BOTTLE	6	12/1/10 8:26	3.39	17850	United Kingdom
5	536365	84029E	RED WOOLLY HOTTIE WHITE HEART.	6	12/1/10 8:26	3.39	17850	United Kingdom
6	536365	22752	SET 7 BABUSHKA NESTING BOXES	2	12/1/10 8:26	7.65	17850	United Kingdom
7	536365	21730	GLASS STAR FROSTED T-LIGHT HOLDER	6	12/1/10 8:26	4.25	17850	United Kingdom
8	536366	22633	HAND WARMER UNION JACK	6	12/1/10 8:28	1.85	17850	United Kingdom
9	536366	22632	HAND WARMER RED POLKA DOT	6	12/1/10 8:28	1.85	17850	United Kingdom
10	536367	84879	ASSORTED COLOUR BIRD ORNAMENT	32	12/1/10 8:34	1.69	13047	United Kingdom
11	536367	22745	POPPY'S PLAYHOUSE BEDROOM	6	12/1/10 8:34	2.10	13047	United Kingdom
12	536367	22748	POPPY'S PLAYHOUSE KITCHEN	6	12/1/10 8:34	2.10	13047	United Kingdom
13	536367	22749	FELTCRAFT PRINCESS CHARLOTTE DOLL	8	12/1/10 8:34	3.75	13047	United Kingdom
14	536367	22310	IVORY KNITTED MUG COSY	6	12/1/10 8:34	1.65	13047	United Kingdom
15	536367	84969	BOX OF 6 ASSORTED COLOUR TEASPOONS	6	12/1/10 8:34	4.25	13047	United Kingdom
16	536367	22623	BOX OF VINTAGE JIGSAW BLOCKS	3	12/1/10 8:34	4.95	13047	United Kingdom
17	536367	22622	BOX OF VINTAGE ALPHABET BLOCKS	2	12/1/10 8:34	9.95	13047	United Kingdom
18	536367	21754	HOME BUILDING BLOCK WORD	3	12/1/10 8:34	5.95	13047	United Kingdom
19	536367	21755	LOVE BUILDING BLOCK WORD	3	12/1/10 8:34	5.95	13047	United Kingdom
20	536367	21777	RECIPE BOX WITH METAL HEART	4	12/1/10 8:34	7.95	13047	United Kingdom

Showing 1 to 20 of 30 entries

```
> d1 = df[0:30, c(1:8)]
>
```

图 5.47

随后查看下列代码：

```
d1[d1$UnitPrice > 3, "Description"] <- "Miscellaneous"
```

对应结果如图 5.48 所示。

从图 5.48 中可以看到，单价大于 3 的数据项，其描述内容已修改为"Miscellaneous"。

除了“>”之外，还可使用其他运算符，进而替换其他列中的数值。

假设包含发票号 536365 的各数据项均来自 United States。鉴于包含了相同的发票号和发票日期，因而可使用二者之一选择列，如下所示：

```
d1[d1$InvoiceNo == 536365, "Country"] = "United States"
```

对应结果如图 5.49 所示。

	InvoiceNo	StockCode	Description	Quantity	InvoiceDate	UnitPrice	CustomerID	Country
1	536365	85123A	WHITE HANGING HEART T-LIGHT HOLDER	6	12/1/10 8:26	2.55	17850	United Kingdom
2	536365	71053	Miscellaneous	6	12/1/10 8:26	3.39	17850	United Kingdom
3	536365	84406B	CREAM CUPID HEARTS COAT HANGER	8	12/1/10 8:26	2.75	17850	United Kingdom
4	536365	84029G	Miscellaneous	6	12/1/10 8:26	3.39	17850	United Kingdom
5	536365	84029E	Miscellaneous	6	12/1/10 8:26	3.39	17850	United Kingdom
6	536365	22752	Miscellaneous	2	12/1/10 8:26	7.65	17850	United Kingdom
7	536365	21730	Miscellaneous	6	12/1/10 8:26	4.25	17850	United Kingdom
8	536366	22633	HAND WARMER UNION JACK	6	12/1/10 8:28	1.85	17850	United Kingdom
9	536366	22632	HAND WARMER RED POLKA DOT	6	12/1/10 8:28	1.85	17850	United Kingdom
10	536367	84879	ASSORTED COLOUR BIRD ORNAMENT	32	12/1/10 8:34	1.69	13047	United Kingdom
11	536367	22745	POPPY'S PLAYHOUSE BEDROOM	6	12/1/10 8:34	2.10	13047	United Kingdom
12	536367	22748	POPPY'S PLAYHOUSE KITCHEN	6	12/1/10 8:34	2.10	13047	United Kingdom
13	536367	22749	Miscellaneous	8	12/1/10 8:34	3.75	13047	United Kingdom
14	536367	22310	IVORY KNITTED MUG COSY	6	12/1/10 8:34	1.65	13047	United Kingdom
15	536367	84969	Miscellaneous	6	12/1/10 8:34	4.25	13047	United Kingdom
16	536367	22623	Miscellaneous	3	12/1/10 8:34	4.95	13047	United Kingdom
17	536367	22622	Miscellaneous	2	12/1/10 8:34	9.95	13047	United Kingdom
18	536367	21754	Miscellaneous	3	12/1/10 8:34	5.95	13047	United Kingdom
19	536367	21755	Miscellaneous	3	12/1/10 8:34	5.95	13047	United Kingdom
20	536367	21777	Miscellaneous	4	12/1/10 8:34	7.95	13047	United Kingdom

Showing 1 to 20 of 30 entries

```
> d1 = df[0:30, c(1:8)]
> d1[d1$UnitPrice > 3, "Description"] <- "Miscellaneous"
>
```

图 5.48

	InvoiceNo	StockCode	Description	Quantity	InvoiceDate	UnitPrice	CustomerID	Country
1	536365	85123A	WHITE HANGING HEART T-LIGHT HOLDER	6	12/1/10 8:26	2.55	17850	United States
2	536365	71053	Miscellaneous	6	12/1/10 8:26	3.39	17850	United States
3	536365	84406B	CREAM CUPID HEARTS COAT HANGER	8	12/1/10 8:26	2.75	17850	United States
4	536365	84029G	Miscellaneous	6	12/1/10 8:26	3.39	17850	United States
5	536365	84029E	Miscellaneous	6	12/1/10 8:26	3.39	17850	United States
6	536365	22752	Miscellaneous	2	12/1/10 8:26	7.65	17850	United States
7	536365	21730	Miscellaneous	6	12/1/10 8:26	4.25	17850	United States
8	536366	22633	HAND WARMER UNION JACK	6	12/1/10 8:28	1.85	17850	United Kingdom
9	536366	22632	HAND WARMER RED POLKA DOT	6	12/1/10 8:28	1.85	17850	United Kingdom
10	536367	84879	ASSORTED COLOUR BIRD ORNAMENT	32	12/1/10 8:34	1.69	13047	United Kingdom
11	536367	22745	POPPY'S PLAYHOUSE BEDROOM	6	12/1/10 8:34	2.10	13047	United Kingdom
12	536367	22748	POPPY'S PLAYHOUSE KITCHEN	6	12/1/10 8:34	2.10	13047	United Kingdom
13	536367	22749	Miscellaneous	8	12/1/10 8:34	3.75	13047	United Kingdom
14	536367	22310	IVORY KNITTED MUG COSY	6	12/1/10 8:34	1.65	13047	United Kingdom
15	536367	84969	Miscellaneous	6	12/1/10 8:34	4.25	13047	United Kingdom
16	536367	22623	Miscellaneous	3	12/1/10 8:34	4.95	13047	United Kingdom
17	536367	22622	Miscellaneous	2	12/1/10 8:34	9.95	13047	United Kingdom
18	536367	21754	Miscellaneous	3	12/1/10 8:34	5.95	13047	United Kingdom
19	536367	21755	Miscellaneous	3	12/1/10 8:34	5.95	13047	United Kingdom
20	536367	21777	Miscellaneous	4	12/1/10 8:34	7.95	13047	United Kingdom

Showing 1 to 20 of 30 entries

图 5.49

需要注意的是，这里使用了“=”，而非“<-”。在当前上下文中，二者均表示为赋值操作，因而可视作等同。

5.4　本 章 小 结

本章讨论了如何使用 R 语言执行数据分析。此外，还介绍了 R 语言和 Hadoop 间的各种集成方案。

第 6 章将讨论 Apache Spark，并在批处理模型的基础上以此执行大数据分析。

第 6 章　Apache Spark 批处理分析

本章主要讨论 Apache Spark，以及如何根据批处理模型进行大数据分析。Spark SQL 是位于 Spark Core 之上的一个组件，可用于查询结构化数据。它正在成为事实上的工具，并逐渐取代了 Hive，成为 Hadoop 批量分析的一种选择方案。

除此之外，本章还将学习如何使用 Spark 并针对结构化数据进行分析（相应地，包含任意文本的文档则是非结构化数据的一个例子，以及需要转换为结构化形式的其他一些数据格式）。我们将考查 DataFrame/数据集的基础知识，以及 SparkSQL API 如何简化结构化数据的查询任务，同时兼顾其健壮性特征。

此外，本章还将介绍数据集，并讨论数据集、DataFrame 和 RDD 间的差异。具体来说，本章主要包含以下主题：

- SparkSQL 和 DataFrame。
- DataFrame 和 SQL API。
- DataFrame 模式。
- 数据集和编码器。
- 加载和保存数据。
- 聚合。
- 连接。

6.1　SparkSQL 和 DataFrame

在 Apache Spark 之前，当在大型数据集上执行 SQL 查询操作时，一般采用 Apache Hive 技术予以实现。Apache Hive 本质上将 SQL 查询转换为 MapReduce，在不需要编写 Java 和 Scala 复杂代码的情况下，可简化多种大数据分析工作。

随着 Apache Spark 的出现，在大数据范围内进行分析方面，存在一个范式转变。Spark SQL 在 Apache Spark 的分布式计算功能之上提供了一个 SQL 层，并以此简化应用。实际上，Spark SQL 可用作在线分析处理数据库。Spark SQL 的工作方式可描述为：将 SQL 语句解析为抽象语法树（AST），随后将该计划转换为逻辑执行计划，并将该逻辑计划优化为可执行的物理计划，如图 6.1 所示。

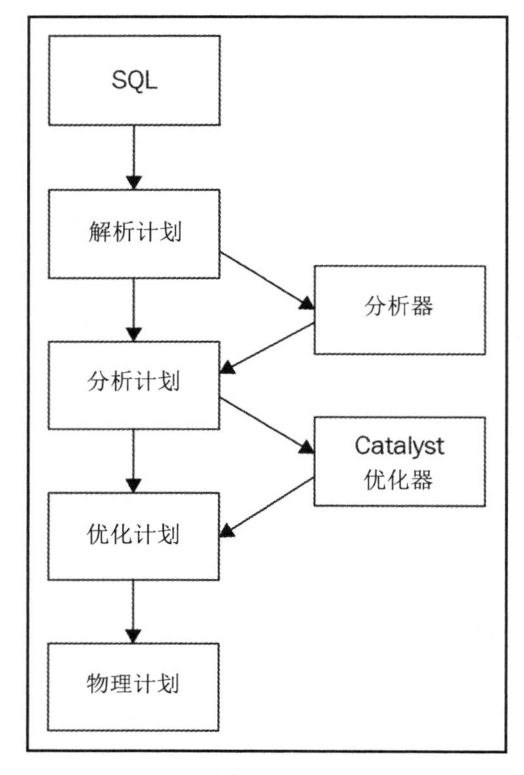

图 6.1

最终执行过程采用了底层的 DataFrame API。通过简单地使用 SQL 接口（无须了解内部情况），对于任何用户来说，都可方便地使用 DataFrame API。本章主要考查 DataFrame API，并通过相关示例展示 Spark SQL，进而比较 API 的不同使用方式。因此，DataFrame API 可视作 Spark SQL 的底层内容。此外，本章还将利用各种技术介绍 DataFrame 的构建方式，包括 SQL 查询，以及 DataFrame 上的各种操作。

DataFrame 表示为弹性分布式数据集（RDD）之上的抽象层，用于处理优化后的高层函数，其高效性通过 Project Tungsten 得以充分体现。

Project Tungsten 可视作 Spark 引擎的最大变化，主要体现在 CPU 的性能以及 Spark 应用程序的内存方面。该项目包含以下内容：

- 内存管理和二进制处理。
- 缓存计算。
- 代码生成。

提示：

关于 Project Tungsten 的更多信息，读者访问 https://databricks.com/blog/2015/04/28/project-tungsten-bringing-spark-closer-to-bare-metal.html。

我们可将数据集视为 RDD 上高效的表，其中包含了优化后的数据的二进制表达。这里，二进制表达通过编码器实现，并将各类对象序列化为二进制结果。与 RDD 相比，这将体现更好的性能。由于 DataFrame 在内部使用了 RDD，因此，DataFrame/数据集的分布方式实际上与 RDD 相似，因而也是一类分布式数据集。显然，这意味着数据集是不可变的。

图 6.2 显示了数据的二进制表达方式。

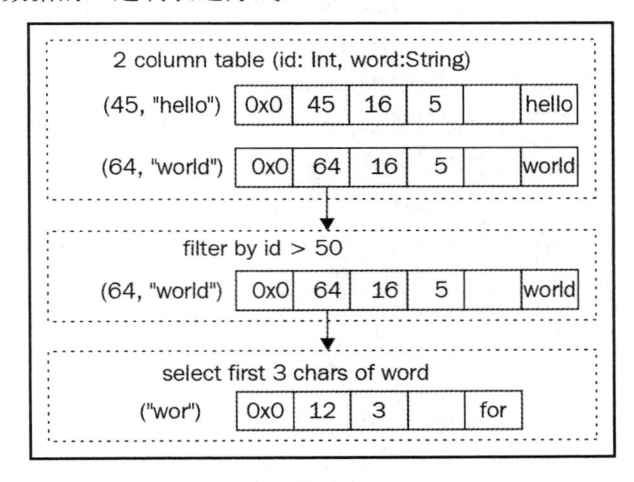

图 6.2

Spark 1.6 中加入了数据集，并在 DataFrame 上提供了强类型机制。实际上，自 Spark 2.0 以来，DataFrame 即简单地表示为数据集（Dataset）的别名。

注意：

http://spark.apache.org/sql/中将 DataFrame 类型定义为 Dataset[Row]，这意味着，大多数 API 均可与数据集和 DataFrame.type DataFrame = Dataset[Row]实现良好的协同工作。

从概念上讲，DataFrame 类似于关系数据库中的表。因此，DataFrame 包含了数据行，且每行由多个列构成。需要注意的是，类似于 RDD，DataFrame 同样不可变。DataFrame 的这种不可变属性意味着，每次转换或操作都会创建一个新的 DataFrame。

下面进一步讨论 DataFrame 及其与 RDD 间的差别。如前所述，RDD 表示为 Apache

Spark 中数据操控的底层 API。DataFrame 则在 EDD 之上创建，同时抽象了 RDD 的底层内部工作机制，且仅展示高层 API——除了易于使用之外，还提供了丰富的功能项。DataFrame 是按照 Python panda 包、R、Julia 等语言中的类似概念加以创建的。

如前所述，DataFrame 将 SQL 代码和特定域的语言表达转换为优化后的执行计划，并在 Spark Core API 之上运行，以便 SQL 语句执行广泛的操作。DataFrame 支持多种不同类型的数据源和操作，其中包括大多数数据库中所有的 SQL 操作类型，例如连接、分组、聚合以及窗口函数。

Spark SQL 与 Hive 查询语言十分类似，其原因在于：Spark 提供了与 Apache Hive 之间的天然适配器，采用 Apache Hive 工作的用户可方便地将其知识运用于 Spark SQL 中，从而进一步缩短了过渡期。实际上，DataFrame 依赖于之前所谈到的表这一概念。

另外，表操作与 Apache Hive 的工作方式也十分类似。实际上，Apache Spark 中的表操作与 Apache Hive 的表处理方式较为相似。对于 DataFrame 表，可将其注册为 DataFrame 表，并利用 Spark SQL 语句对数据进行操作，进而替代 DataFrame API。

DataFrame 依赖于 catalyst 优化器以及 Tungsten 所带来的性能改进，下面简要介绍一下 catalyst 优化器的工作方式。catalyst 优化器从输入的 SQL 中创建解析后的逻辑执行计划，并于随后通过查看全部属性和 SQL 语句中所用的列分析逻辑执行计划。一旦创建了经分析后的逻辑执行计划，catalyst 优化器将尝试进一步优化该执行计划，即整合多项操作，并重组当前逻辑，进而获取更好的性能。

注意：

为了更好地理解 catalyst 优化器，可将其视为一般意义上的逻辑优化器，并可对过滤器和转换这一类操作重新排序；某些时候，还可将多项操作整合为单一操作，进而最小化 worker 节点间混洗的数据量。例如，当在不同的数据集间执行连接操作时，catalyst 优化器可决定广播较小的数据集。另外，catalyst 优化器还可计算 DataFrame 列和分区的统计结果，进而改善执行速度。

如果数据分区上存在转换和过滤器，那么过滤数据和应用转换的顺序对操作的整体性能影响很大。作为最终的优化结果，这将生成优化后的逻辑执行计划，然后将其转换为物理执行计划。

显然，多项物理执行计划可能会执行相同的 SQL 语句，进而生成同一结果。基于开销优化和估算结果，开销优化逻辑负责确定、选取较好的物理执行计划。与之前的版本相比（例如 Spark 1.6 或更早的版本），Tungsten 是 Spark 2.x 性能改进背后的另一个关键因素。Tungsten 实现了对内存管理和其他性能的整体改进。其中，大多数重要的内存管理改进措施采用了二进制对象编码，并在堆外和堆内内存中对其加以引用。因此，Tungsten 支持堆内存应用，即采用二进制编码机制对所有对象进行编码。相应地，二进制编码对象所占用的内存空间则要少得多。

除此之外，Project Tungsten 还进一步改进了混洗性能。具体来说，数据一般通过 DataFrameReader 加载至 DataFrame 中，并通过 DataFrameWriter 从 DataFrame 中予以保存。

6.2　DataFrame API 和 SQL API

DataFrame 可通过多种方式创建，具体如下：

- 执行 SQL 查询，加载 Parquet、JSON、CSV、Text、Hive、JDBC 等外部数据。
- 将 RDD 转换为 DataFrame。
- 加载 CSV 文件。

此处将考查大量的 statesPopulation.csv 文件，并于随后将其作为 DataFrame 加载。

表 6.1 显示了 CSV 文件包含的美国各州的人口数量（2010—2016 年）。

表 6.1

State	Year	Population
Alabama	2010	4785492
Alaska	2010	714031
Arizona	2010	6408312
Arkansas	2010	2921995
California	2010	37332685

由于 CSV 文件包含了文件头，因而可利用隐式模式检测快速地将数据加载至 DataFrame 中，如下所示：

```
scala> val statesDF = spark.read.option("header",
"true").option("inferschema", "true").option("sep",
",").csv("statesPopulation.csv")
statesDF: org.apache.spark.sql.DataFrame = [State: string, Year: int ...
1 more field]
```

待 DataFrame 加载完毕后，即可执行模式检测，如下所示：

```
scala> statesDF.printSchema
root
|-- State: string (nullable = true)
```

```
|-- Year: integer (nullable = true)
|-- Population: integer (nullable = true)
```

提示：

这里，option("header","true").option("inferschema", "true").option("sep", ",")通知 Spark，CSV 文件包含了文件头。其中，逗号分隔符用于分隔字段/列，也可以隐式地对模式进行推断。

DataFrame 将解析逻辑执行计划，对其进行分析和优化，最终运行物理执行计划。DataFrame 上的 explain 显示了执行计划，如下所示：

```
scala> statesDF.explain(true)
== Parsed Logical Plan ==
Relation[State#0,Year#1,Population#2] csv
== Analyzed Logical Plan ==
State: string, Year: int, Population: int
Relation[State#0,Year#1,Population#2] csv
== Optimized Logical Plan ==
Relation[State#0,Year#1,Population#2] csv
== Physical Plan ==
*FileScan csv [State#0,Year#1,Population#2] Batched: false, Format: CSV,
Location: InMemoryFileIndex[file:/Users/salla/states.csv],
PartitionFilters: [], PushedFilters: [], ReadSchema:
struct<State:string,Year:int,Population:int>
```

DataFrame 还可注册为一个表名，随后可像关系数据库那样输入 SQL 语句，如下所示：

```
scala> statesDF.createOrReplaceTempView("states")
```

一旦 DataFrame 定义为结构化的 DataFrame 或表，即可运行相关命令并对数据进行操作，如下所示：

```
scala> statesDF.show(5)
scala> spark.sql("select * from states limit 5").show
+----------+----+----------+
| State|Year|Population|
+----------+----+----------+
| Alabama|2010| 4785492|
| Alaska|2010| 714031|
| Arizona|2010| 6408312|
| Arkansas|2010| 2921995|
|California|2010| 37332685|
+----------+----+----------+
```

在上述代码片段中，我们使用了 SQL 语句，并通过 spark.sql API 执行该语句。

提示：

Spark SQL 被简单地转换为 DataFrame API 并用以执行，而 SQL 只是易于使用的 DSL。

当对 DataFrame 使用 sort 操作时，可通过任意列对 DataFrame 中的行进行排序。此处，我们利用 Population 实现了降序排列；行则通过 Population 实现了降序排列，如下所示：

```
scala> statesDF.sort(col("Population").desc).show(5)
scala> spark.sql("select * from states order by Population desc limit
5").show
+----------+----+----------+
| State|Year|Population|
+----------+----+----------+
|California|2016| 39250017|
|California|2015| 38993940|
|California|2014| 38680810|
|California|2013| 38335203|
|California|2012| 38011074|
+----------+----+----------+
```

使用 groupBy，可以按任何列对 DataFrame 进行分组。下列代码通过 State 对行进行分组，随后针对每一个 State 累加 Population 计数。

```
scala> statesDF.groupBy("State").sum("Population").show(5)
scala> spark.sql("select State, sum(Population)
from states group by State
limit 5").show
+---------+---------------+
| State|sum(Population)|
+---------+---------------+
| Utah| 20333580|
| Hawaii| 9810173|
|Minnesota| 37914011|
| Ohio| 81020539|
| Arkansas| 20703849|
+---------+---------------+
```

使用 agg，可在 DataFrame 的列中执行不同的操作，例如计算某个列的 min、max、avg。除此之外，还可同时执行上述操作并对列进行重命名，进而满足相关用例的要求，如下所示：

```
scala>
statesDF.groupBy("State").agg(sum("Population").alias("Total")).show(5)
scala> spark.sql("select State, sum(Population) as Total from states group
by State limit 5").show
+---------+--------+
|    State|   Total|
+---------+--------+
|     Utah|20333580|
|   Hawaii| 9810173|
|Minnesota|37914011|
|     Ohio|81020539|
| Arkansas|20703849|
+---------+--------+
```

自然状态下，逻辑越复杂，执行计划也随之变得更加复杂。下面深入考查执行 groupBy 和 agg API 操作的计划。下列代码显示了 groupBy 子句，以及每个 State 人口总量的执行计划。

```
scala>
statesDF.groupBy("State").agg(sum("Population").alias("Total")).explain
(true)
== Parsed Logical Plan ==
'Aggregate [State#0], [State#0, sum('Population) AS Total#31886]
+- Relation[State#0,Year#1,Population#2] csv
== Analyzed Logical Plan ==
State: string, Total: bigint
Aggregate [State#0], [State#0, sum(cast(Population#2 as bigint)) AS
Total#31886L]
+- Relation[State#0,Year#1,Population#2] csv
== Optimized Logical Plan ==
Aggregate [State#0], [State#0, sum(cast(Population#2 as bigint)) AS
Total#31886L]
+- Project [State#0, Population#2]
+- Relation[State#0,Year#1,Population#2] csv
== Physical Plan ==
*HashAggregate(keys=[State#0], functions=[sum(cast(Population#2 as
bigint))], output=[State#0, Total#31886L])
+- Exchange hashpartitioning(State#0, 200)
+- *HashAggregate(keys=[State#0], functions=[partial_sum(cast(Population#2
as bigint))], output=[State#0, sum#31892L])
+- *FileScan csv [State#0,Population#2] Batched: false, Format: CSV,
Location: InMemoryFileIndex[file:/Users/salla/states.csv],
```

```
PartitionFilters: [], PushedFilters: [], ReadSchema:
struct<State:string,Population:int>
```

DataFrame 可实现链式操作，其间，执行过程可充分利用（开销）优化结果，即 Tungsten 所带来的性能改进，以及 catalyst 优化器。除此之外，还可在单一语句中实现链式操作，其中涉及基于 State 列的分组操作、Population 值的求和，以及基于列求和结果的 DataFrame 排序，如下所示：

```
scala>
statesDF.groupBy("State").agg(sum("Population").alias("Total")).sort(col
("Total").desc).show(5)
scala> spark.sql("select State, sum(Population) as Total from states group
by State order by Total desc limit 5").show
+----------+---------+
| State| Total|
+----------+---------+
|California|268280590
| Texas|185672865|
| Florida|137618322|
| New York|137409471|
| Illinois| 89960023|
+----------+---------+
```

上述链式操作包含了多项转换和操作，如图 6.3 所示。

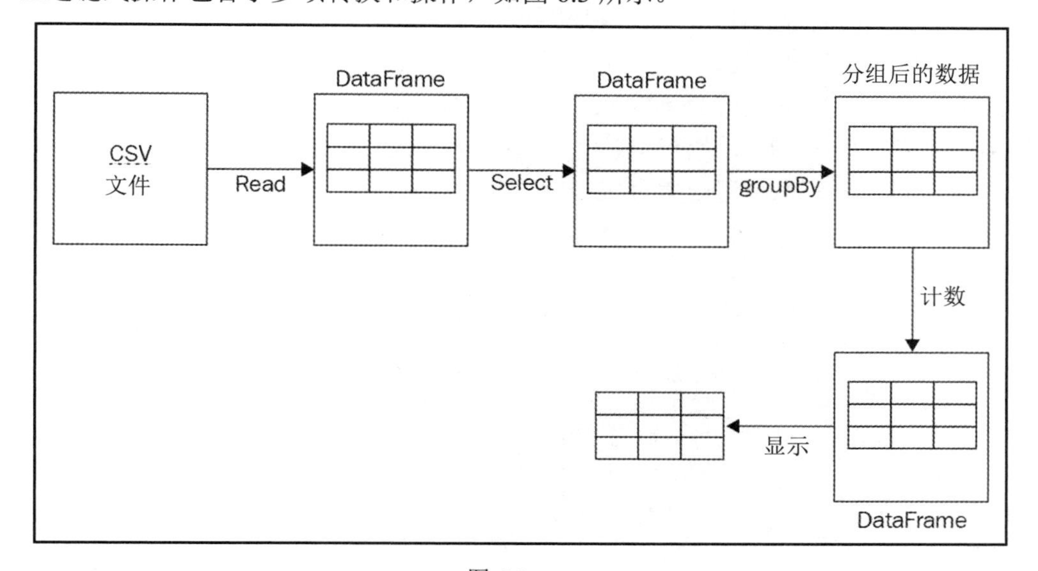

图 6.3

另外，还可同时创建多项聚合操作，如下所示：

```
scala> statesDF.groupBy("State").agg(
min("Population").alias("minTotal"),
max("Population").alias("maxTotal"),
avg("Population").alias("avgTotal"))
.sort(col("minTotal").desc).show(5)
scala> spark.sql("select State, min(Population) as minTotal,
max(Population) as maxTotal, avg(Population) as avgTotal from states group
by State order by minTotal desc limit 5").show
+----------+--------+--------+--------------------+
| State|minTotal|maxTotal| avgTotal|
+----------+--------+--------+--------------------+
|California|37332685|39250017|3.8325798571428575E7|
| Texas|25244310|27862596| 2.6524695E7|
| New York|19402640|19747183| 1.962992442857143E7|
| Florida|18849098|20612439|1.9659760285714287E7|
| Illinois|12801539|12879505|1.2851431857142856E7|
+----------+--------+--------+--------------------+
```

6.2.1　旋转

为了创建更适合执行多项汇总和聚合的不同视图，转换表的最佳方法之一是旋转。对此，可以通过列值并使每个值都成为实际的列来实现这一点。

下面将通过 Year 旋转 DataFrame 中的行，并对结果进行检查。最终结果涵盖了 Year 列中的数值，其中，每个值都形成了一个新的列。这样做的最终结果是，我们不仅可以查看 Year 列，还可以使用创建的 Per Year 列按 Year 进行汇总和聚合，如下所示：

```
scala> statesDF.groupBy("State").pivot("Year").sum("Population").show(5)
+---------+--------+--------+--------+--------+--------+--------+
| State| 2010| 2011| 2012| 2013| 2014| 2015| 2016|
+---------+--------+--------+--------+--------+--------+--------+
| Utah| 2775326| 2816124| 2855782| 2902663| 2941836| 2990632| 3051217|
| Hawaii| 1363945| 1377864| 1391820| 1406481| 1416349| 1425157| 1428557|
|Minnesota| 5311147| 5348562| 5380285| 5418521| 5453109| 5482435| 5519952|
|Ohio|11540983|11544824|11550839|11570022|11594408|11605090|11614373|
| Arkansas| 2921995| 2939493| 2950685| 2958663| 2966912| 2977853| 2988248|
+---------+--------+--------+--------+--------+--------+--------+
```

6.2.2　过滤器

DataFrame 也支持过滤器。通过过滤 DataFrame 行，还可用于生成新的 DataFrame。Filter 可对数据执行较为重要的转换，进而将 DataFrame 缩减至当前用例。

下面考查 DataFrame 过滤机制的执行计划，且仅查看 California 的状态，如下所示：

```
scala> statesDF.filter("State == 'California'").explain(true)
== Parsed Logical Plan ==
'Filter ('State = California)
+- Relation[State#0,Year#1,Population#2] csv
== Analyzed Logical Plan ==
State: string, Year: int, Population: int
Filter (State#0 = California)
+- Relation[State#0,Year#1,Population#2] csv
== Optimized Logical Plan ==
Filter (isnotnull(State#0) && (State#0 = California))
+- Relation[State#0,Year#1,Population#2] csv
== Physical Plan ==
*Project [State#0, Year#1, Population#2]
+- *Filter (isnotnull(State#0) && (State#0 = California))
+- *FileScan csv [State#0,Year#1,Population#2] Batched: false, Format:
CSV, Location: InMemoryFileIndex[file:/Users/salla/states.csv],
PartitionFilters: [], PushedFilters: [IsNotNull(State),
EqualTo(State,California)], ReadSchema:
struct<State:string,Year:int,Population:int>
```

根据上述执行计划，下面执行 filter 命令，如下所示：

```
scala> statesDF.filter("State == 'California'").show
+----------+----+----------+
|     State|Year|Population|
+----------+----+----------+
|California|2010|  37332685|
|California|2011|  37676861|
|California|2012|  38011074|
|California|2013|  38335203|
|California|2014|  38680810|
|California|2015|  38993940|
|California|2016|  39250017|
+----------+----+----------+
```

6.2.3　用户定义的函数

用户定义的函数（UDF）定义了基于列的新函数，并扩展了 Spark SQL 的功能项。当 Spark 的内建函数无法处理某些场景时，UDF 将变得十分有用。

注意：

udf()在内部调用 UserDefinedFunction 类，该类将于内部依次调用 ScalaUDF。

UDF 简单地将 State 列值转换为大写。首先，需要在 Scala 中构建函数，如下所示：

```
import org.apache.spark.sql.functions._
scala> val toUpper: String => String = _.toUpperCase
toUpper: String => String = <function1>
```

随后需要将所生成的函数封装至 udf 中，进而构建 UDF，如下所示：

```
scala> val toUpperUDF = udf(toUpper)
toUpperUDF: org.apache.spark.sql.expressions.UserDefinedFunction =
UserDefinedFunction(<function1>,StringType,Some(List(StringType)))
```

利用所生成的 udf，可将 State 列转换为大写内容，如下所示：

```
scala> statesDF.withColumn("StateUpperCase",
toUpperUDF(col("State"))).show(5)
+----------+----+----------+--------------+
| State|Year|Population|StateUpperCase|
+----------+----+----------+--------------+
| Alabama|2010| 4785492| ALABAMA|
| Alaska|2010| 714031| ALASKA|
| Arizona|2010| 6408312| ARIZONA|
| Arkansas|2010| 2921995| ARKANSAS|
|California|2010| 37332685| CALIFORNIA|
+----------+----+----------+--------------+
```

6.3　模式——数据的结构

模式可视为一种数据结构的描述，可呈现为隐式或显式状态。由于 DataFrame 在内部基于 RDD，因而存在两种主要方式可将现有的 RDD 转换为数据集，具体如下：

- 利用反射推断 RDD 的模式。

- 在编程接口的帮助下，可通过现有的 RDD 展示某种模式，并将 EDD 转换为基于模式的数据集。

6.3.1 隐式模式

下面的示例尝试将逗号分隔的数值（CSV）文件加载至 DataFrame 中。若文本文件包含了一个文件头，那么，read API 可以此对模式进行推断。除此之外，还可指定分隔符，进而分隔文本文件行（可选）。

接下来，将读取 csv 文件并从文件头中推断对应的模式，同时使用逗号作为分隔符。此外，此处还将展示 schema 命令和 printSchema 命令的用法，进而验证输入文件的模式，如下所示：

```
scala> val statesDF = spark.read.option("header", "true")
 .option("inferschema", "true")
 .option("sep", ",")
 .csv("statesPopulation.csv")
statesDF: org.apache.spark.sql.DataFrame = [State: string, Year: int ... 1
more field]
scala> statesDF.schema
res92: org.apache.spark.sql.types.StructType = StructType(
StructField(State,StringType,true),
StructField(Year,IntegerType,true),
StructField(Population,IntegerType,true))
scala> statesDF.printSchema
root
|-- State: string (nullable = true)
|-- Year: integer (nullable = true)
|-- Population: integer (nullable = true)
```

6.3.2 显式模式

该模式使用 StructType 予以描述，并表示为 StructField 对象集合。

注意：

StructType 和 StructField 隶属于 StructType and StructField 包；例如，DataTypes（例如 IntegerType 和 StringType）同样属于 org.apache.spark.sql.types 包。

通过此类导入语句，即可自定义相应的显式模式。

首先须导入下列类：

```
scala> import org.apache.spark.sql.types.{StructType, IntegerType,
StringType}
import org.apache.spark.sql.types.{StructType, IntegerType, StringType}
```

下面定义一个模式，其中包含两个列/字段和一个整数，后跟一个字符串，如下所示：

```
scala> val schema = new StructType().add("i", IntegerType).add("s",
StringType)
schema: org.apache.spark.sql.types.StructType =
StructType(StructField(i,IntegerType,true),StructField(s,StringType,true))
```

对于刚刚生成的 schema，其输出结果如下所示：

```
scala> schema.printTreeString
root
|-- i: integer (nullable = true)
|-- s: string (nullable = true)
```

此外还可输出 JSON，其中使用了 prettyJson 函数，如下所示：

```
scala> schema.prettyJson
res85: String =
{
"type" : "struct",
"fields" : [ {
"name" : "i",
"type" : "integer",
"nullable" : true,
"metadata" : { }
}, {
"name" : "s",
"type" : "string",
"nullable" : true,
"metadata" : { }
} ]
}
```

注意，所有的 Spark SQL 数据类型均位于 org.apache.spark.sql.types 包中，并可通过下列导入语句对其进行访问：

```
import org.apache.spark.sql.types._
```

6.3.3　编码器

针对复杂的数据类型，Spark 2.x 支持多种模式定义方式。这里，首先考查一个简单的示例。其间，当使用 Encoders 时，需要通过 import 语句导入 Encoders，如下所示。

```
import org.apache.spark.sql.Encoders
```

下列示例将定义完毕的元组作为数据类型，并用于数据集 API 中。

```
scala> Encoders.product[(Integer, String)].schema.printTreeString
root
|-- _1: integer (nullable = true)
|-- _2: string (nullable = true)
```

上述代码稍显复杂且经常使用，因而可将其定义为一个类，以供后续操作使用。

相应地，Record 类包含了两个字段，即 Integer 和 String，如下所示：

```
scala> case class Record(i: Integer, s: String)
defined class Record
```

通过 Encoders，可方便地在类之上构建一个模式，进而可方便地使用各种 API，如下所示：

```
scala> Encoders.product[Record].schema.printTreeString
root
|-- i: integer (nullable = true)
|-- s: string (nullable = true)
```

所有的 Spark SQL 数据类型均位于 org.apache.spark.sql.types 包中，并可通过下列导入语句对其进行访问：

```
import org.apache.spark.sql.types._
```

此处应在代码中使用 DataTypes 对象，以便生成复杂的 Spark SQL 类型，例如数组和映射，如下所示：

```
scala> import org.apache.spark.sql.types.DataTypes
import org.apache.spark.sql.types.DataTypes
scala> val arrayType = DataTypes.createArrayType(IntegerType)
arrayType: org.apache.spark.sql.types.ArrayType =
ArrayType(IntegerType,true)
```

表 6.2 列出了 SparkSQL API 所支持的数据类型。

表 6.2

数 据 类 型	Scala 中的值类型	访问或创建数据类型的 API
ByteType	Byte	ByteType
ShortType	Short	ShortType
IntegerType	Int	IntegerType
LongType	Long	LongType
FloatType	Float	FloatType
DoubleType	Double	DoubleType
DecimalType	java.math.BigDecimal	DecimalType
StringType	String	StringType
BinaryType	Array[Byte]	BinaryType
BooleanType	Boolean	BooleanType
TimestampType	java.sql.Timestamp	TimestampType
DateType	java.sql.Date	DateType
ArrayType	scala.collection.Seq	ArrayType(elementType,[containsNull])
MapType	scala.collection.Map	MapType(keyType,valueType,[valueContainsNull]) 。注意，valueContainsNull 的默认值为 TRUE
StructType	org.apache.spark.sql.Row	StructType(fields)。注意，字段表示为 StructFields 的 Seq。另外，此处不支持包含相同名称的两个字段

6.4　加载数据集

Spark SQL 可通过 DataFrameReader 接口从外部存储系统中读取数据，例如文件、Hive 表和 JDBC 数据库。

API 调用格式定义为 spark.read.inputtype，如下所示：

- Parquet。
- CSV。
- Hive 表。
- JDBC。
- ORC。
- Text。
- JSON。

下面通过一组简单示例将 CSV 文件读入 DataFrame 中，如下所示：

```
scala> val statesPopulationDF = spark.read.option("header",
"true").option("inferschema", "true").option("sep",
",").csv("statesPopulation.csv")
statesPopulationDF: org.apache.spark.sql.DataFrame = [State: string, Year:
int ... 1 more field]
scala> val statesTaxRatesDF = spark.read.option("header",
"true").option("inferschema", "true").option("sep",
",").csv("statesTaxRates.csv")
statesTaxRatesDF: org.apache.spark.sql.DataFrame = [State: string, TaxRate:
double]
```

6.5　保存数据集

Spark SQL 可通过 DataFrameWriter 接口从外部存储系统中读取数据，例如文件、Hive 表和 JDBC 数据库。

API 调用格式定义为 dataframe.write.outputtype，如下所示：

- Parquet。
- ORC。
- Text。
- Hive 表
- JSON。
- CSV。
- JDBC

下面通过一组简单示例将 DataFrame 写入或保存至 CSV 文件中，如下所示：

```
scala> statesPopulationDF.write.option("header",
"true").csv("statesPopulation_dup.csv")
scala> statesTaxRatesDF.write.option("header",
"true").csv("statesTaxRates_dup.csv")
```

6.6　聚　　合

聚合表示为基于某种条件的数据采集方法，并在数据上执行分析操作。聚合对于理解所有尺寸的数据非常重要，因为对于大多数用例来说，仅仅拥有数据的原始记录并不是那么有用。

注意：

假设我们可定义一个表，其中包含了 5 年内世界每所城市每天的温度记录值。

如果查看表 6.3 以及同一数据的聚合视图，显然，原始记录无法帮助我们理解其中的数据。表 6.3 以表的形式列出了原始数据。

表 6.3

City	Date	Temperature
Boston	12/23/2016	32
New York	12/24/2016	36
Boston	12/24/2016	30
Philadelphia	12/25/2016	34
Boston	12/25/2016	28

表 6.4 显示了每所城市的平均温度。

表 6.4

City	Average Temperature
Boston	30 - (32 + 30 + 28)/3
New York	36
Philadelphia	34

6.6.1　聚合函数

在 org.apache.spark.sql.functions 包中的相关函数的帮助下，可执行聚合操作。除此之外，还可生成自定义聚合函数，也称作用户定义的聚合函数（UDAF）。

注意：

每个分组操作返回一个 RelationalGroupedDataset，并可在此之上指定聚合。

我们将加载示例数据，并显示所有聚合函数的不同类型，如下所示：

```
val statesPopulationDF = spark.read.option("header", "true").
 option("inferschema", "true").
 option("sep", ",").csv("statesPopulation.csv")
```

1. count

count 是一个最基本的聚合函数，并针对指定的列简单地计算行数量。countDistinct

则定义为 count 的扩展，同时还可消除重复内容。

count API 包含多种实现，实际所用的 API 取决于特定的用例，如下所示：

```
def count(columnName: String): TypedColumn[Any, Long]
 Aggregate function: returns the number of items in a group.
def count(e: Column): Column
 Aggregate function: returns the number of items in a group.
def countDistinct(columnName: String, columnNames: String*): Column
 Aggregate function: returns the number of distinct items in a group.
def countDistinct(expr: Column, exprs: Column*): Column
 Aggregate function: returns the number of distinct items in a group.
```

下面考查一些在 DataFrame 上调用 count 和 countDistinct 的例子，并输出行计数结果，如下所示：

```
import org.apache.spark.sql.functions._
scala> statesPopulationDF.select(col("*")).agg(count("State")).show
scala> statesPopulationDF.select(count("State")).show
+------------+
|count(State)|
+------------+
| 350|
+------------+
scala> statesPopulationDF.select(col("*")).agg(countDistinct("State")).show
scala> statesPopulationDF.select(countDistinct("State")).show
+---------------------+
|count(DISTINCT State)|
+---------------------+
| 50|
```

2. first

first 用于获取 RelationalGroupedDataset 中的第一条记录。

first API 同样包含多种实现，实际所用的 API 取决于特定的用例，如下所示：

```
def first(columnName: String): Column
 Aggregate function: returns the first value of a column in a group.
def first(e: Column): Column
 Aggregate function: returns the first value in a group.
def first(columnName: String, ignoreNulls: Boolean): Column
 Aggregate function: returns the first value of a column in a group.
def first(e: Column, ignoreNulls: Boolean): Column
 Aggregate function: returns the first value in a group.
```

下面考查 DataFrame 上的 first 调用，并输出第一行内容，如下所示：

```
import org.apache.spark.sql.functions._
 scala> statesPopulationDF.select(first("State")).show
+--------------------+
|first(State, false)|
+--------------------+
| Alabama|
+--------------------+
```

3. last

last 获取 RelationalGroupedDataset 中的最后一条记录。

last API 包含了多种实现，实际所用的 API 取决于特定的用例，如下所示：

```
def last(columnName: String): Column
 Aggregate function: returns the last value of the column in a group.
def last(e: Column): Column
 Aggregate function: returns the last value in a group.
def last(columnName: String, ignoreNulls: Boolean): Column
 Aggregate function: returns the last value of the column in a group.
def last(e: Column, ignoreNulls: Boolean): Column
 Aggregate function: returns the last value in a group.
```

下面考查 DataFrame 上的 last 调用，并输出最后一行内容，如下所示：

```
import org.apache.spark.sql.functions._
 scala> statesPopulationDF.select(last("State")).show
 +-------------------+
 |last(State, false)|
 +-------------------+
 | Wyoming|
 +-------------------+
```

4. approx_count_distinct

如果需要对不同的记录进行近似计数，那么，approx_count_distinct 则是一种更快的方法，而不是执行精确计数，这通常会涉及大量的混洗和其他操作。

approx_count_distinct API 包含多种实现，实际所用的 API 取决于特定的用例，如下所示：

```
def approx_count_distinct(columnName: String, rsd: Double): Column
 Aggregate function: returns the approximate number of distinct items in a
 group.
```

```
def approx_count_distinct(e: Column, rsd: Double): Column
 Aggregate function: returns the approximate number of distinct items in a
 group.
def approx_count_distinct(columnName: String): Column
 Aggregate function: returns the approximate number of distinct items in a
 group.
def approx_count_distinct(e: Column): Column
 Aggregate function: returns the approximate number of distinct items in a
 group.
```

下面考查 DataFrame 上的 approx_count_distinct 调用示例，并输出 DataFrame 的近似计数结果，如下所示：

```
import org.apache.spark.sql.functions._
 scala>
statesPopulationDF.select(col("*")).agg(approx_count_distinct("State")).show
 +----------------------------+
 |approx_count_distinct(State)|
 +----------------------------+
 | 48|
 +----------------------------+
 scala> statesPopulationDF.select(approx_count_distinct("State", 0.2)).show
 +----------------------------+
 |approx_count_distinct(State)|
 +----------------------------+
 | 49|
 +----------------------------+
```

5. min

min 表示为 DataFrame 中某一个列的最小列值。如果打算计算某个城市的最低温度，则可使用 min。

min API 包含了多种实现，实际所用的 API 取决于特定的用例，如下所示：

```
def min(columnName: String): Column
 Aggregate function: returns the minimum value of the column in a group.
def min(e: Column): Column
 Aggregate function: returns the minimum value of the expression in a group.
```

下面考查 DataFrame 上的 min 调用示例，并输出最小的 Population，如下所示：

```
import org.apache.spark.sql.functions._
 scala> statesPopulationDF.select(min("Population")).show
 +---------------+
```

```
|min(Population)|
+---------------+
| 564513|
+---------------+
```

6. max

max 表示为 DataFrame 中某个列的最大列值。如果打算计算某个城市的最高温度，则可使用 max。

max API 包含多种实现，实际所用的 API 取决于特定的用例，如下所示：

```
def max(columnName: String): Column
 Aggregate function: returns the maximum value of the column in a group.
def max(e: Column): Column
 Aggregate function: returns the maximum value of the expression in a group.
```

下面考查 DataFrame 上 max 的调用示例，并输出最大 Population，如下所示：

```
import org.apache.spark.sql.functions._
 scala> statesPopulationDF.select(max("Population")).show
+---------------+
 |max(Population)|
 +---------------+
 | 39250017|
 +---------------+
```

7. avg

数值的平均值计算方式可描述为：累加全部值并将对应结果除以数值的数量。

注意：

例如，1，2，3 的平均值为(1+2+3)/3=2。

avg API 包含多种实现，实际所用 denAPI 取决于特定的用例，如下所示：

```
def avg(columnName: String): Column
 Aggregate function: returns the average of the values in a group.
def avg(e: Column): Column
 Aggregate function: returns the average of the values in a group.
```

下面考查 DataFrame 上 avg 的调用示例，并输出 population 的平均值，如下所示：

```
import org.apache.spark.sql.functions._
 scala> statesPopulationDF.select(avg("Population")).show
```

```
+------------------+
| avg(Population)|
+------------------+
|6253399.371428572|
+-----------------+
```

8. sum

sum 用于计算列值的求和结果。作为一种备选方案，sumDistinct 仅可用于累加不同值。sumAPI 包含多种实现，实际所用的 API 取决于特定的用例，如下所示：

```
def sum(columnName: String): Column
 Aggregate function: returns the sum of all values in the given column.
def sum(e: Column): Column
 Aggregate function: returns the sum of all values in the expression.
def sumDistinct(columnName: String): Column
 Aggregate function: returns the sum of distinct values in the expression
def sumDistinct(e: Column): Column
 Aggregate function: returns the sum of distinct values in the expression.
```

下面考查 DataFrame 上 sum 调用示例，并输出 Population 的求和结果（全部值），如下所示：

```
import org.apache.spark.sql.functions._
scala> statesPopulationDF.select(sum("Population")).show
 +---------------+
 |sum(Population)|
 +---------------+
 | 2188689780|
 +---------------+
```

9. kurtosis

kurtosis 是一种量化分布形状差异的方法，从平均值和方差的角度看，它们可能非常相似，但实际上是不同的。

kurtosis API 包含多种实现，实际所用的 API 取决于特定的用例，如下所示：

```
def kurtosis(columnName: String): Column
 Aggregate function: returns the kurtosis of the values in a group.
def kurtosis(e: Column): Column
 Aggregate function: returns the kurtosis of the values in a group.
```

下面考查 DataFrame 上 kurtosis 的调用示例（在 Population 列上），如下所示：

```
import org.apache.spark.sql.functions._
scala> statesPopulationDF.select(kurtosis("Population")).show
+--------------------+
|kurtosis(Population)|
+--------------------+
| 7.727421920829375|
+--------------------+
```

10. skewness

skewness 围绕平均值或中位数计算数据值的非对称性。

skewness 包含了多种实现，实际所用的 API 取决于特定的用例，如下所示：

```
def skewness(columnName: String): Column
 Aggregate function: returns the skewness of the values in a group.
def skewness(e: Column): Column
 Aggregate function: returns the skewness of the values in a group.
```

下面考查 DataFrame 上 skewness 的调用示例（在 Population 列上），如下所示：

```
import org.apache.spark.sql.functions._
 scala> statesPopulationDF.select(skewness("Population")).show
 +--------------------+
 |skewness(Population)|
 +--------------------+
 | 2.5675329049100024|
 +--------------------+
```

11. 方差

方差表示为每个值与平均值之差的平方的平均数。

var API 包含多种实现，实际所用的 API 取决于特定的用例，如下所示：

```
def var_pop(columnName: String): Column
 Aggregate function: returns the population variance of the values in a
group.
def var_pop(e: Column): Column
 Aggregate function: returns the population variance of the values in a
group.
def var_samp(columnName: String): Column
 Aggregate function: returns the unbiased variance of the values in a group.
def var_samp(e: Column): Column
 Aggregate function: returns the unbiased variance of the values in a group.
```

下面考查 DataFrame 上 var_pop 的调用示例，并计算 Population 的方差，如下所示：

```
import org.apache.spark.sql.functions._
scala> statesPopulationDF.select(var_pop("Population")).show
 +--------------------+
 | var_pop(Population)|
 +--------------------+
 |4.948359064356177E13|
 +--------------------+
```

12. 标准偏差

标准偏差表示为方差的平方根。

stddev API 包含多种实现，实际所用的 API 取决于特定的用例，如下所示：

```
def stddev(columnName: String): Column
 Aggregate function: alias for stddev_samp.
def stddev(e: Column): Column
 Aggregate function: alias for stddev_samp.
def stddev_pop(columnName: String): Column
 Aggregate function: returns the population standard deviation of the
 expression in a group.
def stddev_pop(e: Column): Column
 Aggregate function: returns the population standard deviation of the
 expression in a group.
def stddev_samp(columnName: String): Column
 Aggregate function: returns the sample standard deviation of the
expression in a group.
def stddev_samp(e: Column): Column
Aggregate function: returns the sample standard deviation of the expression
in a group.
```

下面考查 DataFrame 上 stddev 的调用示例，并输出 vde 标准偏差，如下所示：

```
import org.apache.spark.sql.functions._
scala> statesPopulationDF.select(stddev("Population")).show
 +-----------------------+
 |stddev_samp(Population)|
 +-----------------------+
 | 7044528.191173398|
 +-----------------------+
```

13. 协方差

协方差是两个随机变量间联合变化的度量过程。

covar 包含多种实现，实际所用的 API 取决于特定的用例，如下所示：

```
def covar_pop(columnName1: String, columnName2: String): Column
 Aggregate function:returns the population covariance for two columns.
def covar_pop(column1: Column, column2: Column): Column
 Aggregate function:returns the population covariance for two columns.
def covar_samp(columnName1: String, columnName2: String): Column
 Aggregate function:returns the sample covariance for two columns.
def covar_samp(column1: Column, column2: Column): Column
 Aggregate function:returns the sample covariance for two columns.
```

下面考查 DataFrame 上 covar_pop 的调用示例，进而计算 Year 和 Population 列间的协方差，如下所示：

```
import org.apache.spark.sql.functions._
scala> statesPopulationDF.select(covar_pop("Year", "Population")).show
 +---------------------------+
 |covar_pop(Year, Population)|
 +---------------------------+
 | 183977.56000006935|
 +---------------------------+
```

14. groupBy

数据分析中较为常见的任务是将数据分组至不同的类别中，并于随后在对应的数据分组中执行计算。

下面尝试在 DataFrame 上运行 groupBy，并输出每个 State 的聚合计数结果，如下所示：

```
scala> statesPopulationDF.groupBy("State").count.show(5)
 +---------+-----+
| State|count|
 +---------+-----+
 | Utah| 7|
 | Hawaii| 7|
 |Minnesota| 7|
 | Ohio| 7|
 | Arkansas| 7|
 +---------+-----+
```

除此之外，还可在 groupBy 的基础上使用之前讨论的聚合函数，例如 min、max、avg、stddev 等，如下所示：

```
import org.apache.spark.sql.functions._
 scala> statesPopulationDF.groupBy("State").agg(min("Population"),
```

```
avg("Population")).show(5)
+---------+--------------+--------------------+
| State|min(Population)| avg(Population)|
+---------+--------------+--------------------+
| Utah| 2775326| 2904797.1428571427|
| Hawaii| 1363945| 1401453.2857142857|
|Minnesota| 5311147| 5416287.285714285|
| Ohio| 11540983|1.1574362714285715E7|
| Arkansas| 2921995| 2957692.714285714|
+---------+--------------+--------------------+
```

15. rollup

rollup 是一类多维聚合，用于执行层次或嵌套计算。例如，如果打算针对每个 State 和 Year 分组以及每个 State 显示记录的数量（在所有的年份中汇总，无论 Year 是什么，都能得到每个 State 的总计结果），则可使用 rollup，如下所示：

```
scala> statesPopulationDF.rollup("State", "Year").count.show(5)
+------------+----+-----+
| State|Year|count|
+------------+----+-----+
|South Dakota|2010| 1|
| New York|2012| 1|
| California|2014| 1|
| Wyoming|2014| 1|
| Hawaii|null| 7|
+------------+----+-----+
```

16. cube

cube 是一类多维聚合，用于执行层次或嵌套计算，这一点与 rollup 类似；二者间的差别在于，cube 针对所有维度均执行相同的操作。例如，如果打算针对每个 State 和 Year 分组以及每个 State 显示记录的数量（在所有的年份中汇总，无论 Year 是什么，都能得到每个 State 的总计结果），则可使用 cube，如下所示：

```
scala> statesPopulationDF.cube("State", "Year").count.show(5)
+------------+----+-----+
| State|Year|count|
+------------+----+-----+
|South Dakota|2010| 1|
| New York|2012| 1|
| null|2014| 50|
| Wyoming|2014| 1|
```

```
| Hawaii|null| 7|
+------------+----+-----+
```

6.6.2 窗口函数

窗口函数可在数据窗口上执行聚合操作，而不是全部数据或某些过滤后的数据。窗口函数的应用示例如下所示：

- ❑ 累计求和。
- ❑ 对于相同的键，相对于上一个值的 Delta。
- ❑ 加权移动平均值。

通过执行简单的计算，我们可指定一个窗口，并查看 T-1、T 和 T+1 行。除此之外，还可在最近/最新的 10 个值上指定一个窗口，如图 6.4 所示。

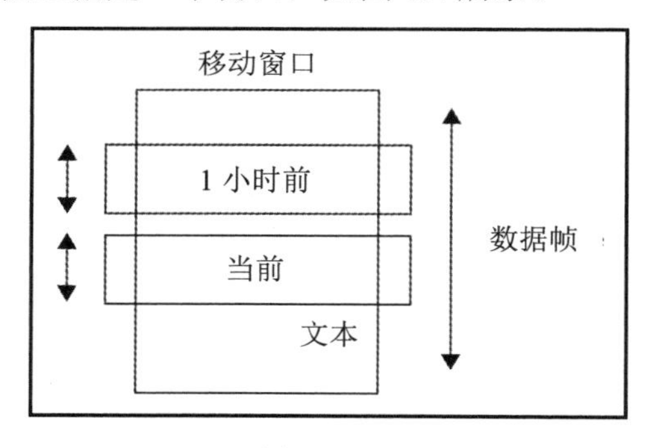

图 6.4

针对 Window 规范的 API 需要使用到 3 个属性，即 partitionBy()、orderBy()和 rowsBetween()。partitionBy 将数据分块至 partitionBy()指定的分区/分组中。orderBy()则用于对每个数据分区中的数据进行排序。

rowsBetween()指定了窗口帧，或者是移动窗口的跨度，进而执行相关计算。

当使用 Windows 函数时，需要导入相关数据包，如下所示：

```
import org.apache.spark.sql.expressions.Window
import org.apache.spark.sql.functions.col
import org.apache.spark.sql.functions.max
```

接下来，可针对经 Population 排序以及 State 划分的分区创建一个窗口规范。除此之外，还需要指定所考查所有的行，直到当前行为窗口的部分内容为止，如下所示：

```
val windowSpec = Window
 .partitionBy("State")
 .orderBy(col("Population").desc)
 .rowsBetween(Window.unboundedPreceding, Window.currentRow)
```

随后计算 Windows 规范的排名。只要对应行落入至所指定的 Windows 中，最终结果表示为添加至每行中的排名（行号）。在该示例中，可选择通过 State 进行划分，随后按降序对每个 State 行进行排序。因此，每个 State 行包含其自身的排名号，如下所示：

```
import org.apache.spark.sql.functions._
 scala> statesPopulationDF.select(col("State"), col("Year"),
 max("Population").over(windowSpec),
 rank().over(windowSpec)).sort("State",
 "Year").show(10)
 +-------+----+----------------------------------------------------------
-
 ----------------------------------------------------------+------------
-
 -----------------------------------------------------------------------
-
 ----------------------------------+
| State|Year|max(Population) OVER (PARTITION BY State ORDER BY Population
 DESC NULLS LAST ROWS BETWEEN UNBOUNDED PRECEDING AND CURRENT ROW)|RANK()
 OVER (PARTITION BY State ORDER BY Population DESC NULLS LAST ROWS BETWEEN
 UNBOUNDED PRECEDING AND CURRENT ROW)|
 +-------+----+----------------------------------------------------------
-
 ----------------------------------------------------------+------------
-
 -----------------------------------------------------------------------
-
 ----------------------------------+
|Alabama|2010| 4863300| 6|
 |Alabama|2011| 4863300| 7|
 |Alabama|2012| 4863300| 5|
 |Alabama|2013| 4863300| 4|
 |Alabama|2014| 4863300| 3|
```

6.6.3　ntiles

ntiles 是一类较为常见的窗口上的聚合操作，一般用于将输入的数据集划分为 n 部分。例如，如果希望通过 State（之前所讨论的窗口规范）划分 statesPopulationDF、通过

population 进行排序，并于随后将数据集划分为两部分内容，则可在 windowspec 上使用 ntile，如下所示：

```
import org.apache.spark.sql.functions._
scala> statesPopulationDF.select(col("State"), col("Year"),
 ntile(2).over(windowSpec), rank().over(windowSpec)).sort("State",
 "Year").show(10)
+-------+----+------------------------------------------------------------
 -----------------------------------------------------------+----------
-
 ---------------------------------------------------------------------
-
 --------------------------+
| State|Year|ntile(2) OVER (PARTITION BY State ORDER BY Population DESC
 NULLS LAST ROWS BETWEEN UNBOUNDED PRECEDING AND CURRENT ROW)|RANK() OVER
 (PARTITION BY State ORDER BY Population DESC NULLS LAST ROWS BETWEEN
 UNBOUNDED PRECEDING AND CURRENT ROW)|
 +-------+----+-----------------------------------------------------------
-
 -----------------------------------------------------------+----------
-
 ---------------------------------------------------------------------
-
 --------------------------+
 |Alabama|2010| 2| 6|
 |Alabama|2011| 2| 7|
 |Alabama|2012| 2| 5|
 |Alabama|2013| 1| 4|
 |Alabama|2014| 1| 3|
 |Alabama|2015| 1| 2|
 |Alabama|2016| 1| 1|
 | Alaska|2010| 2| 7|
 | Alaska|2011| 2| 6|
 | Alaska|2012| 2| 5|
 +-------+----+-----------------------------------------------------------
-
 -----------------------------------------------------------+----------
-
 -------------------------------------------------------------
```

这里使用了 Windows 函数以及 ntile()，并将每个 State 行划分为两个相等的部分。

提示：

ntile()函数的常见应用是计算数据科学模型中所用的十分位数（decile）。

6.7　连　　接

在传统的数据库中，连接往往用于将一个事务表与另一个查找表连接，以生成更完整的视图。例如，对于客户 ID 排序的在线事务表，以及包含客户所在城市和客户 ID 的另一个表，可通过连接生成城市排序后的事务报告内容。

其中，事务表定义了 3 个列，即 CustomerID、Purchased item 以及客户所交付的 Price Paid，如表 6.5 所示。

表 6.5

CustomerID	Purchased Item	Price Paid
1	Headphones	25.00
2	Watch	20.00
3	Keyboard	20.00
1	Mouse	10.00
4	Cable	10.00
3	Headphones	30.00

客户信息表定义了两列，即 CustomerID 和 City，如表 6.6 所示。

表 6.6

CustomerID	City
1	Boston
2	New York
3	Philadelphia
4	Boston

利用客户信息表连接事务表将生成如表 6.7 所示的视图。

表 6.7

CustomerID	Purchased Item	Price Paid	City
1	Headphone	25.00	Boston
2	Watch	100.00	New York
3	Keyboard	20.00	Philadelphia
1	Mouse	10.00	Boston
4	Cable	10.00	Boston
3	Headphones	30.00	Philadelphia

随后可使用连接后的视图生成由 City 排序后的 Total Sale Price，如表 6.8 所示。

表 6.8

City	#Items	Total Sale Price
Boston	3	45.00
Philadelphia	2	50.00
New York	1	100.00

连接是 Spark SQL 中的重要功能，进而可生成前述两个组合在一起的数据集。当然，Spark 不仅仅是用于生成信息报告，还可处理以拍字节（Peta）规模的数据，进而解决实时流式用例、机器学习算法或者是数据分析等问题。为了实现这一类目标，Spark 提供了所需的 API。

两个数据集之间的常见连接操作一般使用左、右数据集的一个或多个键进行，然后将键集合上的条件表达式计算为布尔表达式。如果布尔表达式的结果返回 TRUE，则表明连接成功；否则，连接后的 DataFrame 将不会包含对应的连接结果。相应地，join API 包含 6 种不同的实现，具体如下：

```
join(right: Dataset[_]): DataFrame
 Condition-less inner join
 join(right: Dataset[_], usingColumn: String): DataFrame
 Inner join with a single column
 join(right: Dataset[_], usingColumns: Seq[String]): DataFrame
 Inner join with multiple columns
 join(right: Dataset[_], usingColumns: Seq[String], joinType: String):
 DataFrame
Join with multiple columns and a join type (inner, outer,....)
 join(right: Dataset[_], joinExprs: Column): DataFrame
 Inner Join using a join expression
join(right: Dataset[_], joinExprs: Column, joinType: String): DataFrame
 Join using a Join expression and a join type (inner, outer, ...)
```

我们将通过其中的一种 API 理解 join API 的应用方式。当然，读者也可根据具体用例选择使用其他 API。

```
def join(right: Dataset[_], joinExprs: Column, joinType: String):
 DataFrame
Join with another DataFrame using the given join expression
 right: Right side of the join.
joinExprs: Join expression.
 joinType : Type of join to perform. Default is inner join
// Scala:
```

```
import org.apache.spark.sql.functions._
import spark.implicits._
df1.join(df2, $"df1Key" === $"df2Key", "outer")
```

稍后还将对连接操作进行详细讨论。

6.7.1　连接的内部工作机制

连接的工作方式可描述为：利用多个连接器对 DataFrame 划分进行操作。然而，实际操作以及后续的性能问题取决于连接的类型，以及所连接的数据集的本质。下面将对不同的连接类型加以讨论。

6.7.2　混洗连接

两个较大的数据集之间的连接会涉及混洗连接，其中，左、右数据集的划分横跨执行器。混洗过程的代价相对高昂，但重要的是，混洗可对逻辑进行分析，以确保划分和混洗分布以最优方式完成。

图 6.5 显示了混洗连接的内部工作方式。

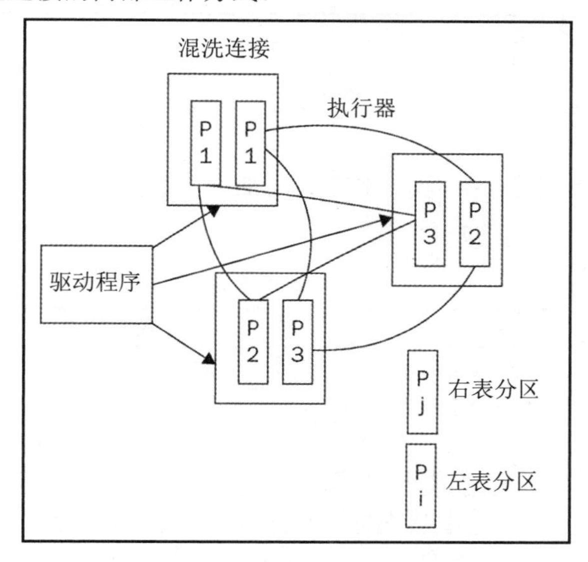

图 6.5

6.7.3　广播连接

在一个大数据集和一个小数据集之间，通过将小数据集广播到所有执行器来执行的连接称为广播连接。

广播连接的内部工作方式如图 6.6 所示。

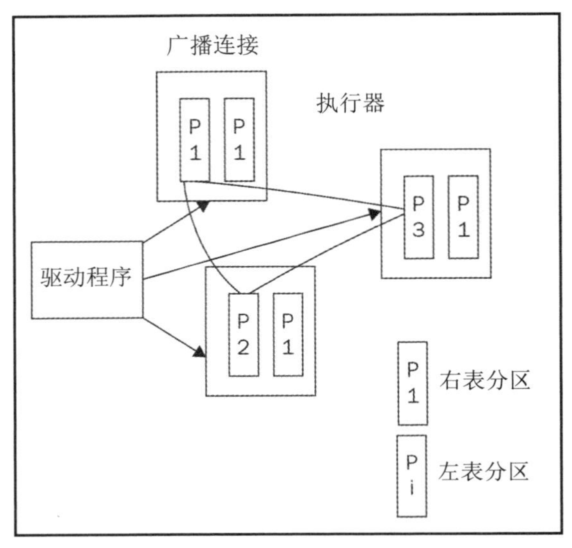

图 6.6

6.7.4 连接类型

表 6.9 显示了不同的连接类型。需要注意的是，当连接两个数据集时，不同的选择策略将在输出和性能上有所差异。

表 6.9

连 接 类 型	描　　述
内部连接	内部连接比较左、右数据集中的各行，仅当二者均包含非 NULL 值时，才组合其中的匹配行
外连接、全连接、全外连接	全外连接生成左、右表中的所有行。如果希望保留两个表中的全部行，可采用全外连接。当一个表中存在匹配项时，全外连接返回所有行
左反连接	左反连接仅生成未出现于右表中，且存在于左表中的数据行
左连接、左外连接	左外连接生成左表中的全部行，以及左、右两表中的公共行（内连接）。如果右表中不存在对应行，则填充 NULL
左半连接	左半连接仅生成左表中的行，当且仅当这些行位于右表中
右连接、右外连接	右外连接生成右表中的所有行，以及左、右表中的公共行（内连接）。如果左表中不存在对应行，则填充 NULL

下面通过一些示例数据集考查不同连接类型的工作方式。

```
scala> val statesPopulationDF = spark.read.option("header",
 "true").option("inferschema", "true").option("sep",
 ",").csv("statesPopulation.csv")
 statesPopulationDF: org.apache.spark.sql.DataFrame = [State: string,
Year:int ... 1 more field]
scala> val statesTaxRatesDF = spark.read.option("header",
 "true").option("inferschema", "true").option("sep",
 ",").csv("statesTaxRates.csv")
 statesTaxRatesDF: org.apache.spark.sql.DataFrame = [State: string,
TaxRate:
 double]
scala> statesPopulationDF.count
 res21: Long = 357
scala> statesTaxRatesDF.count
 res32: Long = 47
%sql
 statesPopulationDF.createOrReplaceTempView("statesPopulationDF")
 statesTaxRatesDF.createOrReplaceTempView("statesTaxRatesDF")
```

6.7.5 内部连接

当 State 在两个数据集中均为非 NULL 时，内部连接生成 statesPopulationDF 和 statesTaxRatesDF 中的数据行，如图 6.7 所示。

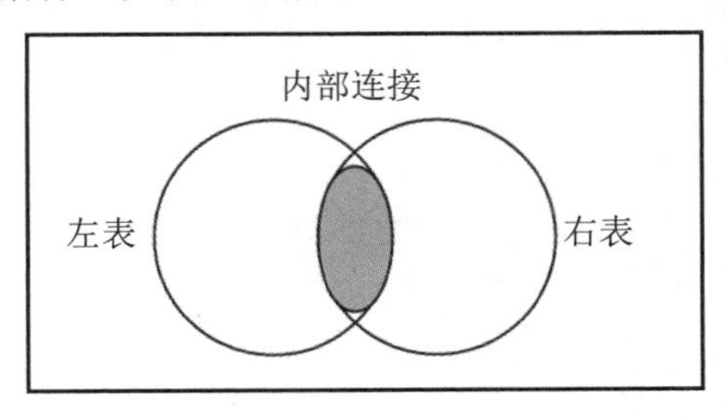

图 6.7

下列代码展示了如何通过 State 列连接两个数据集。

```
val joinDF = statesPopulationDF.join(statesTaxRatesDF,
 statesPopulationDF("State") === statesTaxRatesDF("State"), "inner")
%sql
 val joinDF = spark.sql("SELECT * FROM statesPopulationDF INNER JOIN
```

```
 statesTaxRatesDF ON statesPopulationDF.State = statesTaxRatesDF.State")
scala> joinDF.count
 res22: Long = 329
scala> joinDF.show
 +--------------------+----+----------+--------------------+-----+
 | State|Year|Population| State|TaxRate|
+--------------------+----+----------+--------------------+-----+
 | Alabama|2010| 4785492| Alabama| 4.0|
 | Arizona|2010| 6408312| Arizona| 5.6|
 | Arkansas|2010| 2921995| Arkansas| 6.5|
 | California|2010| 37332685| California| 7.5|
 | Colorado|2010| 5048644| Colorado| 2.9|
 | Connecticut|2010| 3579899| Connecticut| 6.35|
```

另外，还可在 joinDF 上运行 explain()，进而查看当前执行计划，如下所示：

```
scala> joinDF.explain
 == Physical Plan ==
*BroadcastHashJoin [State#570], [State#577], Inner, BuildRight
 :- *Project [State#570, Year#571, Population#572]
 : +- *Filter isnotnull(State#570)
 : +- *FileScan csv [State#570,Year#571,Population#572] Batched: false,
Format: CSV, Location: InMemoryFileIndex[file:/Users/salla/spark-2.1.0-
binhadoop2.7/
 statesPopulation.csv], PartitionFilters: [], PushedFilters:
 [IsNotNull(State)], ReadSchema:
 struct<State:string,Year:int,Population:int>
 +- BroadcastExchange HashedRelationBroadcastMode(List(input[0, string,
true]))
 +- *Project [State#577, TaxRate#578]
 +- *Filter isnotnull(State#577)
 +- *FileScan csv [State#577,TaxRate#578] Batched: false, Format: CSV,
Location: InMemoryFileIndex[file:/Users/salla/spark-2.1.0-binhadoop2.7/
 statesTaxRates.csv], PartitionFilters: [],
PushedFilters:[IsNotNull(State)], ReadSchema:
struct<State:string,TaxRate:double>
```

6.7.6　左外连接

左外连接生成 statesPopulationDF 中的全部行，包括 statesPopulationDF 和 statesTaxRatesDF 中的公共行，如图 6.8 所示。

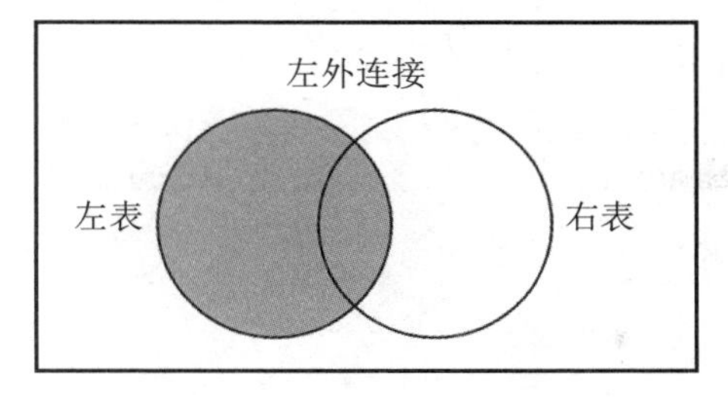

图 6.8

下列代码通过 State 列连接两个数据集。

```
val joinDF = statesPopulationDF.join(statesTaxRatesDF,
 statesPopulationDF("State") === statesTaxRatesDF("State"), "leftouter")
%sql
 val joinDF = spark.sql("SELECT * FROM statesPopulationDF LEFT OUTER JOIN
statesTaxRatesDF ON statesPopulationDF.State = statesTaxRatesDF.State")
 scala> joinDF.count
 res22: Long = 357
 scala> joinDF.show(5)
 +----------+----+----------+----------+-------+
 | State|Year|Population| State|TaxRate|
 +----------+----+----------+----------+-------+
 | Alabama|2010| 4785492| Alabama| 4.0|
 | Alaska|2010| 714031| null| null|
 | Arizona|2010| 6408312| Arizona| 5.6|
 | Arkansas|2010| 2921995| Arkansas| 6.5|
 |California|2010| 37332685|California| 7.5|
 +----------+----+----------+----------+-------+
```

6.7.7　右外连接

右外连接生成 statesTaxRatesDF 中的全部行，包括 statesPopulationDF 和 statesTaxRatesDF 中的公共行，如图 6.9 所示。

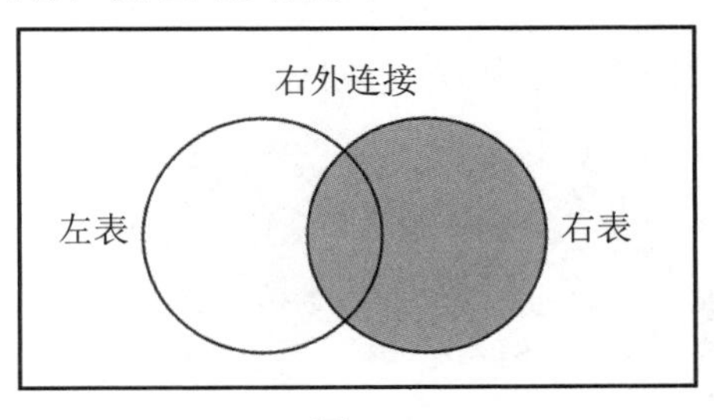

图 6.9

下列代码展示了通过 State 列连接的两个数据集。

```
val joinDF = statesPopulationDF.join(statesTaxRatesDF,
 statesPopulationDF("State") === statesTaxRatesDF("State"), "rightouter")
%sql
 val joinDF = spark.sql("SELECT * FROM statesPopulationDF RIGHT OUTER JOIN
 statesTaxRatesDF ON statesPopulationDF.State = statesTaxRatesDF.State")
scala> joinDF.count
 res22: Long = 323
scala> joinDF.show
 +--------------------+----+----------+--------------------+-----+
 | State|Year|Population| State|TaxRate|
 +--------------------+----+----------+--------------------+-----+
 | Colorado|2011| 5118360| Colorado| 2.9|
 | Colorado|2010| 5048644| Colorado| 2.9|
 | null|null| null|Connecticut| 6.35|
 | Florida|2016| 20612439| Florida| 6.0|
 | Florida|2015| 20244914| Florida| 6.0|
 | Florida|2014| 19888741| Florida| 6.0|
```

6.7.8　全外连接

全外连接生成 statesPopulationDF 和 statesTaxRatesDF 中的所有行，如图 6.10 所示。

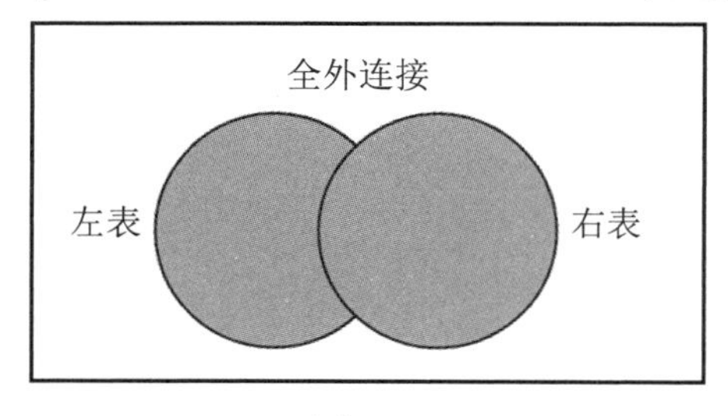

图 6.10

下列代码展示了通过 State 列连接两个数据集。

```
val joinDF = statesPopulationDF.join(statesTaxRatesDF,
 statesPopulationDF("State") === statesTaxRatesDF("State"), "fullouter")
%sql
 val joinDF = spark.sql("SELECT * FROM statesPopulationDF FULL OUTER JOIN
 statesTaxRatesDF ON statesPopulationDF.State = statesTaxRatesDF.State")
scala> joinDF.count
 res22: Long = 351
```

```
scala> joinDF.show
+-----------------+----+----------+--------------------+-------+
| State|Year|Population| State|TaxRate|
+-----------------+----+----------+--------------------+-------+
| Delaware|2010| 899816| null| null|
| Delaware|2011| 907924| null| null|
| West Virginia|2010| 1854230| West Virginia| 6.0|
| West Virginia|2011| 1854972| West Virginia| 6.0|
| Missouri|2010| 5996118| Missouri| 4.225|
| null|null| null| Connecticut| 6.35|
```

6.7.9 左反连接

当且仅当 statesTaxRatesDF 中不存在对应行时，左反连接仅生成 statesPopulationDF 中的数据行，如图 6.11 所示。

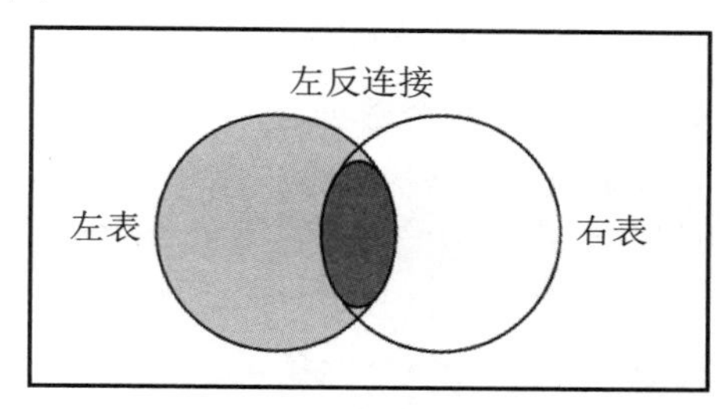

图 6.11

下列代码展示了通过 State 列连接两个数据集。

```
val joinDF = statesPopulationDF.join(statesTaxRatesDF,
 statesPopulationDF("State") === statesTaxRatesDF("State"), "leftanti")
 %sql
 val joinDF = spark.sql("SELECT * FROM statesPopulationDF LEFT ANTI JOIN
 statesTaxRatesDF ON statesPopulationDF.State = statesTaxRatesDF.State")
scala> joinDF.count
res22: Long = 28
 scala> joinDF.show(5)
 +--------+----+----------+
 | State|Year|Population|
 +--------+----+----------+
 | Alaska|2010| 714031|
 |Delaware|2010| 899816|
 | Montana|2010| 990641|
 | Oregon|2010| 3838048|
```

```
| Alaska|2011| 722713|
+--------+----+----------+
```

6.7.10 左半连接

仅当 statesTaxRatesDF 中存在对应行时，左半连接仅生成 statesPopulationDF 中的数据行，如图 6.12 所示。

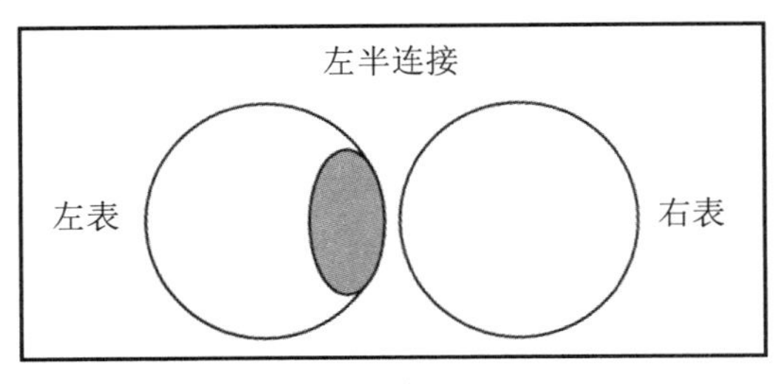

图 6.12

下列代码展示了通过 State 列连接两个数据集。

```
val joinDF = statesPopulationDF.join(statesTaxRatesDF,
 statesPopulationDF("State") === statesTaxRatesDF("State"), "leftsemi")
 %sql

val joinDF = spark.sql("SELECT * FROM statesPopulationDF LEFT SEMI JOIN
 statesTaxRatesDF ON statesPopulationDF.State = statesTaxRatesDF.State")

scala> joinDF.count

res22: Long = 322
 scala> joinDF.show(5)
 +----------+----+----------+
 | State|Year|Population|
 +----------+----+----------+
 | Alabama|2010| 4785492|
 | Arizona|2010| 6408312|
 | Arkansas|2010| 2921995|
 |California|2010| 37332685|
 | Colorado|2010| 5048644|
 +----------+----+----------+
```

6.7.11 交叉连接

交叉连接匹配左、右数据集中的每一行数据，进而生成笛卡儿叉积，如图 6.13 所示。

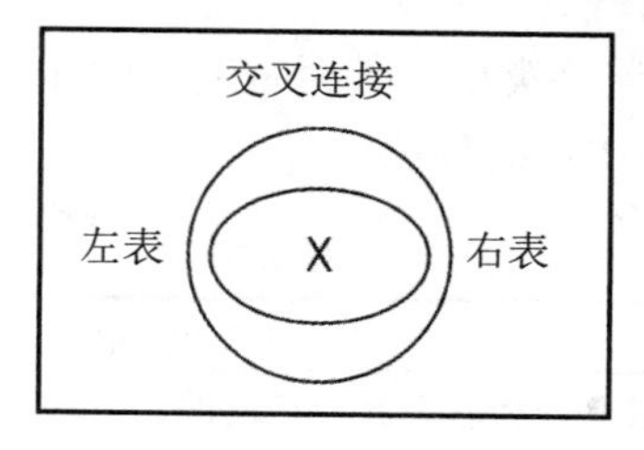

图 6.13

下列代码通过 State 列连接两个数据集。

```
scala> val joinDF=statesPopulationDF.crossJoin(statesTaxRatesDF)
 joinDF: org.apache.spark.sql.DataFrame = [State: string, Year: int ... 3
 more fields]
%sql
val joinDF = spark.sql("SELECT * FROM statesPopulationDF CROSS JOIN
 statesTaxRatesDF")
 scala> joinDF.count
res46: Long = 16450
 scala> joinDF.show(10)
 +-------+----+----------+----------+-------+
 | State|Year|Population| State|TaxRate|
 +-------+----+----------+----------+-------+
 |Alabama|2010| 4785492| Alabama| 4.0|
 |Alabama|2010| 4785492| Arizona| 5.6|
 |Alabama|2010| 4785492| Arkansas| 6.5|
 |Alabama|2010| 4785492| California| 7.5|
 |Alabama|2010| 4785492| Colorado| 2.9|
 |Alabama|2010| 4785492|Connecticut| 6.35|
 |Alabama|2010| 4785492| Florida| 6.0|
 |Alabama|2010| 4785492| Georgia| 4.0|
 |Alabama|2010| 4785492| Hawaii| 4.0|
 |Alabama|2010| 4785492| Idaho| 6.0|
 +-------+----+----------+----------+-------+
```

提示：

此外，还可使用包含交叉 joinType 的连接，而非调用交叉连接 API：statesPopulationDF.join(statesTaxRatesDF,statesPopulationDF("State").isNotNull, "cross").count。

6.7.12　连接的操作性能

所选择的连接类型将直接影响到连接的性能，其原因在于，连接操作需要在执行期间混洗数据。因此，不同的连接类型，甚至是连接顺序，都应予以重视。angle 编写连接代码时，表 6.10 列出了需要注意的事项。

表 6.10

连 接 类 型	性 能 提 示
内连接	内连接要求左表和右表具有相同的列。如果左侧或右侧包含键的一个或多个副本，连接会迅速膨胀为某种笛卡儿连接，这将占用较长的计算时间
交叉连接	交叉连接利用右表中的各行匹配左表中的每一行，并生成笛卡儿叉积。需要注意的是，由于交叉连接是性能最差的连接方式，因而仅在某些特殊用例中加以使用
外连接、全连接和全外连接	全外连接将生成连接子句的左、右两侧表中的全部行（包括匹配和不匹配的内容）。当需要保留两个表中的全部行时，可采用这一连接方式。若某一个表中存在匹配，全外连接将返回所有行。如果表中包含了较少的公共行，最终结果将较为庞大，且性能也会受到一定程度的影响
左反连接	左反连接可描述为：仅生成未出现于右表且来自左表中的行。当希望保留左表中的行，同时这些行未出现于右表中时，即可采用这种方案。该方案具有较好的性能——仅需整体考查一个表，而另一个表仅检查连接条件
左连接、左外连接	除了两个表公共部分（内连接）之外，左外连接还生成左表中的全部行。如果公共行较少，则最终结果相对庞大，因而会对性能产生影响
左半连接	左半连接仅生成左表中的行，当且仅当这些行位于右表中。这与之前讨论的左反连接刚好相反，且不包含右表中的数值。由于仅考查一个表，且另一个表仅执行条件连接检查，因而左半连接具有较好的性能
右连接、右外连接	右外连接生成右表中的全部行，以及两个表中的公共行（内连接）。使用该连接将得到右表中的全部行，以及左、右表中的公共行。如果左表中不存在对应行，则填充 NULL。当采用右外连接时，其性能类似于左外连接

6.8 本 章 小 结

本章讨论了与 DataFrame 相关的基本内容，以及 SparkSQL 如何在 DataFrame 之上提供了相应的 SQL 接口。DataFrame 的强大之处主要体现在：执行时间与基于 RDD 的计算相比明显降低。除此之外，该层还包含了简单的 SQL 接口，其功能也得到了进一步的提升。另外，本章还考查了各种 API（进而可创建、操控 DataFrame），以及聚合的高级特性，包括 groupBy、Window、rollup 和 cube。最后，我们还学习了与连接数据集相关的概念以及各种连接类型，例如内连接、外连接、交叉连接等。

第 7 章将步入本书最令人激动的部分——实时数据处理和分析。

第 7 章　Apache Spark 实时数据分析

本章将介绍 Apache Spark 的流式处理模型，以及如何构建流式实时分析应用程序。本章将着重讨论 Spark Streaming，并展示基于 Spark API 的数据流处理方式。

特别地，读者将学习如何处理 Twitter 中的消息，并通过多种方式处理实时数据流。本章主要涉及以下主题：

- 数据流简介。
- Spark Streaming。
- 离散化数据流。
- 有状态和无状态转换。
- 检查点技术。
- 其他流式平台操作（例如 Apache Kafka）。
- Structured Streaming。

7.1 数据流

在现代社会，越来越多的人通过互联网相互联系。随着智能手机的出现，这一趋势变得越发明显，今天，智能手机已经成为我们生活的一部分，例如浏览社交媒体、在线订餐、在线出租车服务等，我们发现自己比以往任何时候都更加依赖于互联网。未来，这种依赖性只会变得更加强烈。随着这一变化，在数据聚合方面，开发任务也变得更加繁重。随着互联网的蓬勃发展，数据处理的本质发生了变化。任何时候，当 App 或服务通过手机进行访问时，即会产生实时数据处理。鉴于应用程序质量的重要性，公司被迫改进数据处理过程。其间，范式也随之发生了变化。目前正在研究和使用的一种范式是，在高层基础设施上使用高可伸缩、实时（或尽可能接近实时）的处理引擎。这一类引擎应具备快速的特性，且能够适应各种变化和故障。基本上讲，数据处理过程应尽可能接近实时状态且不受干扰。

大多数处于监测状态下的系统，均会以不确定但连续的事件流的形式生成大量数据。与其他数据处理系统一样，挑战来自采集、存储和数据处理等方面。实时需求使得问题变得越加复杂。当采集、处理这一类不确定的数据流时，需要使用到一种具有高可伸缩性的体系结构以及系统迭代，例如 Flink、AMQ、Storm 和 Spark。更新、更现代的系统具有高效和灵活等特征，这也意味着，与以往相比，公司能够更容易、更高效地实现其自身的目标。新技术的发展也使得我们能够使用各种来源的数据，并对其进行处理。所有这些都具有最小的延迟。

当客户通过智能手机订购披萨时，可以通过信用卡支付并以快递方式送达。另外，一些公交系统支持在地图上实时查看公交车辆的行驶状态。其他例子还包括，我们可使用手机访问附近的星巴克咖啡店，并购买一杯咖啡。

通过查看预计的到达时间，我们可对到达机场的路线进行规划。如果公交车短时间内无法到达，这可能会对飞行计划造成不利的影响。对此，我们可以改乘出租车。万一交通堵塞，以致无法准时到达机场，我们可重新预订或取消航班。

为了进一步理解数据的实时处理方式，我们必须首先了解流式架构的基础知识。对于实时流式架构来说，以高速率收集大量数据是非常重要的，同时还应确保数据被正确地处理。

图 7.1 显示了通用的流式处理系统。其中，生产者将事件置入消息系统中，而消费者从消息系统中读取数据。

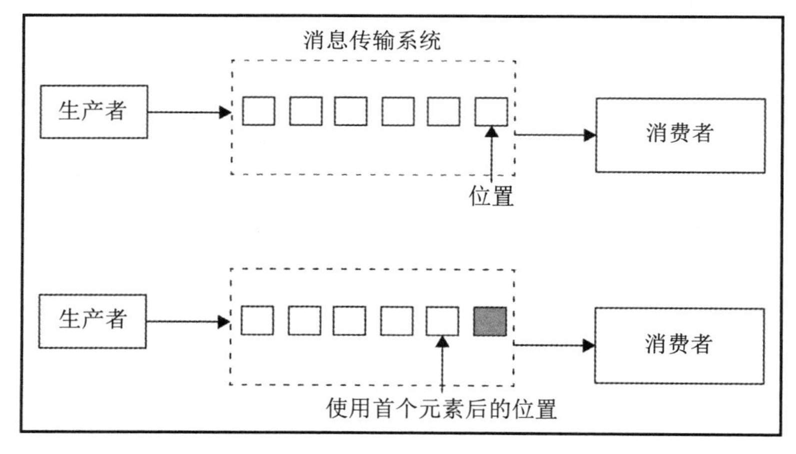

图 7.1

实时流式数据可根据下列 3 个范式进行处理：

- “至少一次”处理。
- “最多一次”处理。
- “仅一次”处理。

其中，“仅一次”处理是最为理想的解决方案，但在各种场合下，这一处理方式一般难以实现。对于这一类相对复杂的标准实现，我们需要对其属性做出一定程度上的让步。

7.1.1　“至少一次”处理

在“至少一次”处理范式中，最后接收的事件位置在该事件被处理后被保存。在发生故障时，使用者仍然能够读取和重新处理旧事件。但是，此处不能假定接收到的事件从未处理或部分处理过，因此，再次调用前一个事件之后，可能会出现重复的结果。这也体现了“至少一次”处理的具体含义。

对于任何应用程序来说，该范式均较为理想，并可用于更新计时器或计量器以显示当前数值。然而，累加、计数器或其他依赖于聚合类型准确性的操作却不适用于上述方案，其主要原因在于，重复事件会导致不正确的结果。

相应地，序列执行下列操作：

- 保存结果。
- 保存游标。

图 7.2 显示了“至少一次”处理范式的操作流程。

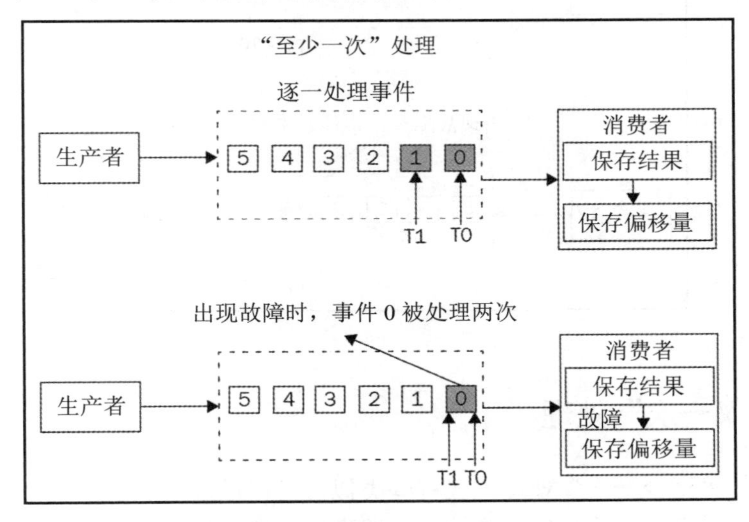

图 7.2

7.1.2　“最多一次”处理

在该范式中，最近一次事件的位置在其处理之前予以保存。如出现故障或消费者重新启动，旧事件并不会被再次读取。但是，该过程可能存在事件丢失的可能性——我们不能假定所有接收到的事件都经过了处理，因而无法再次检索事件，这也体现了该范式的含义：事件要么不被处理，要么最多处理一次。对于需要被更新以显示当前值的计时器或计量器来说，这是一种较为理想的方案。此外，如果准确性并非必需，或者需要使用到全部事件，那么，聚合操作（例如累计和或计数器）均可以正常工作。注意，任何丢失的事件将导致数据丢失或不正确的结果。

相应地，序列执行下列操作：

- ❑ 保存结果。
- ❑ 保存游标。

若存在故障且消费者重新启动，每个事件将会生成一个游标（假设事件在故障出现之前均已被处理），但事件最终可能会丢失，如图 7.3 所示。

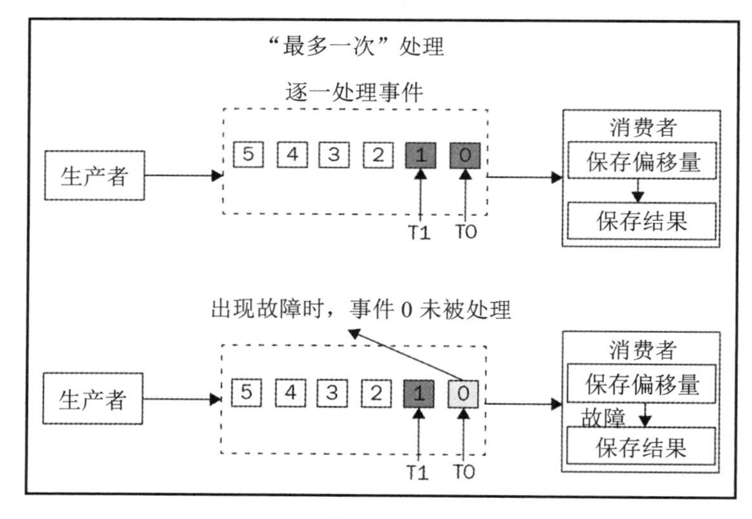

图 7.3

7.1.3 "仅一次"处理

该范式与"至少一次"处理方式有些类似，仅在最近一次事件被处理后对其进行保存。对于故障或消费者重启这一类情况，旧事件可被重新读取和处理。然而，由于无法假设所有事件要么没有被处理，要么只被部分处理，因而可能会存在潜在的重复事件。与"至少一次"处理方式不同，重复事件将被丢弃且不会被处理，最终会形成"仅一次"处理。

相应地，序列执行下列操作：

- ❑ 保存结果。

- ❑ 保存游标。

图 7.4 显示了消费者在故障后重启情形。其间，事件均已被处理，但游标尚未被保存。

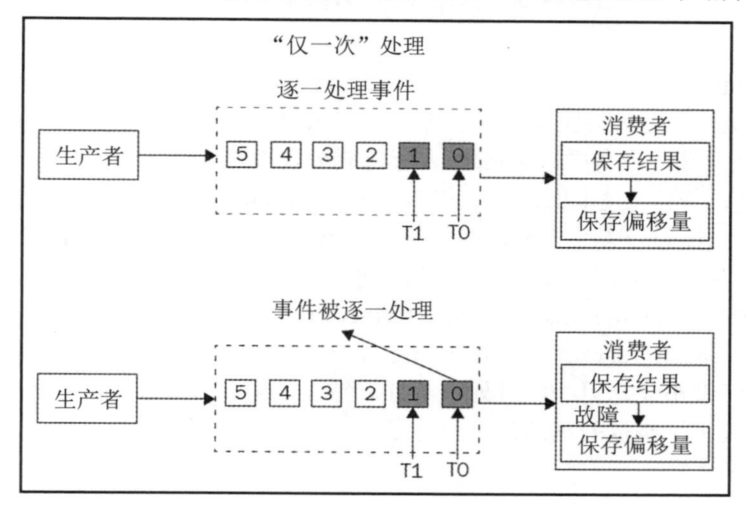

图 7.4

那么，“仅一次”处理方案如何处理重复问题？其中涉及以下处理过程：

- ❑ 幂等（idempotent）更新。
- ❑ 事务更新。

Spark Streaming 在 Spark 2.0 及其更高版本中实现了结构化的流式处理，并支持“仅一次”处理方案。本章稍后将对结构化的流式处理加以讨论。

在幂等更新中，最终结果根据所生成的唯一键或 ID 加以保存。对于重复情形，生成的键或 ID 已处于当前结果中（例如数据库），因而消费者可移除重复内容，且无须对结果进行更新。但是，这一过程可能非常复杂——针对每个事件生成唯一的键并非是一项简单任务，这需要在消费者一端执行额外的处理，除此之外，对于最终结果和游标，数据库可处于独立状态。

在事务更新中，相关结果在开始、提交事务的批次中被保存。因此，在提交事件中，该事件将被成功处理。重复事件可在不更新当前结果的情况下予以删除。但是，与幂等更新相比，其处理过程将会更加复杂——当前，需要存储事务数据。另外，对于最终结果和游标，数据库须保持不变，这也可视为该方案的另一个缺点。

注意：

对于“至少一次”处理和“最多一次”处理方案的使用，对应策略应在考查了所构建的用例之后再做决定，进而保持合理的准确性和性能要求。

7.2　Spark Streaming

Spark Streaming 并不是首先提出的一种流式体系结构。随着时间的推移，业界已涌现出多种技术方案，进而解决各种实时处理需求。Twitter Storm 则是首个较为流行的流式处理技术，并应用于诸多业务中。Spark 涵盖了流式处理库，当前已经成为一种应用广泛的技术，其原因在于：与其他技术相比，Spark Streaming 具备诸多优势，最重要的是它在核心 API 中集成了 Spark Streaming API。不仅如此，Spark Streaming 还进一步与 Spark ML、Spark SQL 以及 GraphX 实现了整合。基于此，Spark 是一种功能强大且兼具多样性的流式处理技术。

关于 Spark Streaming Flink、Heron（Twitter Storm 的后继者）和 Samza，读者可访问 https:// spark.apache.org/docs/2.1.0/ streaming-programming-guide.html 以了解更多信息。例如，它们能够在最小化延迟的同时处理事件。然而，Spark Streaming 可采用微批处理方式处理数据。其中，微批处理的最小尺寸是 500 毫秒。

注意：

在某些场合下，Apache Apex、Flink、Samza、Heron、Gearpump 以及其他新技术均构成了 Spark Streaming 的竞争对手。需要注意的是，当采用逐个事件这一类处理方式时，Spark Streaming 并不适用。

Spark Streaming 的工作方式是按照用户配置的特定时间间隔创建一批事件，然后在另一个指定的时间间隔交付并进行处理。

Spark Streaming 支持多个输入源，从而将最终结果写入多个接收位置处，如图 7.5 所示。

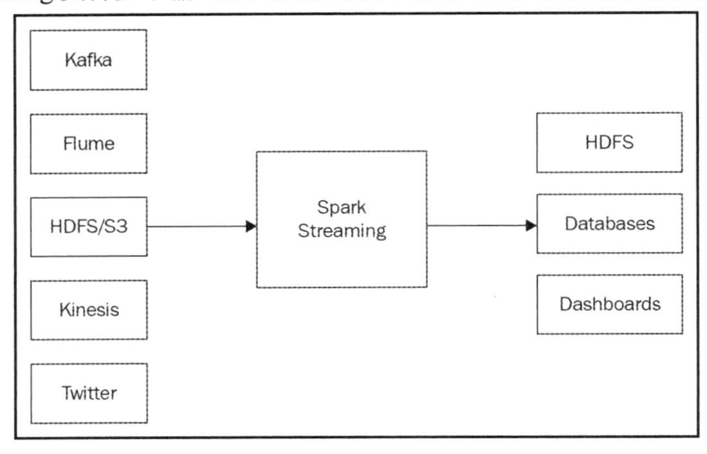

图 7.5

类似于 SparkContext，Spark Streaming 设置了一个 StreamingContext，即流数据的主要入口点。StreamingContext 依赖于 SparkContext；同时，SparkContext 实际上可直接用于流式任务中。StreamingContext 与 SparkContext 十分相似，二者之间唯一的差别在于，StreamingContext 需要程序指定批处理的时间间隔/持续时间，从分钟到毫秒不等，如图 7.6 所示。

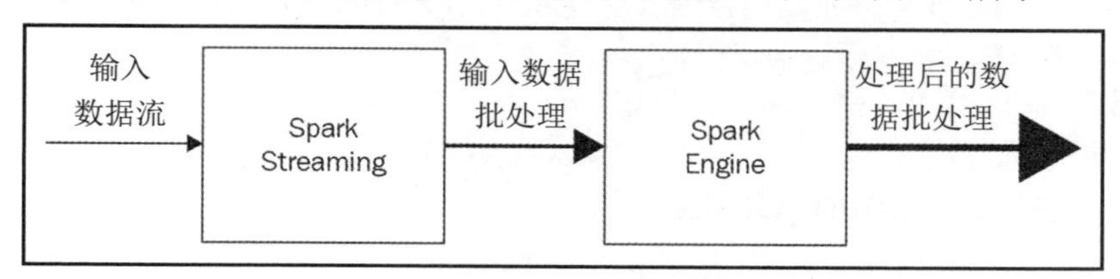

图 7.6

注意：

SparkContext 是主要的入口点。StreamingContext 复用了 SparkContext 中的一部分逻辑（任务调度和资源管理）。

7.2.1 StreamingContext

作为流数据的主要入口点，StreamingContext 处理流式应用程序的相关操作，包括 RDD 的检查点和转换。

7.2.2 创建 StreamingContext

StreamingContext 可通过下列方式创建：

- 利用现有的 SparkContext 创建 StreamingContext，如下所示：

```
StreamingContext(sparkContext: SparkContext, batchDuration:
Duration)
scala> val ssc = new StreamingContext(sc, Seconds(10))
```

- 通过提供所需配置创建 StreamingContext，如下所示：

```
StreamingContext(conf: SparkConf, batchDuration: Duration)
scala> val conf =
newSparkConf().setMaster("local[1]").setAppName("TextStreams")
scala> val ssc = new StreamingContext(conf, Seconds(10))
```

- getOrCreate 函数用于从上一个检查点数据片段重新创建一个 StreamingContext，或者创建一个新的 StreamingContext。如果数据不存在，那么，StreamingContext

则通过调用所提供的 creatingFunc 创建，如下所示：

```
def getOrCreate(
checkpointPath: String,
creatingFunc: () => StreamingContext,
hadoopConf: Configuration = SparkHadoopUtil.get.conf,
createOnError: Boolean = false
): StreamingContext
```

7.2.3　启用 StreamingContext

通过执行 StreamingContext 所定义的数据流，即可启用流式应用程序，如下所示：

```
def start(): Unit
scala> ssc.start()
```

7.2.4　终止 StreamingContext

当 StreamingContext 终止时，所有的处理过程也将结束。我们需要创建一个新的 StreamingContext，对此，需要调用 start()重启应用程序。相应地，存在两种有用的 API 可终止流式处理，具体如下：

- 通过下列方式立即终止数据流的执行（不会等待所接收的数据被处理）：

```
def stop(stopSparkContext: Boolean)
scala> ssc.stop(false)
```

- 停止执行数据流，使用以下选项可处理接收到的数据：

```
def stop(stopSparkContext: Boolean, stopGracefully: Boolean)
scala> ssc.stop(true, true)
```

输入数据流包含多种类型，且均需要创建 StreamingContext。

1．receiverStream

这里，可使用任何用户实现的接收器创建输入流，并可对此自行定义。读者可访问 http://spark.apache.org/docs/latest/streaming-customreceivers.html 以了解更多信息。下列代码显示了部分内容：

```
API declaration for receiverStream:
def receiverStream[T]: ClassTag](receiver: Receiver[T]):
ReceiverInputDStream[T]
```

2. socketTextStream

socketTextStream 使用 TCP 源 hostname:port 创建输入流。数据通过 TCP 套接字接收，所接收的字节被解释为 UTF8，并编码为\n 分隔符行，如下所示：

```
def socketTextStream(hostname: String, port: Int,
storageLevel: StorageLevel = StorageLevel.MEMORY_AND_DISK_SER_2):
ReceiverInputDStream[String]
```

3. rawSocketStream

rawSocketStream 使用网络源 hostname:port 创建输入流，这可视为接收数据时的一种最为高效的方法，如下所示：

```
def rawSocketStream[T: ClassTag](hostname: String, port: Int,
storageLevel: StorageLevel = StorageLevel.MEMORY_AND_DISK_SER_2):
ReceiverInputDStream[T]
```

7.3 fileStream

fileStream 将创建一个输入流，并监测与 Hadoop 兼容的文件系统；同时，还将利用给定的键-值类型和输入格式读取新文件。此处，以“.”开始的文件名将被忽略。当调用一个原子文件重命名函数时，以“.”开始的文件名将被重命名为一个可用的文件名，并可被 fileStream 所选用，进而对其中的内容进行处理，如下所示：

```
def fileStream[K: ClassTag, V: ClassTag, F <: NewInputFormat[K, V]:
ClassTag] (directory: String): InputDStream[(K, V)]
```

7.3.1 textFileStream

textFileStream 命令创建输入流，并检测与 Hadoop 兼容的文件系统，其文件读取方式可描述为：将基于键的文本文件读取为 Longwritable；将数值读取为 text；将输入格式读取为 TextInputFormat。其中，任何以“.”开始的文件名均被忽略，如下所示：

```
def textFileStream(directory: String): Dstream[String]
```

7.3.2 binaryRecordsStream

采用 binaryRecordsStream，可创建一个监测与 Hadoop 兼容的文件系统的输入流。其中，任何以“.”开始的文件名均被忽略，如下所示：

```
def binaryRecordsStream(directory: String, recordLength: Int):
Dstream[Array[Byte]]
```

7.3.3　queueStream

使用 queueStream，将从一个 RDD 队列中创建一个输入流。在每个批次中，将处理由队列返回的一个或全部 RDD，如下所示：

```
def queueStream[T: ClassTag](queue: Queue[RDD[T]], oneAtATime: Boolean =
true): InputDStream[T]
```

1．textFileStream 示例

该示例是一个基于 textFileStream 方法的 Spark Streaming。StreamingContext 从 Spark Shell 的 SparkContext (sc)中进行创建，时间间隔为 10 秒。textFileStream 启动后，将监视名为 streamfiles 的目录，并处理该目录中的任何新文件。在当前示例中，将输出 RDD 中的元素数量，如下所示：

```
scala> import org.apache.spark._
scala> import org.apache.spark.streaming._
scala> val ssc = new StreamingContext(sc, Seconds(10))
scala> val filestream = ssc.textFileStream("streamfiles")
scala> filestream.foreachRDD(rdd => {println(rdd.count())})
scala> ssc.start
```

2．twitterStream 示例

该示例展示了如何使用 Spark Streaming 处理 Twitter 消息，具体步骤如下：

（1）打开终端并将目录调整为 spark-2.1.1-bin-hadoop2.7。

（2）在 spark-2.1.1-binhadoop2.7 文件夹中创建 streamouts 文件夹，并于其中安装 Spark。当应用程序运行时，Streamouts 对象将采集消息，并将其转换为文本文件。

（3）将 JAR 下载至当前目录中，对应网址为 http://central.maven.org/maven2/org/apache/bahir/spark-streaming-twitter_2.11/2.1.0/spark-streamingtwitter_2.11-2.1.0.jar、http://central.maven.org/maven2/org/twitter4j/twitter4j-core/4.0.6/twitter4j-core-4.0.6.jar 和 http://central.maven.org/maven2/org/twitter4j/twitter4j-stream/4.0.6/twitter4jstream-4.0.6.jar。

（4）利用 Twitter 集成所需的全部 JAR 启动 spark-shell，此处定义为./bin/spakr-shell–jarstwitter4jstream-4.0.6.jar、twitter4j-core-4.0.6.jar 和 spark-streamingtwitter_2.11-2.1.0.jar。

（5）下列代码用于测试所处理的 Twitter 事件。

```
import org.apache.spark._
import org.apache.spark.streaming._
import org.apache.spark.streaming.twitter._
import twitter4j.auth.OAuthAuthorization
import twitter4j.conf.ConfigurationBuilder
//you can replace the next 4 settings with your own twitter account
settings.
System.setProperty("twitter4j.oauth.consumerKey",
"8wVysSpBc0LGzbwKMRh8hldSm")
System.setProperty("twitter4j.oauth.consumerSecret",
"FpV5MUDWliR6sInqIYIdkKMQEKaAUHdGJkEb4MVhDkh7dXtXPZ")
System.setProperty("twitter4j.oauth.accessToken",
"817207925756358656-
yR0JR92VBdA2rBbgJaF7PYREbiV8VZq")
System.setProperty("twitter4j.oauth.accessTokenSecret",
"JsiVkUItwWCGyOLQEtnRpEhbXyZS9jNSzcMtycn68aBaS")
val ssc = new StreamingContext(sc, Seconds(10))
val twitterStream = TwitterUtils.createStream(ssc, None)
twitterStream.saveAsTextFiles("streamouts/tweets", "txt")
ssc.start()
```

当前，streamouts 文件夹中包含了多个消息输出文本文件。我们可尝试打开此类文件，并查看其中的内容，以确保文件中涵盖相关消息。

7.3.4　离散流

离散流（DStream）表示为 Spark Streaming 之上构建的抽象层。其中，每个 DStream 体现为一个 RDD 序列，并在特定的时间间隔处被创建。随后，DStream 可采用与常规 RDD 类似的方式加以处理，并借助于有向循环图（DAG）执行计划这一类概念。类似于常规 RDD 处理过程，隶属于这一执行计划中的任何转换和操作均在 DStream 场合下被处理，如图 7.7 所示。

DStream 根据时间间隔将较长的数据流划分为较小的块，并作为 RDD 处理每个块。这一类微批处理以独立方式操作，且均为无状态微批处理。这里，假设批处理的时间间隔为 5 秒，且事件呈实时状态；同时，微批处理作为 RDD 被进一步处理。关于 Spark Streaming，需要注意的是，用于处理微批处理事件的 API 调用集成到 Spark API 中，以便与体系结构的其余部分集成。当微批处理创建完毕后，将变为一个 RDD，并支持基于 Spark APIDE 无缝处理，如图 7.8 所示。

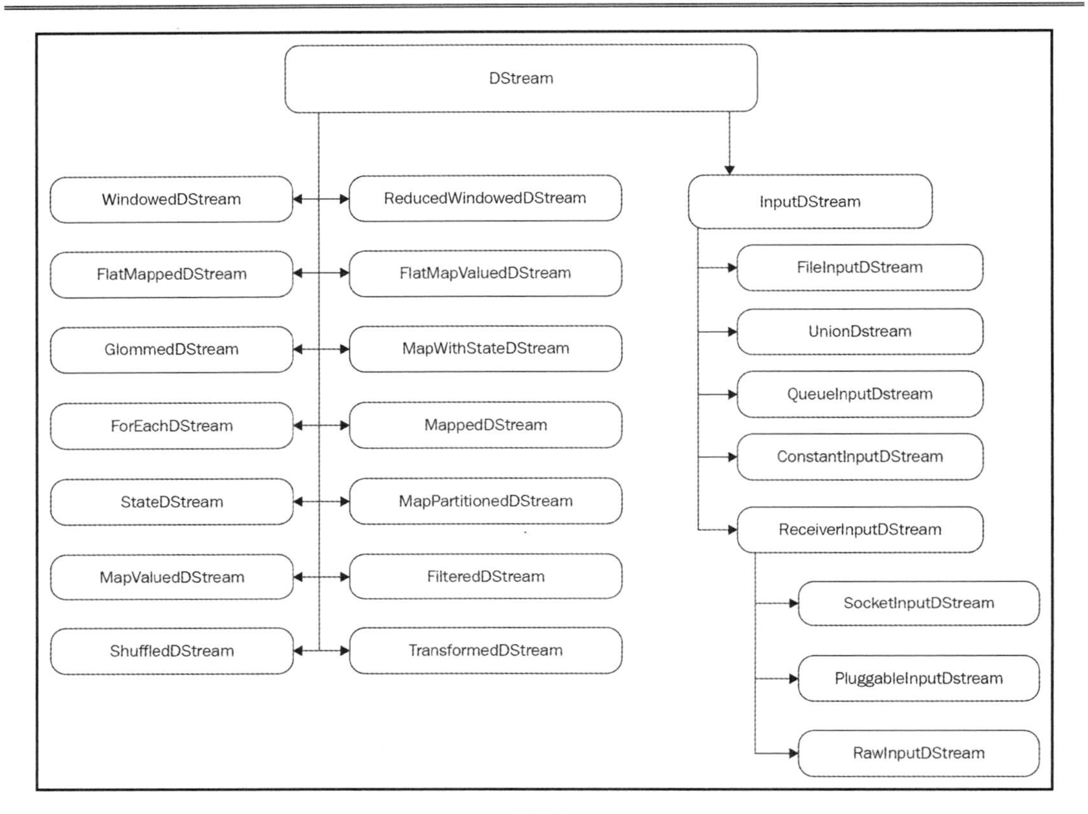
DStream
WindowedDStream
ReducedWindowedDStream
InputDStream
FlatMappedDStream
FlatMapValuedDStream
FileInputDStream
UnionDstream
GlommedDStream
MapWithStateDStream
QueueInputDstream
ForEachDStream
MappedDStream
ConstantInputDStream
StateDStream
MapPartitionedDStream
ReceiverInputDStream
MapValuedDStream
FilteredDStream
SocketInputDStream
ShuffledDStream
TransformedDStream
PluggableInputDstream
RawInputDStream

图 7.7

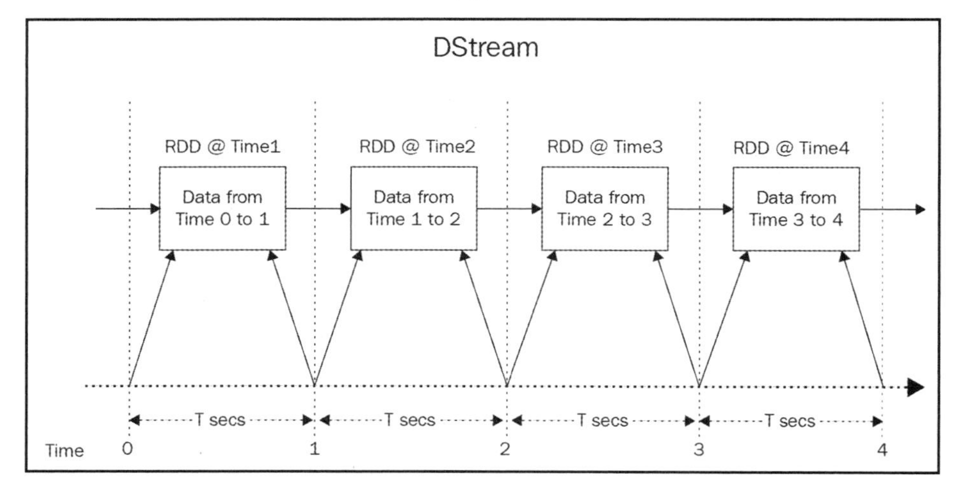
DStream
RDD @ Time1
RDD @ Time2
RDD @ Time3
RDD @ Time4
Data from Time 0 to 1
Data from Time 1 to 2
Data from Time 2 to 3
Data from Time 3 to 4
T secs
T secs
T secs
T secs
Time
0
1
2
3
4

图 7.8

下列代码定义了 DStream 类：

```
class DStream[T: ClassTag] (var ssc: StreamingContext)
//hashmap of RDDs in the DStream
var generatedRDDs = new HashMap[Time, RDD[T]]()
```

其中创建了 StreamingContext，并每隔 5 秒生成一个微批处理，进而创建一个与 Spark Core API RDD 类似的 RDD。这一类数据流中的 RDD 可采用与其他 RDD 类似的方式进行处理。

构建一个数据流应用程序的相关步骤如下：

（1）从 SparkContext 中创建 StreamingContext。

（2）从某个数据流中创建 DStream。

（3）上下文环境提供了可用于 RDD 的转换和操作。

（4）调用 treamingContext 上的 start 启用流应用程序。Spark Streaming 应用程序将以实时方式进行处理。

注意：

一旦启用了 Spark Streaming 应用程序，则无法进一步添加操作。处于终止状态的 StreamingContext 也无法重新启动，必须创建一个新的 StreamingContext。

下列示例展示了如何生成一个流作业，并以此访问 Twitter。

（1）从 SparkContext 中创建一个 StreamingContext，如下所示：

```
scala> val ssc = new StreamingContext(sc, Seconds(5))
ssc: org.apache.spark.streaming.StreamingContext =
org.apache.spark.streaming.StreamingContext@8ea5756
```

（2）从 StreamingContext 中创建一个 DStream，如下所示：

```
scala> val twitterStream = TwitterUtils.createStream(ssc, None)
twitterStream:
org.apache.spark.streaming.dstream.ReceiverInputDStream[twitter4j.
Status] =
org.apache.spark.streaming.twitter.TwitterInputDStream@46219d14
```

（3）提供可用于每个独立 RDD 的转换和操作，如下所示：

```
val aggStream = twitterStream
.flatMap(x => x.getText.split(" ")).filter(_.startsWith("#"))
.map(x => (x, 1))
.reduceByKey(_ + _)
```

（4）调用 StreamingContext 上的 start 启用流应用程序，如下所示：

```
ssc.start()
//to stop just call stop on the StreamingContext
ssc.stop(false)
```

在步骤（2）中，我们创建了 ReceiverInputDStream 类型的 DStream，表示为一个抽象类，定义了须在 worker 节点上启动接收器才能接收外部数据的 InputDStream。

下面这个例子展示了从 Twitter Stream 中接收到的信息：

```
class InputDStream[T: ClassTag](_ssc: StreamingContext) extends
DStream[T](_ssc)
class ReceiverInputDStream[T: ClassTag](_ssc: StreamingContext) extends
InputDStream[T](_ssc)
```

运行 twitterStream 上的 flatMap()转换将生成一个 FlatMappedDStream，如下所示：

```
scala> val wordStream = twitterStream.flatMap(x => x.getText().split(" "))
wordStream: org.apache.spark.streaming.dstream.DStream[String] =
org.apache.spark.streaming.dstream.FlatMappedDStream@1ed2dbd5
```

7.4 转　　换

DStream 上的转换与 Spark Core RDD 十分类似。DStream 由 RDD 构成，因而转换应用于每个 RDD 上，并针对每个 RDD 生成一个转换后的 RDD，进而创建转换后的 DStream。每个转换均会创建一个特定的 DStream 派生类。

相应地，存在多个 DStream 类并针对某项功能予以定义。例如，映射转换、窗口函数、Reduce 操作以及不同的 InputStream 类型均采用了不同的 DStream 派生类加以实现。

表 7.1 显示了可能的转换类型。

表 7.1

转　换	具 体 含 义
map(func)	针对 DStream 的每个元素使用 transformation 函数，并返回行的 DStream
filter(func)	过滤 DStream 的记录，并返回一个新的 DStream
repartition(numPartitions)	创建分区并重新分布数据，进而调整并行机制
union(otherStream)	组合两个源 DStream 中的元素，并返回一个新的 DStream
count()	计算源 DStream 的每个 RDD 中的元素数量，并返回新的 DStream
reduce(func)	在源 DStream 的每个 RDD 上使用 reduce 函数，并返回一个新的 DStream
countByValue()	计算每个 Key 的频率，并返回(Key, Long)对的新 DStream

续表

转　　换	具 体 含 义
reduceByKey(func, [numTasks])	通过源 DStream 的 RDD 中的 Key 聚合数据，并返回一个新的(Key, Value)对的 DStream
join(otherStream, [numTasks])	连接两个(K, V)和(K, W)对的 DStream，并返回一个新的(K, (V, W))对的 DStream，同时整合两个 DStream 中的数值
cogroup(otherStream, [numTasks])	当调用(K, V)和(K, W)对的 DStream 时，cogroup 转换将返回一个(K, Seq(V),Seq(W))元组的新 DStream
transform(func)	在源 DStream 的每个 RDD 上应用转换函数，并返回一个新的 DStream
updateStateByKey(func)	通过将既定函数应用于键的之前状态和键的新值，更新每个键的状态。通常用于维护状态机

7.4.1　窗口操作

Spark Streaming 支持窗口处理，并可在事件的滑动窗口上应用转换。其中，滑动窗口在在特定的间隔上被创建。

每次窗口滑过一个 DStream 时，落入窗口规范中的源 RDD 经整合后将生成一个窗口 DStream，如图 7.9 所示。该窗口须包含以下两个参数。

- 窗口长度：指定了所考查的间隔长度。
- 滑动间隔：窗口的生成间隔。

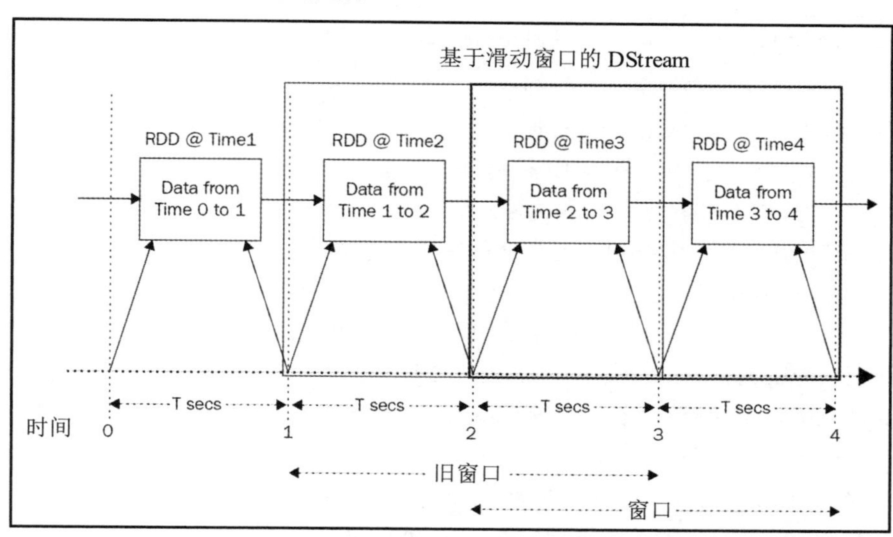

图 7.9

注意：

窗口长度和滑动间隔应为块间隔的倍数。

表 7.2 列出了常见的转换。

表 7.2

转　　换	具 体 含 义
window(windowLength, slideInterval)	在源 DStream 上创建一个窗口，并作为新的 DStream 予以返回
countByWindow(windowLength,slideInterval)	通过滑动窗口返回 DStream 值的元素计数
reduceByWindow(func,windowLength,slideInterval)	在创建了长度为 windowLength 的滑动窗口后，针对源 DStream 的每个元素应用 reduce 函数，进而返回一个新的 DStream
reduceByKeyAndWindow(func,windowLength,slideInterval, [numTasks])	在应用于源 DStream 的 RDD 的窗口中，通过 Key 聚合数据，并返回(Key, Value)对的新 DStream。该计算过程由 func 函数提供
reduceByKeyAndWindow(func, invFunc,windowLength, slideInterval,[numTasks])	窗口 w 应用于源 DStream 的 RDD 的间隔，并返回(Key, Value)对的新 DStream。该函数与上述 reduceByKeyAndWindow 函数间的主要差别在于 invFunc（在滑动窗口开始时即完成计算）
countByValueAndWindow(windowLength,slideInterval, [numTasks])	计算每个 Key 的频率，并返回滑动窗口中的、(Key,Long)对的新 DStream

下面再次考查前述 Twitter Stream 示例，当前目标是每隔 5 秒输出消息流中常用的 5 个单词，窗口长度为 15 秒且每隔 10 秒滑动。当运行对应代码时，需要执行下列步骤：

（1）打开终端，并将目录调整为 spark-2.1.1-bin-hadoop2.7。

（2）在 spark-2.1.1-bin-hadoop2.7 文件夹中创建名为 streamouts 的文件，其中安装了 Spark。当运行应用程序时，streamouts 文件夹中包含了全部消息文本文件。

（3）下载 JAR 并将其置于当前目录中，对应网址为 http://central.maven.org/maven2/org/apache/bahir/spark-streaming-twitter_2.11/2.1.0/spark-streaming-twitter_2.11-2.1.0.jar、http://central.maven.org/maven2/org/twitter4j/twitter4j-core/4.0.6/twitter4j-core-4.0.6.jar 和 http://central.maven.org/maven2/org/twitter4j/twitter4j-stream/4.0.6/twitter4j-stream-4.0.6.jar。

（4）利用消息集成所需的 JAR 启动 Spark Shell，即./bin/spark-shell--jars twitter4jstream-4.0.6.jar、twitter4j-core-4.0.6.jar 和 spark-streamingtwitter_2.11-2.1.0.jar。

（5）下列示例代码用于测试消息处理：

```
import org.apache.log4j.Logger
import org.apache.log4j.Level
Logger.getLogger("org").setLevel(Level.OFF)
import java.util.Date
import org.apache.spark._
import org.apache.spark.streaming._
import org.apache.spark.streaming.twitter._
import twitter4j.auth.OAuthAuthorization
import twitter4j.conf.ConfigurationBuilder
System.setProperty("twitter4j.oauth.consumerKey","8wVysSpBc0LGzbwKMRh8hldSm")
System.setProperty("twitter4j.oauth.consumerSecret",
"FpV5MUDWliR6sInqIYIdkKMQEKaAUHdGJkEb4MVhDkh7dXtXPZ")
System.setProperty("twitter4j.oauth.accessToken",
"817207925756358656-yR0JR92VBdA2rBbgJaF7PYREbiV8VZq")
System.setProperty("twitter4j.oauth.accessTokenSecret",
"JsiVkUItwWCGyOLQEtnRpEhbXyZS9jNSzcMtycn68aBaS")

val ssc = new StreamingContext(sc, Seconds(5))
val twitterStream = TwitterUtils.createStream(ssc, None)
val aggStream = twitterStream
.flatMap(x => x.getText.split(" "))
.filter(_.startsWith("#"))
.map(x => (x, 1))
.reduceByKeyAndWindow(_ + _, _ - _, Seconds(15), Seconds(10), 5)

ssc.checkpoint("checkpoints")
aggStream.checkpoint(Seconds(10))
aggStream.foreachRDD((rdd, time) => {
val count = rdd.count()
if (count > 0) {
val dt = new Date(time.milliseconds)
println(s"\n\n$dt rddCount = $count\nTop 5 words\n")

val top5 = rdd.sortBy(_._2, ascending = false).take(5)
top5.foreach {
case (word, count) =>
println(s"[$word] - $count")
}}})
ssc.start
//wait 60 seconds
ss.stop(false)
```

```
The output is shown on the console every 15 seconds, looking like
the following:
Mon May 29 02:44:50 EDT 2017 rddCount = 1453
Top 5 words
[#RT] - 64
[#de] - 24
[#a] - 15
[#to] - 15
[#the] - 13
Mon May 29 02:45:00 EDT 2017 rddCount = 3312
Top 5 words
[#RT] - 161
[#df] - 47
[#a] - 35
[#the] - 29
[#to] - 29
```

7.4.2 有状态/无状态转换

Spark Streaming 使用了 DStream 这一概念，基本上可视为 RRD 数据微批处理。除此之外，我们还介绍了一些可应用于 DStream 之上的转换。DStream 转换可分为两种类型，即有状态转换和无状态转换。

在无状态转换中，每个数据的微批处理是否被处理与之前的数据批次无关，也就是说，每个批处理完全独立于之前的数据批次。

在有状态转换中，每个数据的微批处理部分或全部取决于之前的数据批次。因此，每个批处理将考查前一次批处理，并在处理过程中使用到相关信息。

1. 无状态转换

通过将转换应用于 DStream 中的每个 RDD，DStream 将被转换为另一个 DStream，如图 7.10 所示。相关示例包括 map()、flatMap()、union()、join()和 reduceBykey()。

2. 有状态转换

有状态转换应用于某个 DStream 上，但依赖于前一次处理状态。相关示例包括 countByValueAndWindow()、reduceByKeyAndWindow()、mapWithState()和 updateStateByKey()。根据定义，所有基于窗口的转换均为有状态转换；我们须跟踪 DStream 的窗口长度和滑动间隔。

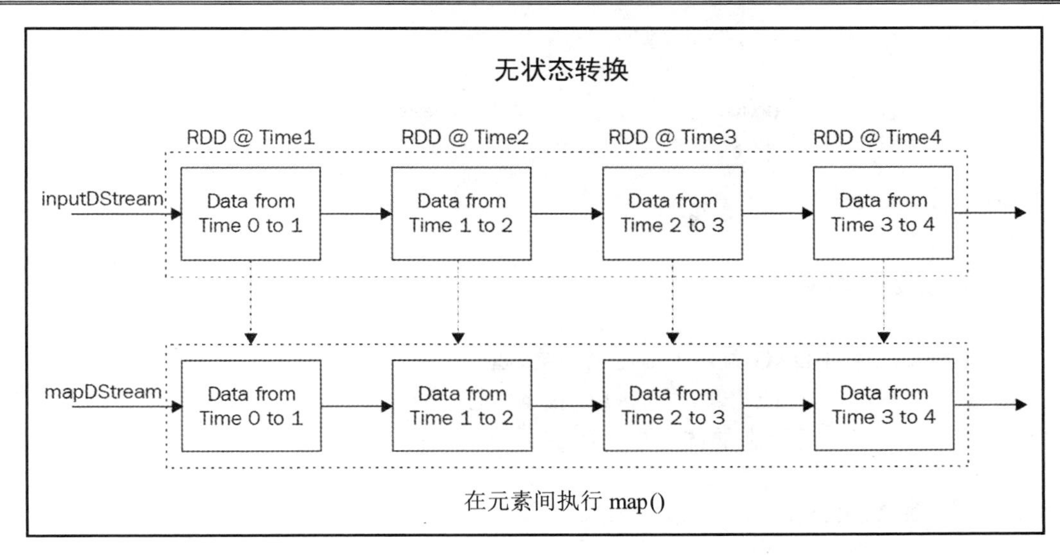

图 7.10

7.5 检查点

正如预期的那样，实时流应用程序将运行很长一段时间，同时保持对故障的弹性，Spark Streaming 实现了一种称为检查点的机制，该机制跟踪一定量的信息，并可从任何故障中予以恢复。对此，存在两种检查点类型，如下所示：

- 元数据检查点。
- 数据检查点。

通过调用 StreamingContext 上的 checkpoint()即可启用检查点，如下所示：

```
def checkpoint(directory: String)
```

这指定了存储检查点数据的目录。需要注意的是，此处须采用基于容错机制的文件系统，例如 HDFS。

当检查点目录设置完毕后，任何 DStream 都可以根据时间间隔对其进行检查。在前述 Twitter 示例中，每隔 10 秒将对每个 DStream 予以检查，如下所示：

```
val ssc = new StreamingContext(sc, Seconds(5))
val twitterStream = TwitterUtils.createStream(ssc, None)
val wordStream = twitterStream.flatMap(x => x.getText().split(" "))
val aggStream = twitterStream
```

```
.flatMap(x => x.getText.split(" ")).filter(_.startsWith("#"))
.map(x => (x, 1))
.reduceByKeyAndWindow(_ + _, _ - _, Seconds(15), Seconds(10), 5)
ssc.checkpoint("checkpoints")
aggStream.checkpoint(Seconds(10))
wordStream.checkpoint(Seconds(10))
```

7.5.1　元数据检查点

元数据检查点将 DAG 表示的、定义数据流操作的信息保存至 HDFS 中，从而在发生故障时恢复 DAG，以便应用程序重新启动。随后，驱动程序重启并从 HDFS 中读取全部元数据，在恢复崩溃前状态的同时重新构建 DAG。

7.5.2　数据检查点

数据检查点将 RDD 保存至 HDFS 中。对于流应用程序中的故障，可对 RDD 予以恢复并持续进行处理。除了恢复功能之外，数据检查点还对缓存清空或执行器丢失所导致的 RDD 丢失提供帮助。生成后的 RDD 无须等待 DAG 中父 RDD 被重新处理。

对于以下应用程序，需要启用检查点机制：

- 采用有状态转换时。如果使用 updateStateBykey()或 reduceByKeyAndWindow()函数（及其反函数），那么，需要针对 RDD 检查点给定检查点目录。
- 在运行应用程序时恢复驱动程序故障。元数据检查点有助于恢复工作进度中的信息。

如果不存在有状态转换，应用程序则可在未启用检查点的情况下运行。

注意：

一种可能的情况是，丢失已接收但尚未处理的数据。

需要注意的是，RDD 检查点将每个 RDD 保存至存储中，这将增加基于检查点的 RDD 的批处理时间。因此，为了避免性能损失，需要设置和调整检查点间隔，这对于实时处理来说十分重要。较小的批处理尺寸（例如 1 秒）意味着检查点频繁出现，这可能会降低操作的吞吐量。相反，不频繁的检查点会导致任务尺寸增加，由于队列数据量太大，进而会产生处理延迟问题。

基于 RDD 检查点的有状态转换通常包含默认的检查点间隔（至少 10 秒）。一种较好的方案是，检查点间隔设置为 5～10 个 DStream 滑动间隔。

7.6 驱动程序故障恢复

借助于 StreamingContext.getOrCreate()，可对驱动程序故障进行恢复。如前所述，这将从已存在的检查点处初始化 StreamingContext，或者生成新的 StreamingContext。

此处并不打算实现 createStreamContext0 函数，该函数创建一个 StreamingContext、设置 DStream 并解释消息内容，并生成前 5 个最为常用的主题标签（每隔 15 秒使用一个窗口）。这里并未依次调用 createStreamContext()和 ssc.start()，而是调用了 getOrCreate()——如果检查点存在，那么 StreamingContext 将根据 checkpointDirectory 中的数据予以重建。如果不存在此类目录，或者应用程序首次运行，那么将调用 createStreamContext()，如下所示：

```
val ssc = StreamingContext.getOrCreate(checkpointDirectory,
createStreamContext _)
```

下列代码展示了对应函数的定义方式，以及 getOrCreate()的调用方式。

```
val checkpointDirectory = "checkpoints"
//Creating and setting up a new StreamingContext
def createStreamContext(): StreamingContext = {
val ssc = new StreamingContext(sc, Seconds(5))
val twitterStream = TwitterUtils.createStream(ssc, None)
val wordStream = twitterStream.flatMap(x => x.getText().split(" "))
val aggStream = twitterStream
.flatMap(x => x.getText.split(" ")).filter(_.startsWith("#"))
.map(x => (x, 1))
.reduceByKeyAndWindow(_ + _, _ - _, Seconds(15), Seconds(10), 5)
ssc.checkpoint(checkpointDirectory)
aggStream.checkpoint(Seconds(10))
wordStream.checkpoint(Seconds(10))
aggStream.foreachRDD((rdd, time) => {
val count = rdd.count()
if (count > 0) {
val dt = new Date(time.milliseconds)
println(s"\n\n$dt rddCount = $count\nTop 5 words\n")
val top10 = rdd.sortBy(_._2, ascending = false).take(5)
top10.foreach {
case (word, count) =>
println(s"[$word] - $count")
}
```

```
}
})
ssc
}
//Retrieve StreamingContext from checkpoint data or create a new one
val ssc = StreamingContext.getOrCreate(checkpointDirectory,
createStreamContext _)
```

7.7　与流平台的互操作性（Apache Kafka）

Spark Streaming与Apache Kafka这一当前最为流行的消息平台间实现了良好的集成。其中涉及多种方案，且随着时间的变化，性能和可靠性也得到了不断的完善。

其中包括以下 3 种主要方案：

- 基于接收器的方案。
- Direct Stream 方案。
- Structured Streaming。

7.7.1　基于接收器的方案

第一种集成方案是 Spark 和 Kafka 之间的集成。在基于接收器的方案中，驱动程序启动执行程序上的接收器，然后使用 Kafka 代理的高级 API 提取数据。鉴于事件从 Kafka 代理中获取，因而接收器将偏移量更新到 Zookeeper 中，Kafka 集群也采用了这个方法。这里，重点是采用了预写日志（WAL）功能，这是接收器在 Kafka 采集数据时写入的内容。

如果存在问题，且执行器或接收器须重启抑或丢失，WAL 可用于恢复事件并对其进行处理。最终，这种基于日志的设计方案可提供可靠、一致的结果。

事件的输入 DStream 由每个 Kafka 主题的接收器创建，同时它查询 Kafka 主题、代理和偏移量。考虑到处于登录和运行状态下的接收器，并行机制将变得更加复杂——随着应用程序规模不断壮大，工作负载无法得到适当的分配。另一个问题则是对 HDFS 的依赖，以及写操作的重复性。对于“仅一次”处理范式，也存在可靠性要求，其原因在于，只有幂等方案可正常工作。事务性方法也无法在基于接收器的方案中工作，因为没有一种方法可以访问基于 Zookeeper 或 HDFS 位置的偏移量范围。尽管如此，接收器方案仍是较为通用的方法，它适用于任何消息传递系统，如图 7.11 所示。

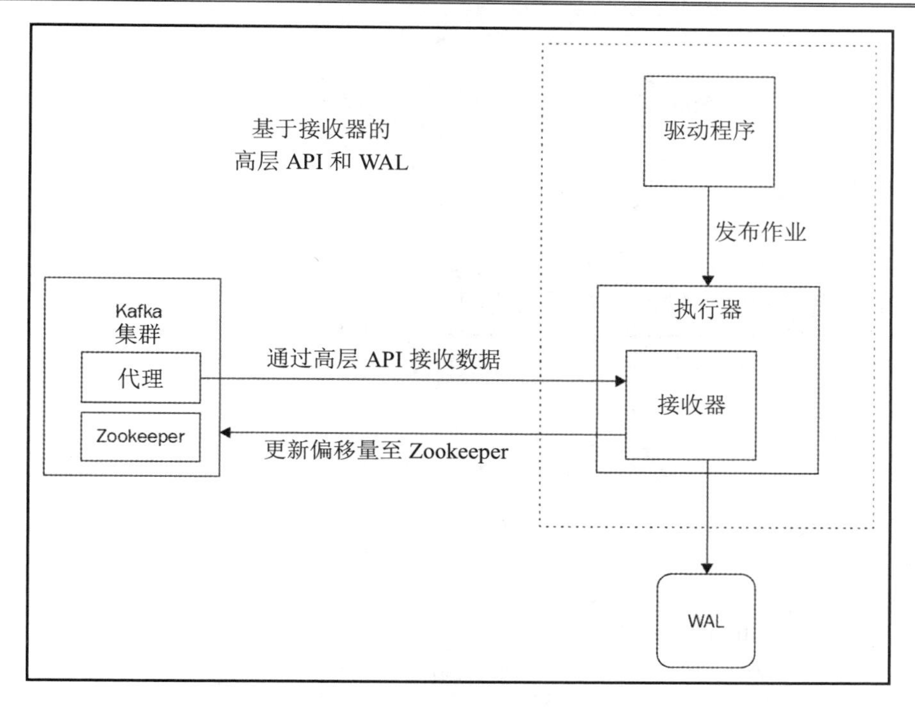

图 7.11

基于接收器的数据流可通过调用 createStream() API 创建，如下所示：

```
def createStream(
ssc: StreamingContext,
//StreamingContext object
zkQuorum: String,
//Zookeeper quorum (hostname:port,hostname:port,..)
groupId: String,
//Group id for the consumer
topics: Map[String, Int],
//Map of (topic_name to numPartitions) to consume
storageLevel: StorageLevel = StorageLevel.MEMORY_AND_DISK_SER_2
//Storage level to use for storing the received objects
(default: StorageLevel.MEMORY_AND_DISK_SER_2)
): ReceiverInputDStream[(String, String)]
//DStream of (Kafka message key,
Kafka message value)
```

下列示例代码显示了如何创建基于接收器的数据流（从 Kafka 代理中提取消息）。

```
val topicMap = topics.split(",").map((_, numThreads.toInt)).toMap
val lines = KafkaUtils.createStream(ssc, zkQuorum, group,
topicMap).map(_._2)
```

7.7.2　Direct Stream

可以创建一个输入流，它不使用接收器直接从 Kafka 代理中提取消息，这可确保每个 Kafka 消息在转换中只包含一次，如图 7.12 所示。

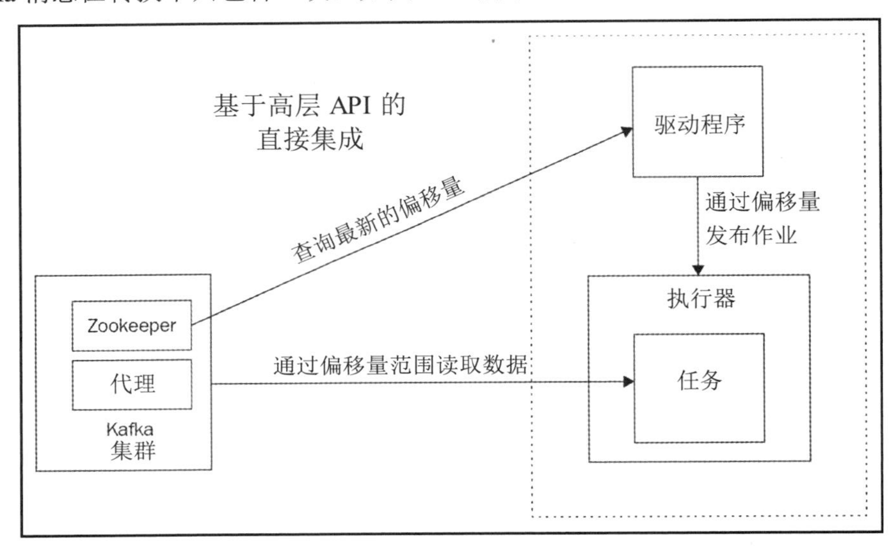

图 7.12

Direct Stream 的属性具体如下。

- 不存在接收器：Direct Stream 不使用接收器，且直接查询 Kafka。
- 偏移量：Direct Stream 不使用 Zookeeper 存储偏移量，任何所使用的偏移量均通过数据流自身被跟踪。每个批处理中使用的偏移量可从生成的 RDD 中访问。
- 故障恢复：须开启数据流上下文中的检查点，进而从驱动程序故障中予以恢复。
- 端到端示意图：数据流可确保所有记录可一次性地被接收和转换，但却无法保证转换后的数据一次性地输出。

Direct Stream 的创建方式如下所示，其中使用了 Kafka Utils 中的 createDirectStream() API。

```
def createDirectStream[
K: ClassTag,
//K type of Kafka message key
```

```
V: ClassTag,
//V type of Kafka message value
KD <: Decoder[K]: ClassTag,
//KD type of Kafka message key decoder
VD <: Decoder[V]: ClassTag,
//VD type of Kafka message value decoder
R: ClassTag
//R type returned by messageHandler
](
ssc: StreamingContext,
//StreamingContext object
kafkaParams: Map[String, String],
/*
kafkaParams Kafka <a
href="http://kafka.apache.org/documentation.html#configuration">
configuration parameters</a>. Requires "metadata.broker.list" or
"bootstrap.servers"
to be set with Kafka broker(s) (NOT Zookeeper servers) specified in
host1:port1,host2:port2 form.
*/
fromOffsets: Map[TopicAndPartition, Long],
//fromOffsets Pertopic/
partition Kafka offsets defining the (inclusive) starting point of
the stream
messageHandler: MessageAndMetadata[K, V] => R
//messageHandler Function
for translating each message and metadata into the desired type
): InputDStream[R]
//DStream of R
```

下列 Direct Stream 示例从 Kafka 主题中获取数据，进而生成 DStream。

```
val topicsSet = topics.split(",").toSet
val kafkaParams : Map[String, String] =
Map("metadata.broker.list" -> brokers,
"group.id" -> groupid )
val rawDstream = KafkaUtils.createDirectStream[String, String,
StringDecoder, StringDecoder](ssc, kafkaParams, topicsSet)
```

Direct Stream API 仅可与 Kafka 协同使用，因而不具备通用性。

7.7.3　Structured Streaming

Apache Spark 2.0+之后加入了结构化的 Structured Streaming，但仍处于开发的测试阶

段，稍后将详细介绍 Structured Streaming 的应用方式。关于 Structured Streaming 中的 Kafka 集成，读者可访问 https://spark.apache.org/docs/latest/structured-streaming-kafka-integration.html 以了解更多信息。

下面的代码片段展示了如何在 Structured Streaming 中使用 Kafka 源。

```
val ds1 = spark
.readStream
.format("kafka")
.option("kafka.bootstrap.servers", "host1:port1,host2:port2")
.option("subscribe", "topic1")
.load()
ds1.selectExpr("CAST(key AS STRING)", "CAST(value AS STRING)")
.as[(String, String)]
```

下面的示例显示了 Kafka 源的使用方式，而非源数据流（如果希望使用批处理分析方案）。

```
val ds1 = spark
.read
.format("kafka")
.option("kafka.bootstrap.servers", "host1:port1,host2:port2")
.option("subscribe", "topic1")
.load()
ds1.selectExpr("CAST(key AS STRING)", "CAST(value AS STRING)")
.as[(String, String)]
```

Structured Streaming 是一种具有容错性、可伸缩的流处理引擎，且构建于 Spark SQL 引擎之上。但是，Structured Streaming 允许在接收的数据中指定事件时间，以便后续数据可被自动关注。需要注意的是，在 Spark 2.1 中，Structured Streaming 仍处于测试阶段，且相关 API 被标记为“experimental”。对此，读者可访问 https://spark.apache.org/docs/latest/structured-streaming-programmingguide.html 以了解更多信息。

Structured Streaming 背后的驱动思想是将数据流视为不断加入其中的无界表。随后，计算和 SQL 查询可应用于该表之上，通常可对批处理数据采取此类做法。例如，Spark SQL 查询将处理无界表。由于 DStream 随时间变化，因而处于增长的数据量将被处理，进而生成结果表。该表可写入外部接收位置处，也称作输出。

下面通过监听本地主机端口 9999 的输入内容，进而考查 Structured Streaming 查询的创建示例。在 Linux 或 macOS 环境下，可利用下列命令在端口 9999 上启动服务器：

```
nc -lk 9999
```

下列示例通过调用 SparkSession 的 readStream API 创建 inputStream，并于随后从各行中析取单词。接下来，单词将被分组，并统计出现次数。最后，相关结果将写入输出流中，如下所示：

```
//Creating stream reading from localhost 999
val inputLines = spark.readStream
.format("socket")
.option("host", "localhost")
.option("port", 9999)
.load()
inputLines: org.apache.spark.sql.DataFrame = [value: string]
// Splitting the inputLines into words
val words = inputLines.as[String].flatMap(_.split(" "))
words: org.apache.spark.sql.Dataset[String] = [value: string]
// Generating running word count
val wordCounts = words.groupBy("value").count()
wordCounts: org.apache.spark.sql.DataFrame = [value: string, count:
bigint]
val query = wordCounts.writeStream
.outputMode("complete")
.format("console")

query:
org.apache.spark.sql.streaming.DataStreamWriter[org.apache.spark.sql.Row] =
org.apache.spark.sql.streaming.DataStreamWriter@4823f4d0
query.start()
```

只要单词输入终端中，查询即被更新，随后将其不断地输出至控制台，如下所示：

```
scala> -------------------------------------------
Batch: 0
-------------------------------------------
+-----+-----+
|value|count|
+-----+-----+
| dog| 1|
+-----+-----+
-------------------------------------------
Batch: 1
-------------------------------------------
+-----+-----+
|value|count|
+-----+-----+
```

```
| dog| 1|
| cat| 1|
+-----+-----+
scala> -------------------------------------------
Batch: 2
-------------------------------------------
+-----+-----+
|value|count|
+-----+-----+
| dog| 2|
| cat| 1|
+-----+-----+
```

7.8 处理事件时间和延迟日期

事件时间是指数据内部的时间。Spark Streaming 将这一时间定义为 DStream 的接收时间。但对于需要使用到事件时间的许多应用程序来说，这已然足够。例如，如果计算消息中出现的主题标签的次数（每分钟），则需要使用数据生成的时间，而不是 Spark 接收事件的时间。

下列示例将监听服务器端口 9999，Timestamp 当前作为输入数据的一部分内容被启用。因此，现在可在无界表上执行窗口操作，如下所示：

```
import java.sql.Timestamp
import org.apache.spark.sql.SparkSession
import org.apache.spark.sql.functions._
// Creating DataFrame that represent the stream of input lines from
connection
to host:port
val inputLines = spark.readStream
.format("socket")
.option("host", "localhost")
.option("port", 9999)
.option("includeTimestamp", true)
.load()
// Splitting the lines into words, retaining timestamps
val words = inputLines.as[(String, Timestamp)].flatMap(line =>
line._1.split(" ").map(word => (word, line._2))
).toDF("word", "timestamp")
// Grouping the data by window and word and computing the count of each
```

```
val windowedCounts = words.withWatermark("timestamp", "10 seconds")
.groupBy(
window($"timestamp", "10 seconds", "10 seconds"), $"word"
).count().orderBy("window")
// Begin executing the query which will print the windowed word counts to
the
console
val query = windowedCounts.writeStream.outputMode("complete")
.format("console")
.option("truncate", "false")

query.start()
query.awaitTermination()
```

7.9　容错示意图

“仅一次”处理范式在传统流式机制中相对复杂，需要使用到外部数据库/存储维护偏移量。另外，Structured Streaming 仍处于变化中，在被广泛使用之前仍需要克服许多问题。

7.10　本章小结

本章讨论了流处理系统、Spark Streaming、Apache Spark 中的 DStream、DAG 和 DStream 系统以及转换操作。除此之外，本章还介绍了窗口流式处理，以及基于 Spark Streaming 的 Twitter 消息处理机制。接下来，我们学习了基于接收器和 Direct Stream 的数据方案（结合 Kafka）。最后，本章还讲解了 Structured Streaming 这一较新的开发技术。据此，我们可以解决诸多挑战性问题，例如故障容错、数据流中的“仅一次”处理方案，以及与消息系统（例如 Kafka）间集成的简化方法，同时保持与其他输入流类型集成的灵活性和可扩展性。

第 8 章将讨论 Apache Flink，这也是 Spark 计算平台中的关键问题。

第 8 章 Apache Flink 批处理分析

本章主要向读者介绍 Apache Flink，并根据批处理模型使用 Flink 进行大数据分析。此外，本章还将考查 DataSet API，对于执行大数据上的批处理分析，它提供了易于使用的方法。

本章主要涉及以下主题：

- Apache Flink 简介。
- 安装 Flink。
- 使用 Scala Shell。
- 使用 Flink 集群 UI。
- 利用 Flink 进行批处理分析。

8.1 Apache Flink 简介

Flink 是一个针对分布式流处理的开源框架，并包含以下特性：

- 提供了准确的结果，即使对无序和延迟数据也是如此。
- Flink 是有状态的且兼具容错特性，当维护“仅一次”应用程序状态时，可无缝地从故障中进行恢复。
- 规模较大，可运行于数千个节点上，具有非常好的吞吐量和延迟特性。

图 8.1 显示了 Apache Flink 应用方式的官方文档截图。

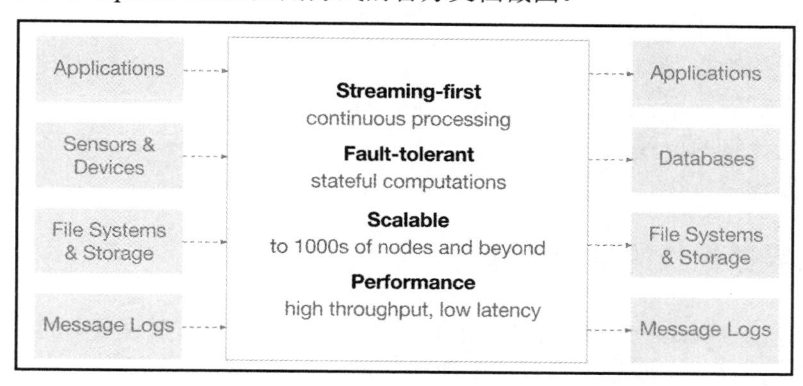

图 8.1

图 8.2 显示了 Apache Flink 框架的另一种观察方式。

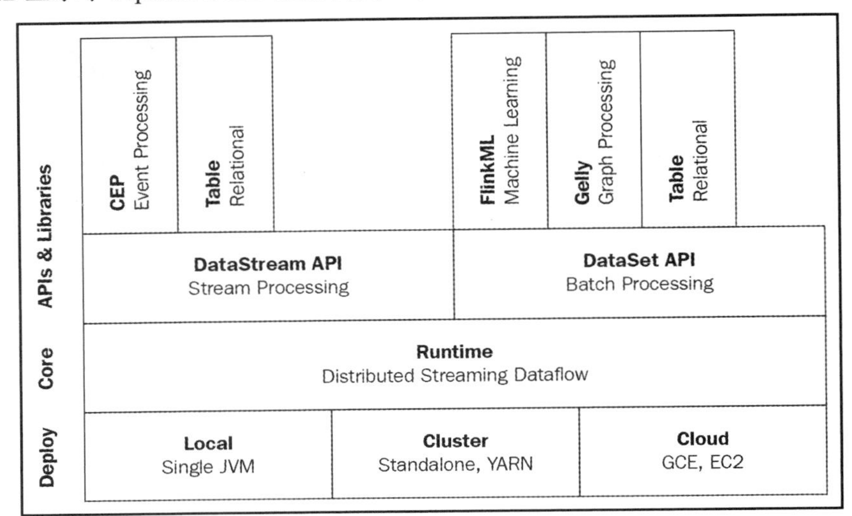

图 8.2

当运行程序的主方法时，所有的 Flink 程序均以延迟方式执行，且并不会直接执行数据加载和转换。相反，每项操作在创建后将添加至程序的计划中。当显式地被执行环境中的 execute()所触发后，相关操作方被真正地予以执行。程序在本地或集群上执行取决于执行环境的类型。相应地，延迟程序可构建更加复杂的程序——Flink 作为一个整体计划单元被执行。

Flink 程序看上去与转换数据集的常规程序并无两样，每个程序均由相同的基本内容构成，如下所示：

- 获取执行环境。
- 加载初始数据。
- 指定对应数据上的转换、聚合和连接。
- 指定计算的放置位置。
- 触发程序执行。

8.1.1　无界数据集的连续处理

在深入讨论 Apache Flink 之前，首先在较高层次上介绍一下数据集的类型，在处理数据的过程中，以及选取处理过程中的执行模型时会涉及这一类问题。这两个概念经常被混淆，因而有必要了解其中的差别。

首先，存在以下两种数据集类型。

- 无界：可持续添加的无限数据集。
- 有界：有限的、无不可改变的数据集。

实际上，许多传统意义上视为有界或批处理的数据集均为无界数据集，无论数据是否存储于 HDFS 上的目录序列，或者是基于日志的系统（例如 Apache Kafka）中，这一结论均成立。

一些无界数据集包括但不仅限于以下形式：

- 与移动或 Web 应用程序交互的终端用户。
- 提供测量结果的物理传感器。
- 金融市场。
- 机器产生的日志数据。

其次，类似于上述两种数据集类型，还存在以下两种执行模型类型。

- 数据流：只要产生数据，即持续执行的处理过程。
- 批处理：在有限的时间内执行并完成的处理过程，随后释放计算资源。

数据集类型和执行模型间的组合并无硬性规定，但最终结果不一定是最优的。例如，批处理执行一直以来应用于无界数据集上，尽管存在一些窗口机制、状态管理以及无序数据等问题。

Flink 依赖于流执行模型，这对于处理无界数据集较为直观：数据流执行过程持续处理连续生成的数据。在准确性和性能方面，数据集类型和执行模型类型之间的协调涵盖了诸多优点。

8.1.2　Flink、数据流模型和有界数据集

在 Apache Flink 中，可使用 DataStream API 与无界数据协同工作，并使用 DataSet API 与有界数据协同工作。Flink 在有界和无界数据集之间指定了较为自然的关系。相应地，有界数据集可视为无界数据集的特例，因而可针对两种数据集类型使用相同的概念。

有界数据集作为有限数据流在 Flink 内部被处理，对于 Flink 的管理方式，有界数据集和无界数据集间仅存在较小的差异。最终，可采用 Flink 对有界和无界数据进行处理，两种 API 运行于同一分布式流式执行引擎上——这是一种简单且功能强大的体系结构。

8.2　安装 Flink

本节将讨论 Apache Flink 的下载和安装过程。

Flink 可运行于 Linux、OS X 和 Windows 环境下。当运行 Flink 时，唯一的条件是需要安装 Java 7.x（或更高版本）。如果用户工作于 Windows 环境下，可访问 https://ci.apache.org/projects/flink/flinkdocs-release-1.4/start/flink_on_windows.html 以查看操作指南，其中详细描述了如何针对本地设置在 Windows 上运行 Flink。

通过下列命令，可查看当前的 Java 版本。

```
java -version
```

对于 Java 8，输出结果如下所示：

```
java version "1.8.0_111"
Java(TM) SE Runtime Environment (build 1.8.0_111-b14)
Java HotSpot(TM) 64-Bit Server VM (build 25.111-b14, mixed mode)
```

1. 下载 Flink

读者可访问 https://flink.apache.org/downloads.html，并下载与对应平台相关的 Apache Flink 二进制文件，如图 8.3 所示。

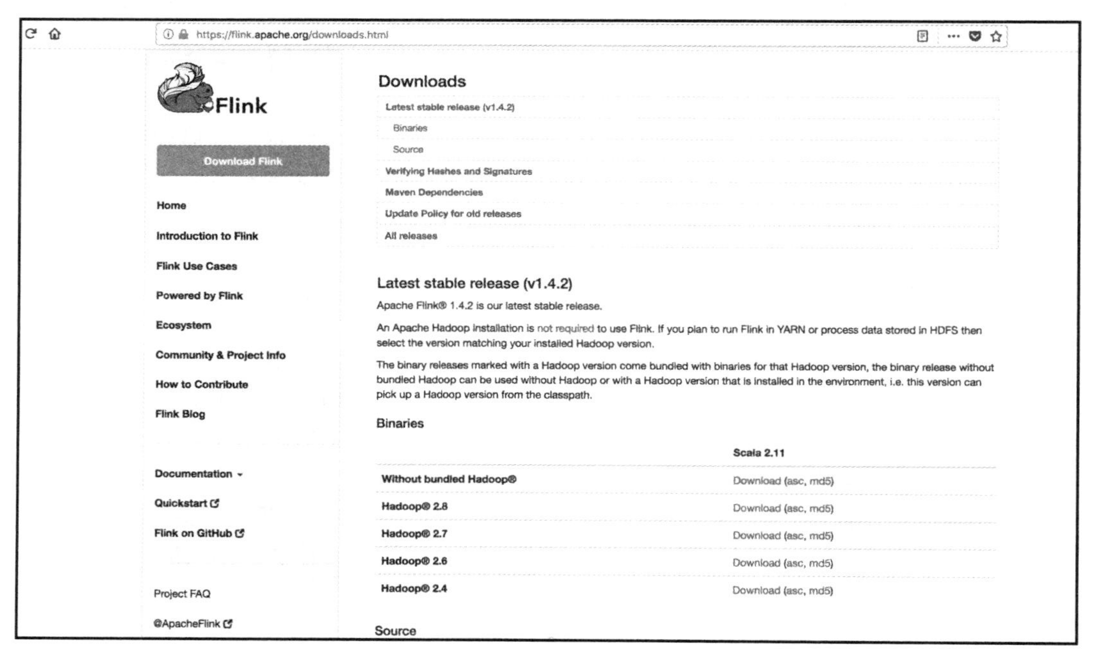

图 8.3

单击 Download 项将下载 Hadoop 2.8，随后可在浏览器中看到相应的下载页面，如图 8.4

所示。

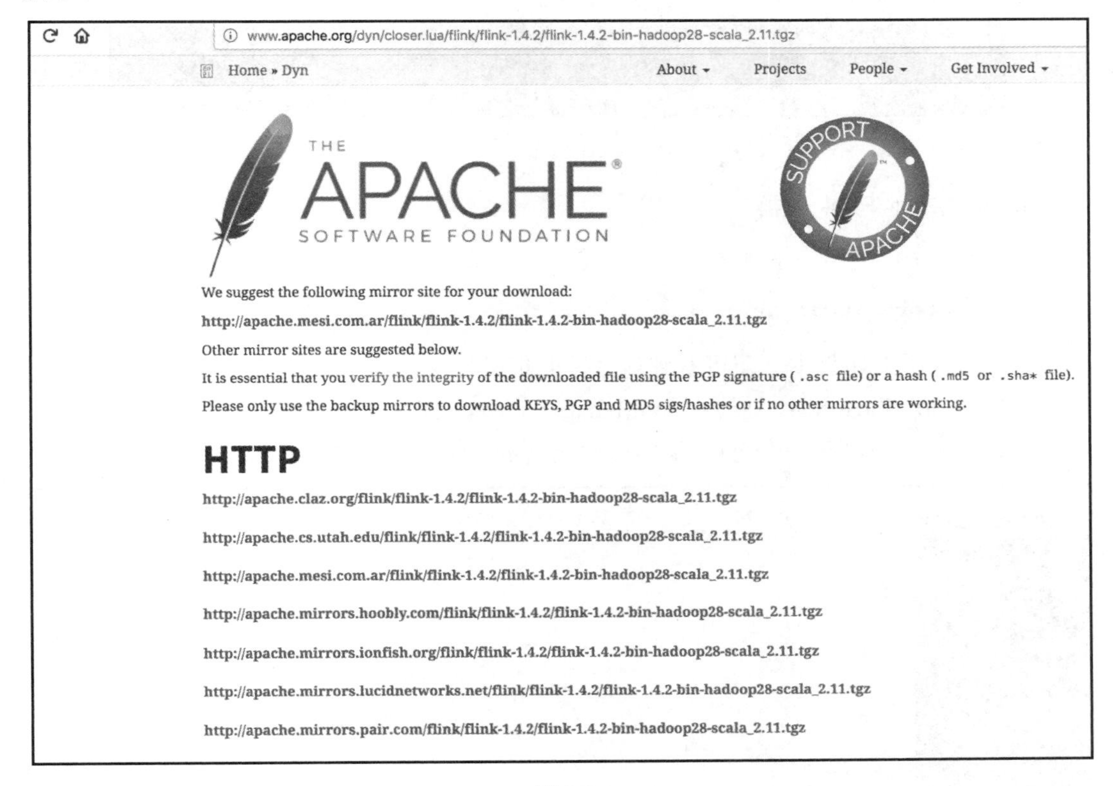

图 8.4

此处下载了 flink-1.4.2-bin-hadoop28-scala_2.11.tgz，这也是最新的版本（在本书撰写时）。

待下载完毕后，析取二进制文件。在 Mac 或 Linux 机器上，可运行 tar 命令，如图 8.5 所示。

```
Moogie:~ sridharalla$ tar -xvzf flink-1.4.2-bin-hadoop28-scala_2.11.tgz
x flink-1.4.2/
x flink-1.4.2/lib/
x flink-1.4.2/lib/flink-dist_2.11-1.4.2.jar
```

图 8.5

2. 安装 Flink

首先需要将当前目录调整为析取 Apache Flink 的位置，如下所示：

```
cd flink-1.4.2
```

相关内容如图 8.6 所示。

图 8.6

3. 启动本地 Flink 集群

通过 bin 文件夹中的下列脚本，即可启用本地集群：

```
./bin/start-local.sh
```

一旦运行了上述脚本，即可看到启用后的集群。

在 http://localhost:8081 处检查 JobManager 的 Web 前端，确保一切处于正常运行状态。这里，Web 前端应显示一个可用的 Task Managers 实例，如图 8.7 所示。

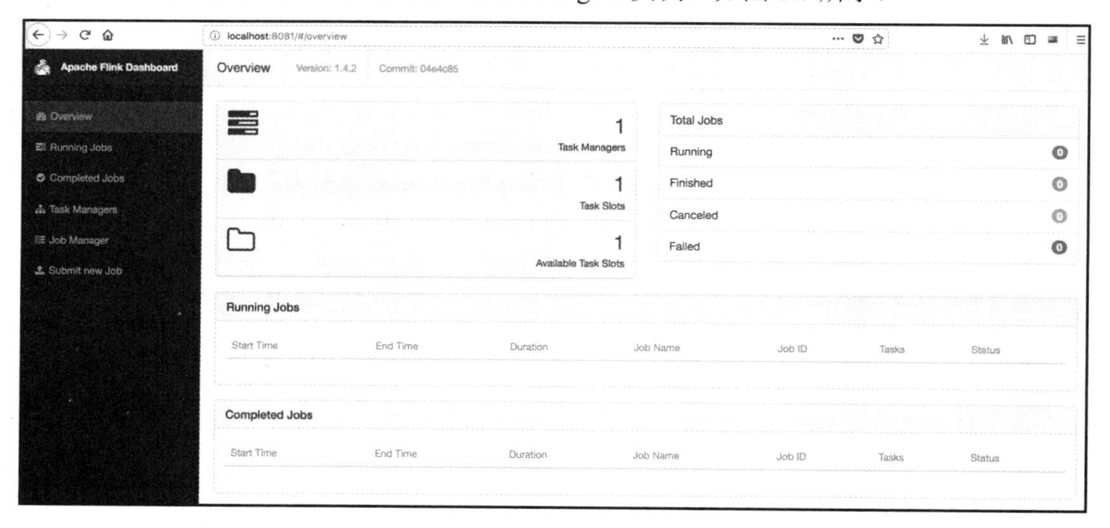

图 8.7

通过检查 logs 目录中的日志文件，还可进一步验证当前系统处于运行状态，如下所示：

```
tail log/flink-*-jobmanager-*.log
```

输出结果如图 8.8 所示。

```
Moogie:flink-1.4.2 sridharalla$ tail log/flink-*-jobmanager-*.log
2018-04-27 13:38:50,400 INFO  org.apache.flink.runtime.jobmanager.JobManager                - Starting JobManager actor
2018-04-27 13:38:50,405 INFO  org.apache.flink.runtime.blob.BlobServer                      - Created BLOB server storage directory /var/folde
rs/vv/5yxz8w1j3js11f94tpj4n2_h0000gn/T/blobStore-02f217cf-9d4d-4ce9-9351-769b543ddbce
2018-04-27 13:38:50,408 INFO  org.apache.flink.runtime.blob.BlobServer                      - Started BLOB server at 0.0.0.0:57213 - max concu
rrent requests: 50 - max backlog: 1000
2018-04-27 13:38:50,492 INFO  org.apache.flink.runtime.jobmanager.JobManager                - Starting JobManager at akka.tcp://flink@localhos
t:6123/user/jobmanager.
2018-04-27 13:38:50,492 INFO  org.apache.flink.runtime.jobmanager.MemoryArchivist           - Started memory archivist akka://flink/user/archi
ve
2018-04-27 13:38:50,502 INFO  org.apache.flink.runtime.jobmanager.JobManager                - JobManager akka.tcp://flink@localhost:6123/user/
jobmanager was granted leadership with leader session ID Some(00000000-0000-0000-0000-000000000000).
2018-04-27 13:38:50,505 INFO  org.apache.flink.runtime.clusterframework.standalone.StandaloneResourceManager  - Trying to associate with JobMa
nager leader akka.tcp://flink@localhost:6123/user/jobmanager
2018-04-27 13:38:50,510 INFO  org.apache.flink.runtime.clusterframework.standalone.StandaloneResourceManager  - Resource Manager associating w
ith leading JobManager Actor[akka://flink/user/jobmanager#-1691988345] - leader session 00000000-0000-0000-0000-000000000000
2018-04-27 13:38:51,561 INFO  org.apache.flink.runtime.clusterframework.standalone.StandaloneResourceManager  - TaskManager 675697f1ebad9bc592
62c69d1fb85311 has started.
2018-04-27 13:38:51,563 INFO  org.apache.flink.runtime.instance.InstanceManager             - Registered TaskManager at 10.0.0.103 (akka.tcp:/
/flink@moogie.local:57215/user/taskmanager) as bba64099cb58b029ccfceb7a4ecd7720. Current number of registered hosts is 1. Current number of al
ive task slots is 1.
```

图 8.8

当使用 Scala Shell 时，可输入下列命令：

```
./bin/start-scala-shell.sh remote localhost 6123
```

对应结果如图 8.9 所示。

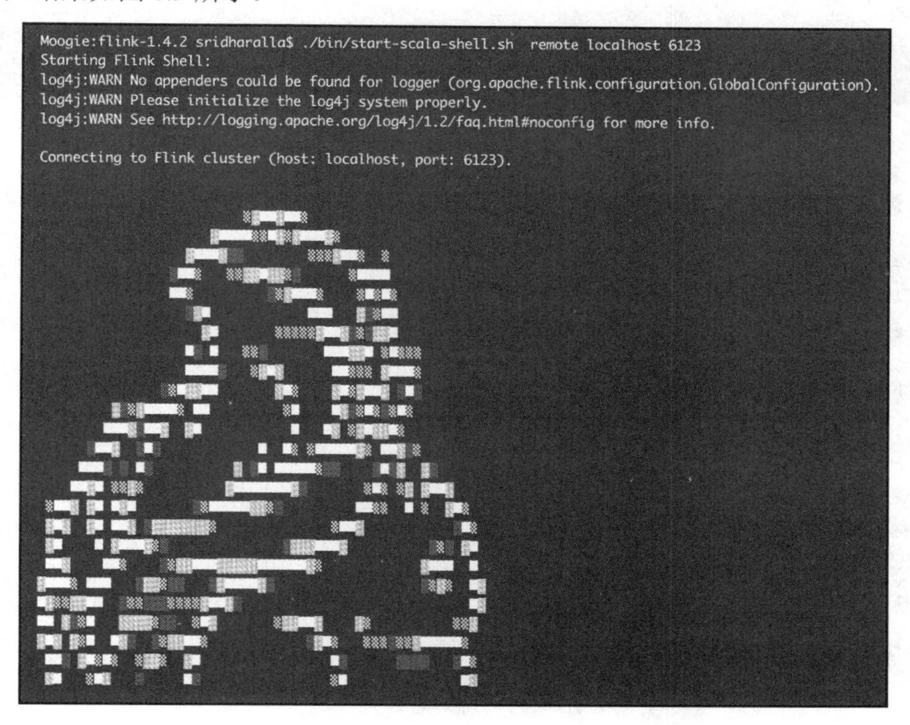

图 8.9

输入下列命令可加载数据：

```
val dataSet = benv
.readTextFile("OnlineRetail.csv")
dataSet.count()
```

对应结果如图 8.10 所示。

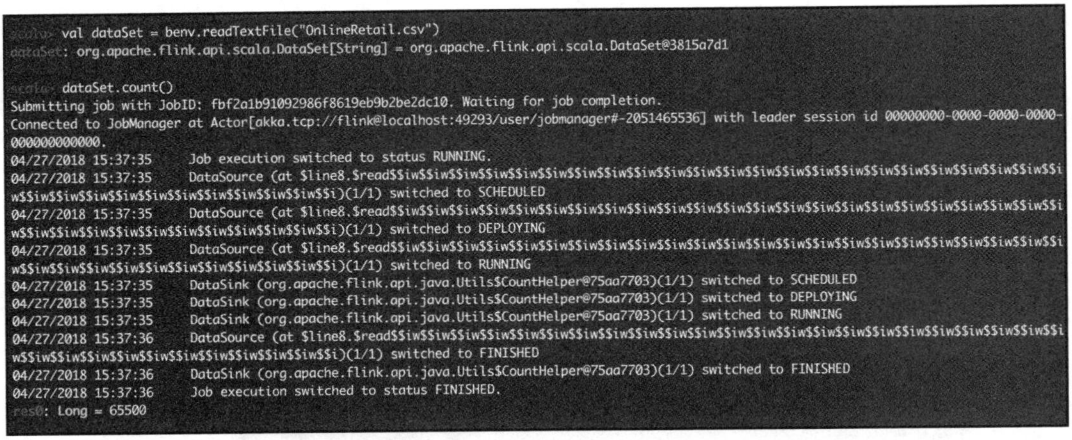

```
scala> val dataSet = benv.readTextFile("OnlineRetail.csv")
dataSet: org.apache.flink.api.scala.DataSet[String] = org.apache.flink.api.scala.DataSet@3815a7d1

scala> dataSet.count()
Submitting job with JobID: fbf2a1b91092986f8619eb9b2be2dc10. Waiting for job completion.
Connected to JobManager at Actor[akka.tcp://flink@localhost:49293/user/jobmanager#-2051465536] with leader session id 00000000-0000-0000-0000-
000000000000.
04/27/2018 15:37:35     Job execution switched to status RUNNING.
04/27/2018 15:37:35     DataSource (at $line8.$read$$iw$$iw$$iw$$iw$$iw$$iw$$iw$$iw$$iw$$iw$$iw$$iw$$iw$$iw$$iw$$iw$$iw$$iw$$iw$$iw$$iw$$iw$$i
w$$iw$$iw$$iw$$iw$$iw$$iw$$iw$$iw$$iw$$iw$$i)(1/1) switched to SCHEDULED
04/27/2018 15:37:35     DataSource (at $line8.$read$$iw$$iw$$iw$$iw$$iw$$iw$$iw$$iw$$iw$$iw$$iw$$iw$$iw$$iw$$iw$$iw$$iw$$iw$$iw$$iw$$iw$$iw$$i
w$$iw$$iw$$iw$$iw$$iw$$iw$$iw$$iw$$iw$$iw$$i)(1/1) switched to DEPLOYING
04/27/2018 15:37:35     DataSource (at $line8.$read$$iw$$iw$$iw$$iw$$iw$$iw$$iw$$iw$$iw$$iw$$iw$$iw$$iw$$iw$$iw$$iw$$iw$$iw$$iw$$iw$$iw$$iw$$i
w$$iw$$iw$$iw$$iw$$iw$$iw$$iw$$iw$$iw$$iw$$i)(1/1) switched to RUNNING
04/27/2018 15:37:35     DataSink (org.apache.flink.api.java.Utils$CountHelper@75aa7703)(1/1) switched to SCHEDULED
04/27/2018 15:37:35     DataSink (org.apache.flink.api.java.Utils$CountHelper@75aa7703)(1/1) switched to DEPLOYING
04/27/2018 15:37:35     DataSink (org.apache.flink.api.java.Utils$CountHelper@75aa7703)(1/1) switched to RUNNING
04/27/2018 15:37:36     DataSource (at $line8.$read$$iw$$iw$$iw$$iw$$iw$$iw$$iw$$iw$$iw$$iw$$iw$$iw$$iw$$iw$$iw$$iw$$iw$$iw$$iw$$iw$$iw$$iw$$i
w$$iw$$iw$$iw$$iw$$iw$$iw$$iw$$iw$$iw$$iw$$i)(1/1) switched to FINISHED
04/27/2018 15:37:36     DataSink (org.apache.flink.api.java.Utils$CountHelper@75aa7703)(1/1) switched to FINISHED
04/27/2018 15:37:36     Job execution switched to status FINISHED.
res0: Long = 65500
```

图 8.10

通过下列代码，可输出数据集的前 5 行内容：

```
dataSet
.first(5)
.print()
```

对应结果如图 8.11 所示。

通过 map()，可执行简单的转换操作，如下所示：

```
dataSet
.map(x => x.split(",")(2))
.first(5)
.print()
```

对应结果如图 8.12 所示。

```
scala> dataSet.first(5).print()
Submitting job with JobID: 2b31324ec79f774ab1c9baaa35b64bb6. Waiting for job completion.
Connected to JobManager at Actor[akka.tcp://flink@localhost:49293/user/jobmanager#-2051465536] with leader session id 00000000-0000-0000-0000-
000000000000.
04/27/2018 15:38:59     Job execution switched to status RUNNING.
04/27/2018 15:38:59     DataSource (at $line8.$read$$iw$$iw$$iw$$iw$$iw$$iw$$iw$$iw$$iw$$iw$$iw$$iw$$iw$$iw$$iw$$iw$$iw$$iw$$iw$$iw$$iw$$i
w$$iw$$iw$$iw$$iw$$iw$$iw$$iw$$iw$$iw$$i)(1/1) switched to SCHEDULED
04/27/2018 15:38:59     DataSource (at $line8.$read$$iw$$iw$$iw$$iw$$iw$$iw$$iw$$iw$$iw$$iw$$iw$$iw$$iw$$iw$$iw$$iw$$iw$$iw$$iw$$iw$$iw$$i
w$$iw$$iw$$iw$$iw$$iw$$iw$$iw$$iw$$iw$$i)(1/1) switched to DEPLOYING
04/27/2018 15:38:59     DataSource (at $line8.$read$$iw$$iw$$iw$$iw$$iw$$iw$$iw$$iw$$iw$$iw$$iw$$iw$$iw$$iw$$iw$$iw$$iw$$iw$$iw$$iw$$iw$$i
w$$iw$$iw$$iw$$iw$$iw$$iw$$iw$$iw$$iw$$i)(1/1) switched to RUNNING
04/27/2018 15:38:59     GroupReduce (GroupReduce at org.apache.flink.api.scala.DataSet.first(DataSet.scala:820))(1/1) switched to SCHEDULED
04/27/2018 15:38:59     GroupReduce (GroupReduce at org.apache.flink.api.scala.DataSet.first(DataSet.scala:820))(1/1) switched to DEPLOYING
04/27/2018 15:38:59     GroupReduce (GroupReduce at org.apache.flink.api.scala.DataSet.first(DataSet.scala:820))(1/1) switched to RUNNING
04/27/2018 15:38:59     DataSink (collect())(1/1) switched to SCHEDULED
04/27/2018 15:38:59     DataSink (collect())(1/1) switched to DEPLOYING
04/27/2018 15:38:59     GroupReduce (GroupReduce at org.apache.flink.api.scala.DataSet.first(DataSet.scala:820))(1/1) switched to FINISHED
04/27/2018 15:38:59     DataSink (collect())(1/1) switched to RUNNING
04/27/2018 15:38:59     DataSink (collect())(1/1) switched to FINISHED
04/27/2018 15:38:59     DataSource (at $line8.$read$$iw$$iw$$iw$$iw$$iw$$iw$$iw$$iw$$iw$$iw$$iw$$iw$$iw$$iw$$iw$$iw$$iw$$iw$$iw$$iw$$iw$$i
w$$iw$$iw$$iw$$iw$$iw$$iw$$iw$$iw$$iw$$i)(1/1) switched to FINISHED
04/27/2018 15:38:59     Job execution switched to status FINISHED.
InvoiceNo,StockCode,Description,Quantity,InvoiceDate,UnitPrice,CustomerID,Country
536365,85123A,WHITE HANGING HEART T-LIGHT HOLDER,6,12/1/10 8:26,2.55,17850,United Kingdom
536365,71053,WHITE METAL LANTERN,6,12/1/10 8:26,3.39,17850,United Kingdom
536365,84406B,CREAM CUPID HEARTS COAT HANGER,8,12/1/10 8:26,2.75,17850,United Kingdom
536365,84029G,KNITTED UNION FLAG HOT WATER BOTTLE,6,12/1/10 8:26,3.39,17850,United Kingdom
```

图 8.11

```
scala> dataSet.map(x => x.split(",")(2)).first(5).print()
Submitting job with JobID: 6513e762d3a224dfa3ba3a611d5a8136. Waiting for job completion.
Connected to JobManager at Actor[akka.tcp://flink@localhost:49293/user/jobmanager#-2051465536] with leader session id 00000000-0000-0000-0000-
000000000000.
04/27/2018 15:39:51     Job execution switched to status RUNNING.
04/27/2018 15:39:51     CHAIN DataSource (at $line8.$read$$iw$$iw$$iw$$iw$$iw$$iw$$iw$$iw$$iw$$iw$$iw$$iw$$iw$$iw$$iw$$iw$$iw$$iw$$iw$
$iw$$iw$$iw$$iw$$iw$$iw$$iw$$iw$$iw$$iw$$iw$$i) -> Map (Map at $line13.$read$$iw$$iw$$iw$$iw$$iw$$iw$$iw$$iw$$iw$$iw$$iw$$iw$$iw$$iw$$
iw$$iw$$iw$$iw$$iw$$iw$$iw$$iw$$iw$$iw$$iw$$iw$$iw$$iw$$iw$$iw$$iw$$iw$.<init>(<console>:65))(1/1) switched to SCHEDULED
04/27/2018 15:39:51     CHAIN DataSource (at $line8.$read$$iw$$iw$$iw$$iw$$iw$$iw$$iw$$iw$$iw$$iw$$iw$$iw$$iw$$iw$$iw$$iw$$iw$$iw$$iw$
$iw$$iw$$iw$$iw$$iw$$iw$$iw$$iw$$iw$$iw$$iw$$i) -> Map (Map at $line13.$read$$iw$$iw$$iw$$iw$$iw$$iw$$iw$$iw$$iw$$iw$$iw$$iw$$iw$$iw$$
iw$$iw$$iw$$iw$$iw$$iw$$iw$$iw$$iw$$iw$$iw$$iw$$iw$$iw$$iw$$iw$$iw$$iw$.<init>(<console>:65))(1/1) switched to DEPLOYING
04/27/2018 15:39:51     CHAIN DataSource (at $line8.$read$$iw$$iw$$iw$$iw$$iw$$iw$$iw$$iw$$iw$$iw$$iw$$iw$$iw$$iw$$iw$$iw$$iw$$iw$$iw$
$iw$$iw$$iw$$iw$$iw$$iw$$iw$$iw$$iw$$iw$$iw$$i) -> Map (Map at $line13.$read$$iw$$iw$$iw$$iw$$iw$$iw$$iw$$iw$$iw$$iw$$iw$$iw$$iw$$iw$$
iw$$iw$$iw$$iw$$iw$$iw$$iw$$iw$$iw$$iw$$iw$$iw$$iw$$iw$$iw$$iw$$iw$$iw$.<init>(<console>:65))(1/1) switched to RUNNING
04/27/2018 15:39:51     GroupReduce (GroupReduce at org.apache.flink.api.scala.DataSet.first(DataSet.scala:820))(1/1) switched to SCHEDULED
04/27/2018 15:39:51     GroupReduce (GroupReduce at org.apache.flink.api.scala.DataSet.first(DataSet.scala:820))(1/1) switched to DEPLOYING
04/27/2018 15:39:51     GroupReduce (GroupReduce at org.apache.flink.api.scala.DataSet.first(DataSet.scala:820))(1/1) switched to RUNNING
04/27/2018 15:39:51     DataSink (collect())(1/1) switched to SCHEDULED
04/27/2018 15:39:51     DataSink (collect())(1/1) switched to DEPLOYING
04/27/2018 15:39:51     GroupReduce (GroupReduce at org.apache.flink.api.scala.DataSet.first(DataSet.scala:820))(1/1) switched to FINISHED
04/27/2018 15:39:51     DataSink (collect())(1/1) switched to RUNNING
04/27/2018 15:39:51     DataSink (collect())(1/1) switched to FINISHED
04/27/2018 15:39:51     CHAIN DataSource (at $line8.$read$$iw$$iw$$iw$$iw$$iw$$iw$$iw$$iw$$iw$$iw$$iw$$iw$$iw$$iw$$iw$$iw$$iw$$iw$$iw$
$iw$$iw$$iw$$iw$$iw$$iw$$iw$$iw$$iw$$iw$$iw$$i) -> Map (Map at $line13.$read$$iw$$iw$$iw$$iw$$iw$$iw$$iw$$iw$$iw$$iw$$iw$$iw$$iw$$iw$$
iw$$iw$$iw$$iw$$iw$$iw$$iw$$iw$$iw$$iw$$iw$$iw$$iw$$iw$$iw$$iw$$iw$$iw$.<init>(<console>:65))(1/1) switched to FINISHED
04/27/2018 15:39:51     Job execution switched to status FINISHED.
Description
WHITE HANGING HEART T-LIGHT HOLDER
WHITE METAL LANTERN
CREAM CUPID HEARTS COAT HANGER
KNITTED UNION FLAG HOT WATER BOTTLE
```

图 8.12

又如：

```
dataSet
.flatMap(x => x.split(","))
```

```
.map(x=> (x,1))
.groupBy(0)
.sum(1)
.first(10)
.print()
```

对应结果如图 8.13 所示。

```
(,25447)
( ",95)
( 1 HANGER ,30)
( 3 TIER,1)
( 4 PURPLE FLOCK DINNER CANDLES,4)
( BACK DOOR ",31)
( BAROQUE",3)
( BILLBOARD FONTS DESIGN",12)
( BIRTHDAY CARD,28)
( BLUE",1)
```

图 8.13

8.3 使用 Flink 集群 UI

当使用 Flink 集群 UI 时，可查看并监测集群中的运行内容，并可进一步深入考查各种作业和任务。其中包括作业状态的检测、取消作业或者调试作业中的任何问题。通过查看标签，还可对代码问题进行诊断和修复。

图 8.14 显示了 Completed Jobs 列表。

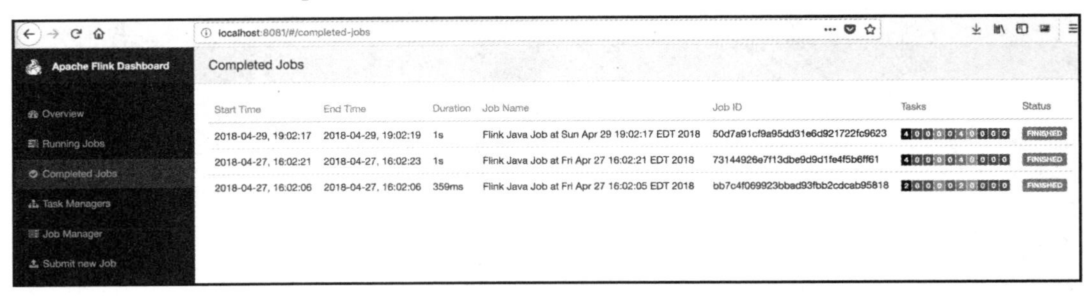

图 8.14

随后，可查看特定的作业，并考查与作业执行相关的详细内容，如图 8.15 所示。

例如，可查看作业的 Timeline 以了解更多信息，如图 8.16 所示。

图 8.17 显示了 Task Managers 作业，以及所有的任务管理器。这将有助于我们理解任务管理器的数量和状态。

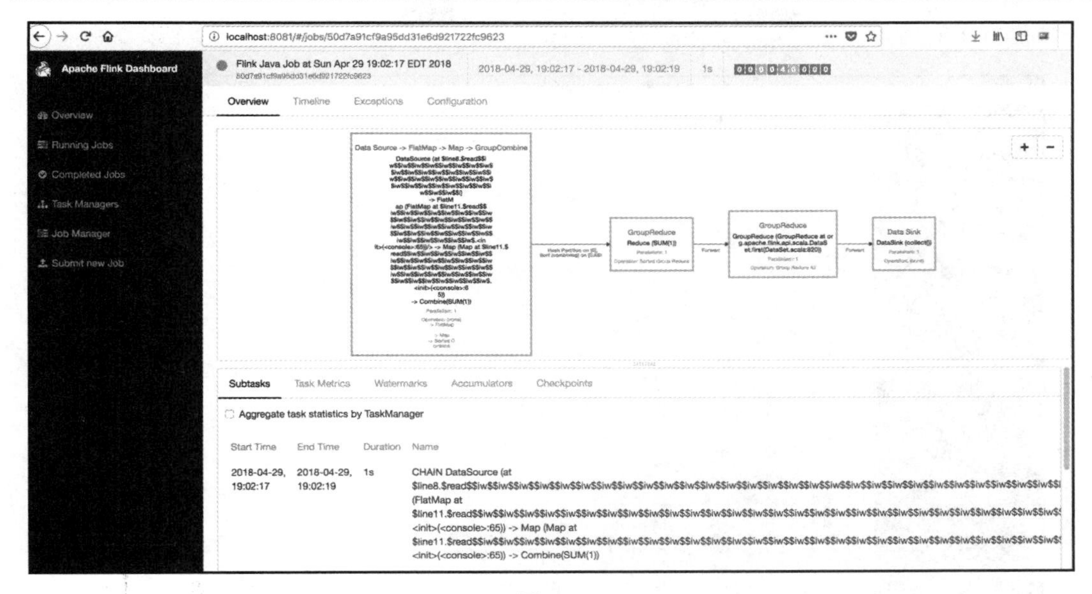

图 8.15

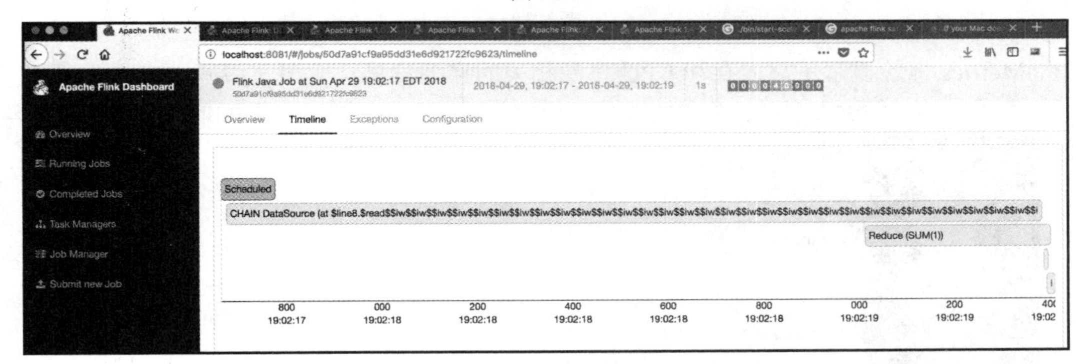

图 8.16

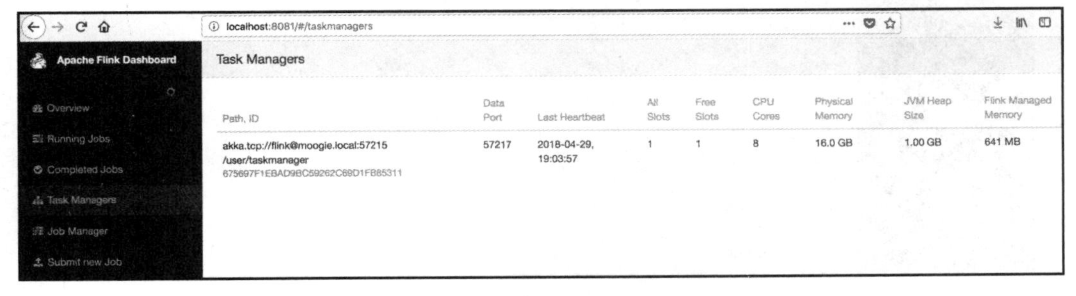

图 8.17

除此之外，还可对 Logs 进行检查，如图 8.18 所示。

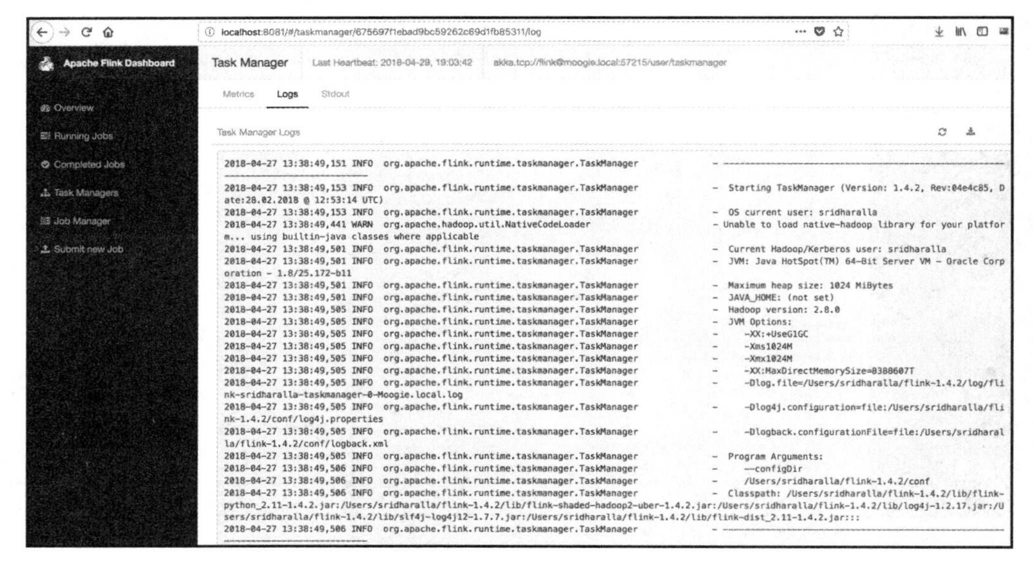

图 8.18

Metrics 标签显示了与内存和 CPU 资源相关的详细信息，如图 8.19 所示。

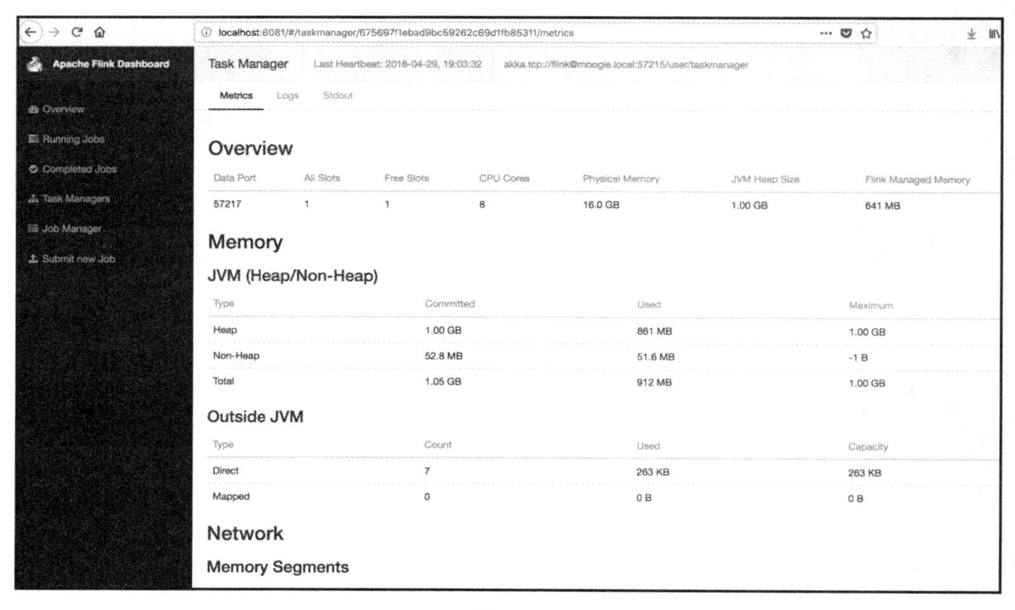

图 8.19

另外，还可作为作业提交 JAR，并代替在 Scala Shell 中编写的所有内容，如图 8.20 所示。

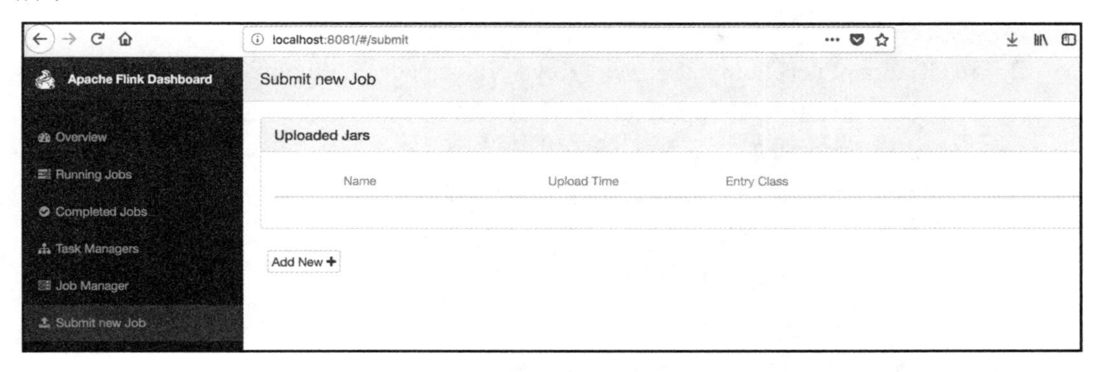

图 8.20

8.4 批处理分析

Apache Flink 中的批处理分析与流式分析较为类似，Flink 使用相同的 API 处理两种类型的分析，因而具有较强的灵活性，以便代码在类中不同的分析类型之间复用。

本节将在 OnlineRetail.csv 提供的样本数据上考查一些分析作业。此外，还将加载 cities.csv 和 temperature.csv 以执行更多的连接操作。

8.4.1 读取文件

Flink 包含多种内建格式，并从普通的文件格式中生成数据集。其中，许多格式在执行环境中具有相应的快捷方法。

1. 文件源

文件源可利用下列 API 进行读取。

- readTextFile(path)/TextInputFormat：逐行读取文件并以字符串的形式返回。
- readTextFileWithValue(path)/TextValueInputFormat：逐行读取文件，并作为 StringValues 将其返回。这里，StringValues 为可变的字符串。
- readCsvFile(path)/CsvInputFormat：解析用逗号（或另一个字符）分隔的字段的文件，并返回元组形式的 DataSet、类对象或 POJO；同时还支持基本的 Java 类型及其 Value 对应的字段类型。

- ❑ readFileOfPrimitives(path, delimiter)/PrimitiveInputFormat：使用给定的分隔符解析换行（或另一个字符序列）分隔的、原始数据类型（例如 String 或 Integer）的文件。
- ❑ readHadoopFile(FileInputFormat, Key, Value,path)/FileInputFormat：创建一个 JobConf，并通过 FileInputFormat、Key 类和 Value 类从指定的路径读取文件，并将其作为 Tuple2<Key, Value>予以返回。
- ❑ readSequenceFile(Key, Value, path)/SequenceFileInputFormat：创建一个 JobConf，并利用 SequenceFileInputFormat 类型、Key 类和 Value 类从指定的路径中读取文件，随后将其作为 Tuple2<Key, Value>返回。

2. 基于集合的源

基于集合的源（例如利润表、数组等数据结构）可利用下列 API 读取。

- ❑ fromCollection(Seq)：从 Seq 中创建一个 DataSet。集合中的全部元素须为相同的类型。
- ❑ fromCollection(Iterator)：从 Iterator 中创建一个 DataSet。该类指定了迭代器返回元素的数据类型。
- ❑ fromElements(elements: _*)：从给定的对象序列中创建一个 DataSet，全部对象须为同一类型。
- ❑ fromParallelCollection(SplittableIterator)：从迭代器中以并行方式创建一个 DataSet。该类指定了迭代器返回元素的数据类型。
- ❑ generateSequence(from, to)：在给定间隔中以并行方式生成一个数字序列。

3. 通用源

通用（自定义）源可利用下列 API 进行读取。

- ❑ readFile(inputFormat, path)/FileInputFormat：接收一种文件输入格式。
- ❑ createInput(inputFormat)/InputFormat：接收通用输入格式。

下面考查其中的一个 API，即 readTextFile()。通过该 API 读取一个文件将把一个文件（本地文本文件、HDFS 文件、Amazon s3 文件等）加载至 DataSet 中。这一 DataSet 包含了载入数据的划分位置，因而可执行 TB 级的数据。

下列代码展示了 OnlineRetail.csv 文件的加载示例。

```
val dataSet = benv.readTextFile("OnlineRetail.csv")
dataSet.first(10).print()
```

这将输出加载后的 DataSet 中的内容，对应结果如下所示：

```
InvoiceNo,StockCode,Description,Quantity,InvoiceDate,UnitPrice,Custome
rID,Country
 536365,85123A,WHITE HANGING HEART T-LIGHT HOLDER,6,12/1/10
8:26,2.55,17850,United Kingdom
 536365,71053,WHITE METAL LANTERN,6,12/1/10 8:26,3.39,17850,United Kingdom
 536365,84406B,CREAM CUPID HEARTS COAT HANGER,8,12/1/10
8:26,2.75,17850,United Kingdom
 536365,84029G,KNITTED UNION FLAG HOT WATER BOTTLE,6,12/1/10
8:26,3.39,17850,United Kingdom
 536365,84029E,RED WOOLLY HOTTIE WHITE HEART.,6,12/1/10
8:26,3.39,17850,United Kingdom
 536365,22752,SET 7 BABUSHKA NESTING BOXES,2,12/1/10
8:26,7.65,17850,UnitedKingdom
 536365,21730,GLASS STAR FROSTED T-LIGHT HOLDER,6,12/1/10
8:26,4.25,17850,United Kingdom
 536366,22633,HAND WARMER UNION JACK,6,12/1/10 8:28,1.85,17850,United
Kingdom
 536366,22632,HAND WARMER RED POLKA DOT,6,12/1/10 8:28,1.85,17850,United
Kingdom
```

不难发现，第一行内容表示为数据头行，因而在分析过程中并无实际用处。通过 filter()函数可过滤掉一行或多行内容。

下列代码显示了文件的加载过程，并移除了第一行内容，最后返回一个 DataSet。

```
val dataSet =benv
    .readTextFile("OnlineRetail.csv")
    .filter(!_.startsWith("InvoiceNo"))
dataSet.first(10).print()
```

这将输出加载后的 DataSet 中的内容，如下所示：

```
 536365,85123A,WHITE HANGING HEART T-LIGHT HOLDER,6,12/1/10
8:26,2.55,17850,United Kingdom
 536365,71053,WHITE METAL LANTERN,6,12/1/10 8:26,3.39,17850,United Kingdom
 536365,84406B,CREAM CUPID HEARTS COAT HANGER,8,12/1/10
8:26,2.75,17850,United Kingdom
 536365,84029G,KNITTED UNION FLAG HOT WATER BOTTLE,6,12/1/10
8:26,3.39,17850,United Kingdom
 536365,84029E,RED WOOLLY HOTTIE WHITE HEART.,6,12/1/10
8:26,3.39,17850,United Kingdom
 536365,22752,SET 7 BABUSHKA NESTING BOXES,2,12/1/10
8:26,7.65,17850,United Kingdom
 536365,21730,GLASS STAR FROSTED T-LIGHT HOLDER,6,12/1/10
```

```
8:26,4.25,17850,United Kingdom
 536366,22633,HAND WARMER UNION JACK,6,12/1/10 8:28,1.85,17850,United
Kingdom
 536366,22632,HAND WARMER RED POLKA DOT,6,12/1/10
8:28,1.85,17850,United Kingdom
 536367,84879,ASSORTED COLOUR BIRD ORNAMENT,32,12/1/10
8:34,1.69,13047,United Kingdom
```

不难发现，下列数据头行被移除：

```
InvoiceNo,StockCode,Description,Quantity,InvoiceDate,UnitPrice,Custome
rID,Country
```

对于载入后的 DataSet，下面讨论其上的更多操作。

8.4.2 转换

通过对原始 DataSet 的每一行使用转换逻辑，转换可将 DataSet 转换为一个新的 DataSet。例如，如果希望移除输入中的第一行（数据头），则可使用 filter()操作。

下列代码使用了两项 filter()操作：首先是移除数据头；随后，确保每一行中包含正确的列数（此处为 8）。

```
val dataSet = benv.readTextFile("OnlineRetail.csv")
    .filter(!_.startsWith("InvoiceNo"))
    .filter(_.split(",").length == 8)

dataSet.map(x => x.split(",")(2))
    .first(10).print()
```

这将输出加载后的 DataSet 中的内容，如下所示：

```
WHITE HANGING HEART T-LIGHT HOLDER
WHITE METAL LANTERN
CREAM CUPID HEARTS COAT HANGER
KNITTED UNION FLAG HOT WATER BOTTLE
RED WOOLLY HOTTIE WHITE HEART.
SET 7 BABUSHKA NESTING BOXES
GLASS STAR FROSTED T-LIGHT HOLDER
HAND WARMER UNION JACK
HAND WARMER RED POLKA DOT
```

类似地，还可输出 DataSet 中的 quantity 列，如下所示：

```
dataSet.map(x => x.split(",")(3))
    .first(10).print()
```

这将输出加载后的 DataSet 中的内容，如下所示：

```
6
6
8
6
6
2
6
6
6
```

同样，还可输出 DataSet 中 description 和 quantity 类构成的元组，如下所示：

```
dataSet.map(x => (x.split(",")(2), x.split(",")(3).toInt))
    .first(10).print()
```

这将输出加载后的 DataSet 中的内容，如下所示：

```
(WHITE HANGING HEART T-LIGHT HOLDER,6)
(WHITE METAL LANTERN,6)
(CREAM CUPID HEARTS COAT HANGER,8)
(KNITTED UNION FLAG HOT WATER BOTTLE,6)
(RED WOOLLY HOTTIE WHITE HEART.,6)
(SET 7 BABUSHKA NESTING BOXES,2)
(GLASS STAR FROSTED T-LIGHT HOLDER,6)
(HAND WARMER UNION JACK,6)
(HAND WARMER RED POLKA DOT,6)
```

表 8.1 对有效的转换进行了适当的总结，读者也可访问 https://ci.apache.org/projects/flink/flink-docs-release-1.4/dev/batch/dataset_transformations.html 以了解相关内容。

表 8.1

转　　换	描　　述
map	接收一个元素并生成一个元素，如下所示： data.map { x => x.toInt }
flatMap	接收一个元素并生成 0、1 或更多个元素，如下所示： data.flatMap { str => str.split(" ") }
mapPartition	在单一函数调用中转换某个并行分区。该函数以迭代器的形式获得分区，并可生成任意数量的结果值。每个分区中的元素数量取决于并行度和上一次操作，如下所示： data.mapPartition { in => in map { (_, 1) } }
filter	计算每个元素的布尔函数，并保留该函数返回 TRUE 的值

续表

转　换	描　述
reduce	针对每个元素计算布尔函数，并保留函数返回 TRUE 的元素。注意：系统假设函数不会修改应用断言的元素；否则将导致错误的结果。相关示例如下所示： data.filter { _ > 1000 }
reduceGroup	将一组元素整合至单一元素中，也就是说，重复地将两个元素整合至一个元素中。其间，reduce 操作可能会应用于全数据集上，或者是分组后的数据集上。对应示例如下所示： data.reduce { _ + _ }
aggregate	将一组数值聚合至单一数值上。聚合函数可视为内建 reduce 函数。其间，聚合操作可应用于全数据集上，或者是一个分组后的数据集上，对应示例如下所示： val input: DataSet[(Int, String, Double)] = // [...] val output: DataSet[(Int, String, Double)] = input.aggregate(SUM, 0).aggregate(MIN, 2) 除此之外，还可针对 min、max、sum 聚合使用快捷语法，如下所示： val input: DataSet[(Int, String, Double)] = // [...] val output: DataSet[(Int, String, Double)] = input.sum(0).min(2)
distinct	对于元素的全部字段，或者是字段的子集，返回数据集的不同元素，并从输入 DataSet 中移除重复项，如下所示： data.distinct()

join	通过创建键相等的全部元素对连接两个数据集。作为可选方案，还可使用 JoinFunction 将元素对转换为单一元素；或者利用 FlatJoinFunction 将元素对转换为任意多个（包括 0 个）元素，如下所示：
	// In this case tuple fields are used as keys. "0" //is the join //field on the first tuple
	// "1" is the join field on the second tuple.
	val result = input1.join(input2).where(0).equalTo(1)
	可通过 Join Hints 指定运行期执行连接的方式，即连接是否采用分区、广播、排序或哈希算法。读者可访问 https://ci.apache.org/projects/flink/flink-docs-release-1.4/dev/batch/dataset_transformations.html#join-algorithmhints 以了解相关信息。如果对此未予指定，系统则尝试估算输出尺寸，并根据相应的结果选取最佳策略，如下所示：
	// This executes a join by broadcasting the first data set
	// using a hash table for the broadcast data
	val result = input1.join(input2, JoinHint.BROADCAST_HASH_FIRST)
	　　　　　　　　.where(0).equalTo(1)
	需要注意的是，join 转换仅适用于等值连接，其他 join 类型则需要通过 OuterJoin 或 CoGroup 予以表示

续表

转　　换	描　　述

OuterJoin	在两个数据集上执行左、右或全连接。外部连接类似于常规（内部）连接，并生成键相等的全部元素对。除此之外，如果另一侧中不存在匹配键，外侧的记录（左、右或全部）将被保留。匹配的元素对（或其他输入的单一元素和 null 值）将被置入 JoinFunction，并将元素对转换为单一元素；抑或置入 FlatJoinFunction，并将元素对转换为任意多个元素（包括 0 个），如下所示：
	val joined = left.leftOuterJoin(right).where(0).equalTo(1) {
	    (left, right) =>
	      val a = if (left == null) "none" else left._1
	      (a, right)
	  }

coGroup	reduce 操作的二维变化版本。coGroup 将一个或多个字段上的每个输入分组，然后将这些分组连接起来。转换函数针对每个分组对进行调用。读者可访问 https://ci.apache.org/projects/flink/flink-docs-release-1.4/dev api_concepts.html#specifying-keys，以了解如何定义 coGroup 键。相关示例如下所示： data1.coGroup(data2).where(0).equalTo(1)
cross	构建两个输入的笛卡儿积（叉积），并生成所有的元素对。作为可选方案，使用 CrossFunction 将元素对转换为单一元素，如下所示： val data1: DataSet[Int] = // [...] val data2: DataSet[String] = // [...] val result: DataSet[(Int, String)] = data1.cross(data2) 注意，Cross 是一类计算密集型操作，即使在大型计算群集上也是如此，建议使用 crossWithTiny()、crossWithHuge()、数据集的大小对当前系统予以提示
union	生成两个数据集的并集，如下所示： data.union(data2)
rebalance	重新平衡数据集的并行分区，以消除数据倾斜。只有 map 转换支持再平衡转换，如下所示： val data1: DataSet[Int] = // [...] val result: DataSet[(Int, String)] = data1.rebalance().map(...)
哈希分区	在给定键上对数据集执行哈希分区。其中，键可以指定为位置键、表达式键和键选择器函数，如下所示： val in: DataSet[(Int, String)] = // [...] val result = in.partitionByHash(0).mapPartition { ... }
范围分区	在给定键上对数据集执行范围分区。其中，键可以指定为位置键、表达式键和键选择器函数，如下所示： val in: DataSet[(Int, String)] = // [...] val result = in.partitionByRange(0).mapPartition { ... }

续表

转　　换	描　　述
自定义分区	手动指定数据上的分区。注意，该方法仅适用于单字段键，如下所示： val in: DataSet[(Int, String)] = // [...] val result = in 　.partitionCustom(partitioner: Partitioner[K], key)
排序分区	以指定的顺序对指定字段上的数据集的所有分区进行本地排序。字段可指定为元组位置或字段表达式。在多个字段上排序则是通过链接 sortPartition()调用完成的，如下所示： val in: DataSet[(Int, String)] = // [...] val result = in.sortPartition(1, Order.ASCENDING).mapPartition { ... }

First-n	返回数据集的前 n 个元素。First-n 适用于规则数据集、分组后的数据集、分组-排序后的数据集。分组键可指定为键-选择器函数、元组位置或类字段，如下所示：
	val in: DataSet[(Int, String)] = // [...]
	// regular data set
	val result1 = in.first(3)
	// grouped data set
	val result2 = in.groupBy(0).first(3)
	// grouped-sorted data set
	val result3 = in.groupBy(0).sortGroup(1,
	Order.ASCENDING).first(3)

8.4.3　groupBy

groupBy 操作通过某些列实现了 DataSet 的行聚合操作。groupBy()接收用于聚合行的列索引。

下列命令通过 Description 进行分组，并输出前 10 条记录。

```
dataSet.map(x => (x.split(",")(2), x.split(",")(3).toInt))
       .groupBy(0)
       .first(10).print()
```

这将生成加载后的 DataSet 内容，如下所示：

```
(WOODLAND DESIGN COTTON TOTE BAG,1)
(WOODLAND DESIGN COTTON TOTE BAG,1)
(WOODLAND DESIGN COTTON TOTE BAG,6)
(WOODLAND DESIGN COTTON TOTE BAG,1)
```

```
(WOODLAND DESIGN COTTON TOTE BAG,2)
(WOODLAND DESIGN COTTON TOTE BAG,1)
(WOODLAND DESIGN COTTON TOTE BAG,6)
(WOODLAND DESIGN COTTON TOTE BAG,1)
(WOODLAND DESIGN COTTON TOTE BAG,1)
(WOODLAND DESIGN COTTON TOTE BAG,12)
(WOODLAND PARTY BAG + STICKER SET,2)
(WOODLAND PARTY BAG + STICKER SET,16)
(WOODLAND PARTY BAG + STICKER SET,1)
(WOODLAND PARTY BAG + STICKER SET,8)
(WOODLAND PARTY BAG + STICKER SET,4)
```

groupBy()的定义如下所示：

```
/**
 * Groups a {@link Tuple} {@link DataSet} using field position keys.
 *
 * <p><b>Note: Field position keys only be specified for Tuple
DataSets.</b>
 *
 * <p>The field position keys specify the fields of Tuples on which the
DataSet is grouped.
 * This method returns an {@link UnsortedGrouping} on which one of the
following grouping transformation
 * can be applied.
 * <ul>
 * <li>{@link UnsortedGrouping#sortGroup(int,
org.apache.flink.api.common.operators.Order)} to get a {@link
SortedGrouping}.
 * <li>{@link UnsortedGrouping#aggregate(Aggregations, int)} to apply an
Aggregate transformation.
 * <li>{@link
UnsortedGrouping#reduce(org.apache.flink.api.common.functions.ReduceFu
nction)} to apply a Reduce transformation.
 * <li>{@link
UnsortedGrouping#reduceGroup(org.apache.flink.api.common.functions.GroupRed
uceFunction)} to apply a GroupReduce transformation.
 * </ul>
 *
 * @param fields One or more field positions on which the DataSet will be
grouped.
 * @return A Grouping on which a transformation needs to be applied to obtain
a transformed DataSet.
```

```
 *
 * @see Tuple
 * @see UnsortedGrouping
 * @see AggregateOperator
 * @see ReduceOperator
 * @see org.apache.flink.api.java.operators.GroupReduceOperator
 * @see DataSet
 */
public UnsortedGrouping<T> groupBy(int... fields) {
return new UnsortedGrouping<>(this, new Keys.ExpressionKeys<>(fields,
getType()));
}
```

8.4.4 聚合

在某些列应用 groupBy()之后，聚合操作将逻辑应用于数据集的分组行。groupBy()接收用户聚合行的列索引；聚合操作接收列索引并执行聚合操作。

下列命令通过 Description 进行分组，并针对每个 Description 添加 Quantities，最后输出前 10 条记录，如下所示：

```
dataSet.map(x => (x.split(",")(2), x.split(",")(3).toInt))
    .groupBy(0)
    .sum(1)
    .first(10).print()
```

这将输出加载后的 DataSet 的内容，如下所示：

```
(,-2117)
(*Boombox Ipod Classic,1)
(*USB Office Mirror Ball,2)
(10 COLOUR SPACEBOY PEN,823)
(12 COLOURED PARTY BALLOONS,102)
(12 DAISY PEGS IN WOOD BOX,62)
(12 EGG HOUSE PAINTED WOOD,16)
(12 IVORY ROSE PEG PLACE SETTINGS,80)
(12 MESSAGE CARDS WITH ENVELOPES,238)
(12 PENCIL SMALL TUBE WOODLAND,444)
```

下列命令通过 Description 进行分组，并针对每个 Description 添加 Quantities，最后输出包含最大 Quantity 的 Description，如下所示：

```
dataSet.map(x => (x.split(",")(2), x.split(",")(3).toInt))
    .groupBy(0)
```

```
    .sum(1)
    .max(1)
    .first(10).print()
```

这将输出加载后的 DataSet 的内容，如下所示：

```
(reverse 21/5/10 adjustment,8189)
```

类似地，下列命令通过 Description 进行分组，并针对每个 Description 添加 Quantities，最后输出包含最小 Quantity 的 Description，如下所示：

```
dataSet.map(x => (x.split(",")(2), x.split(",")(3).toInt))
    .groupBy(0)
    .sum(1)
    .min(1)
    .first(10).print()
```

这将输出加载后的 DataSet 的内容，如下所示：

```
(reverse 21/5/10 adjustment,-7005)
```

sum() API 的定义如下所示：

```
// private helper that allows to set a different call location name
 private AggregateOperator<T> aggregate(Aggregations agg, int field, String
callLocationName) {
 return new AggregateOperator<T>(this, agg, field, callLocationName);
 }
/**
 * Syntactic sugar for aggregate (SUM, field).
 * @param field The index of the Tuple field on which the aggregation
function is applied.
 * @return An AggregateOperator that represents the summed DataSet.
 *
 * @see org.apache.flink.api.java.operators.AggregateOperator
 */
 public AggregateOperator<T> sum (int field) {
 return this.aggregate (Aggregations.SUM, field,
Utils.getCallLocationName());
 }
```

8.4.5　连接

首先读取 cities.csv 文本文件，如下所示：

```
val cities = benv.readTextFile("cities.csv")
```

对应结果如图 8.21 所示。

```
scala> val cities = benv.readTextFile("cities.csv")
cities: org.apache.flink.api.scala.DataSet[String] = org.apache.flink.api.scala.DataSet@bd09a26

scala> cities.first(10).print()
Submitting job with JobID: 4b05ca2711840f76d1a7b7d6799eb781. Waiting for job completion.
Connected to JobManager at Actor[akka.tcp://flink@localhost:6123/user/jobmanager#-61006016] with lead
00000000.
05/21/2018 15:31:45     Job execution switched to status RUNNING.
05/21/2018 15:31:45     DataSource (at $line16.$read$$iw$$iw$$iw$$iw$$iw$$iw$$iw$$iw$$iw$$iw$$iw$$iw$
w$$iw$$iw$$iw$$iw$$iw$$iw$$iw$$iw$$iw$$iw$$)(1/1) switched to SCHEDULED
05/21/2018 15:31:45     DataSource (at $line16.$read$$iw$$iw$$iw$$iw$$iw$$iw$$iw$$iw$$iw$$iw$$iw$$iw$
w$$iw$$iw$$iw$$iw$$iw$$iw$$iw$$iw$$iw$$iw$$)(1/1) switched to DEPLOYING
05/21/2018 15:31:45     DataSource (at $line16.$read$$iw$$iw$$iw$$iw$$iw$$iw$$iw$$iw$$iw$$iw$$iw$$iw$
w$$iw$$iw$$iw$$iw$$iw$$iw$$iw$$iw$$iw$$iw$$)(1/1) switched to RUNNING
05/21/2018 15:31:45     GroupReduce (GroupReduce at org.apache.flink.api.scala.DataSet.first(DataSet.
05/21/2018 15:31:45     GroupReduce (GroupReduce at org.apache.flink.api.scala.DataSet.first(DataSet.
05/21/2018 15:31:45     DataSource (at $line16.$read$$iw$$iw$$iw$$iw$$iw$$iw$$iw$$iw$$iw$$iw$$iw$$iw$
w$$iw$$iw$$iw$$iw$$iw$$iw$$iw$$iw$$iw$$iw$$)(1/1) switched to FINISHED
05/21/2018 15:31:45     GroupReduce (GroupReduce at org.apache.flink.api.scala.DataSet.first(DataSet.
05/21/2018 15:31:45     DataSink (collect())(1/1) switched to SCHEDULED
05/21/2018 15:31:45     DataSink (collect())(1/1) switched to DEPLOYING
05/21/2018 15:31:45     GroupReduce (GroupReduce at org.apache.flink.api.scala.DataSet.first(DataSet.
05/21/2018 15:31:45     DataSink (collect())(1/1) switched to RUNNING
05/21/2018 15:31:45     DataSink (collect())(1/1) switched to FINISHED
05/21/2018 15:31:45     Job execution switched to status FINISHED.
Id,City
1,Boston
2,New York
3,Chicago
4,Philadelphia
5,San Francisco
7,Las Vegas
```

图 8.21

cities.csv 文本文件如下所示：

```
Id,City
1,Boston
2,New York
3,Chicago
4,Philadelphia
5,San Francisco
7,Las Vegas
```

随后读取 temperatures.csv 文件，如下所示：

```
val temp = benv.readTextFile("temperatures.csv")
```

对应结果如图 8.22 所示。

```
scala> val temp = benv.readTextFile("temperatures.csv")
temp: org.apache.flink.api.scala.DataSet[String] = org.apache.flink.api.scala.DataSet@3497ecea

scala> temp.first(10).print()
Submitting job with JobID: b42267930e15fcc3ed7153e84664bef5. Waiting for job completion.
Connected to JobManager at Actor[akka.tcp://flink@localhost:6123/user/jobmanager#-61006016] with 1
00000000.
05/21/2018 15:34:19     Job execution switched to status RUNNING.
05/21/2018 15:34:19     DataSource (at $line22.$read$$iw$$iw$$iw$$iw$$iw$$iw$$iw$$iw$$iw$$iw$$
w$$iw$$iw$$iw$$iw$$iw$$iw$$iw$$iw$$iw$$)(1/1) switched to SCHEDULED
05/21/2018 15:34:19     DataSource (at $line22.$read$$iw$$iw$$iw$$iw$$iw$$iw$$iw$$iw$$iw$$iw$$
w$$iw$$iw$$iw$$iw$$iw$$iw$$iw$$iw$$iw$$)(1/1) switched to DEPLOYING
05/21/2018 15:34:19     DataSource (at $line22.$read$$iw$$iw$$iw$$iw$$iw$$iw$$iw$$iw$$iw$$iw$$
w$$iw$$iw$$iw$$iw$$iw$$iw$$iw$$iw$$iw$$)(1/1) switched to RUNNING
05/21/2018 15:34:19     GroupReduce (GroupReduce at org.apache.flink.api.scala.DataSet.first(DataS
05/21/2018 15:34:19     GroupReduce (GroupReduce at org.apache.flink.api.scala.DataSet.first(DataS
05/21/2018 15:34:19     DataSource (at $line22.$read$$iw$$iw$$iw$$iw$$iw$$iw$$iw$$iw$$iw$$iw$$
w$$iw$$iw$$iw$$iw$$iw$$iw$$iw$$iw$$iw$$)(1/1) switched to FINISHED
05/21/2018 15:34:19     GroupReduce (GroupReduce at org.apache.flink.api.scala.DataSet.first(DataS
05/21/2018 15:34:19     DataSink (collect())(1/1) switched to SCHEDULED
05/21/2018 15:34:19     DataSink (collect())(1/1) switched to DEPLOYING
05/21/2018 15:34:19     GroupReduce (GroupReduce at org.apache.flink.api.scala.DataSet.first(DataS
05/21/2018 15:34:19     DataSink (collect())(1/1) switched to RUNNING
05/21/2018 15:34:19     DataSink (collect())(1/1) switched to FINISHED
05/21/2018 15:34:19     Job execution switched to status FINISHED.
Date,Id,Temperature
2018-01-01,1,21
2018-01-01,2,22
2018-01-01,3,23
2018-01-01,4,24
2018-01-01,5,25
2018-01-01,6,22
2018-01-02,1,23
2018-01-02,2,24
2018-01-02,3,25
```

图 8.22

temperatures.csv 文件如下所示：

```
Date,Id,Temperature
2018-01-01,1,21
2018-01-01,2,22
2018-01-01,3,23
2018-01-01,4,24
2018-01-01,5,25
2018-01-01,6,22
2018-01-02,1,23
2018-01-02,2,24
2018-01-02,3,25
```

下面将 cities.csv 和 temperatures.csv 载入 DataSet 中，并移除文件头，如下所示：

```
val cities = benv.readTextFile("cities.csv")
  .filter(!_.contains("Id,"))
```

```
val temp = benv.readTextFile("temperatures.csv")
   .filter(!_.contains("Id,"))
```

接下来将 DataSet 转换为一个元组 DataSet。其中，第一个 DataSet（即城市 DataSet）将生成<cityId, cityName>元组；第二个 DataSet（即温度 DataSet）生成<cityId, temperature>元组，如下所示：

```
val cities2 = cities.map(x => (x.split(",")(0), x.split(",")(1)))
cities2.first(10).print()
val temp2 = temp.map(x => (x.split(",")(1), x.split(",")(2)))
temp2.first(10).print()
```

1. 内部连接

内部连接需要使用到左、右表以包含相同列。如果左表或右表中包含了重复键或多个键副本，连接操作将会迅速膨胀为笛卡儿连接。相应地，计算时间也将随之增长。内部连接如图 8.23 所示。

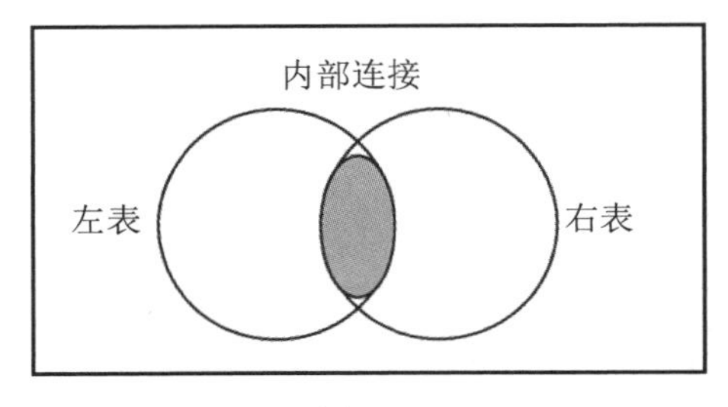

图 8.23

下列代码展示了内部连接的执行方式，进而连接两个元组 DataSet。

```
cities2.join(temp2)
 .where(0)
 .equalTo(0)
 .first(10).print()
```

该作业的输出结果显示了源自两个 DataSet 的元组，其中，cityID 位于两个 DataSet 中。对应结果如下所示：

```
((1,Boston),(1,21))
((2,New York),(2,22))
((3,Chicago),(3,23))
((4,Philadelphia),(4,24))
((5,San Francisco),(5,25))
((1,Boston),(1,23))
((2,New York),(2,24))
```

```
((3,Chicago),(3,25))
((4,Philadelphia),(4,26))
((5,San Francisco),(5,18))
```

如果使用聚合并针对每所城市添加温度值，将得到每所城市的全部温度值，对应代码如下所示：

```
cities2
    .join(temp2)
    .where(0)
    .equalTo(0)
    .map(x=> (x._1._2, x._2._2.toInt))
    .groupBy(0)
    .sum(1)
    .first(10).print()
```

对应结果如下所示：

```
(Boston,111)
(Chicago,116)
(New York,119)
(Philadelphia,116)
(San Francisco,113)
```

当前作业可在 Flink UI 中进行查看，如图 8.24 所示。

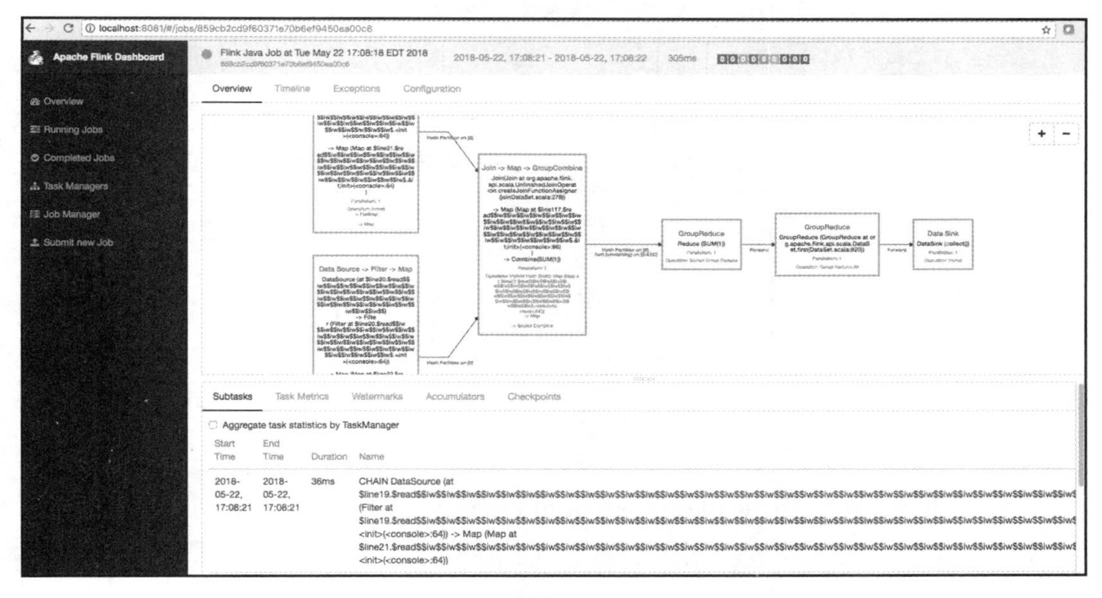

图 8.24

join() API 的定义如下所示：

```
/**
 * Initiates a Join transformation.
 *
 * <p>A Join transformation joins the elements of two
 * {@link DataSet DataSets} on key equality and provides multiple ways *
to combine
 * joining elements into one DataSet.
 *
 * <p>This method returns a {@link JoinOperatorSets} on which one of the
{@code where} methods
 * can be called to define the join key of the first joining (i.e., *
this)DataSet.
 *
 * @param other The other DataSet with which this DataSet is joined.
 * @return A JoinOperatorSets to continue the definition of the Join
 * transformation.
 *
 * @see JoinOperatorSets
 * @see DataSet
 */
public <R> JoinOperatorSets<T, R> join(DataSet<R> other) {
return new JoinOperatorSets<>(this, other);
}
```

2. 左外连接

除了两个表中的公共行之外（内连接），左外连接生成左表中的全部行。如果两个表中的公共行较少，最终结果将十分庞大，因而性能也相对低下。左外连接如图 8.25 所示。

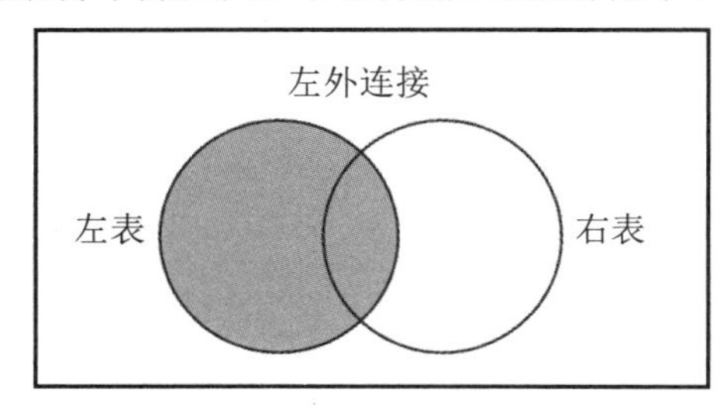

图 8.25

下面将执行左外连接，进而连接两个元组 DataSet，如下所示：

```
cities2
    .leftOuterJoin(temp2)
    .where(0)
```

```
    .equalTo(0) {
        (x,y) => (x, if (y==null) (x._1,0) else (x._1, y._2.toInt))
    }
    .map(x=> (x._1._2, x._2._2.toInt))
    .groupBy(0)
    .sum(1)
    .first(10).print()
```

该作业的输出结果展示了源自两个 DataSet 中的元组。其中，cityID 位于左表或两个 DataSet 中，如下所示：

```
(Boston,111)
(Chicago,116)
(Las Vegas,0) // Las vegas has no records in temperatures DataSet so
is assigned 0
(New York,119)
(Philadelphia,116)
(San Francisco,113)
```

当前作业可在 Flink UI 中进行查看，如图 8.26 所示。

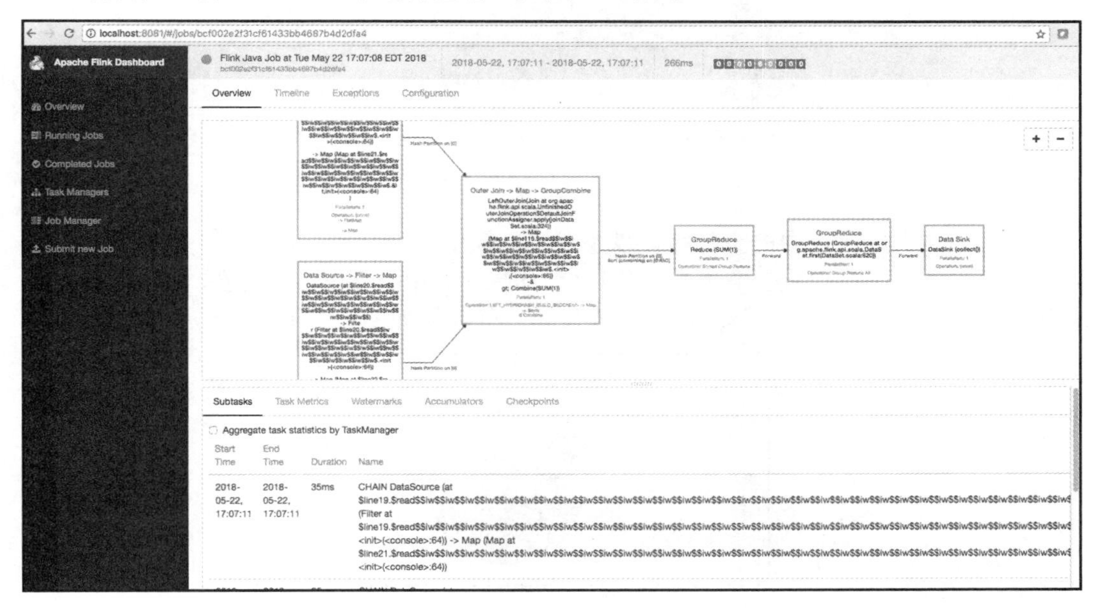

图 8.26

leftOuterJoin() API 的定义如下所示：

```
/**
 * Initiates a Left Outer Join transformation.
```

```
 *
 * <p>An Outer Join transformation joins two elements of two
 * {@link DataSet DataSets} on key equality and provides multiple ways *
to combine
 * joining elements into one DataSet.
 *
 * <p>Elements of the <b>left</b> DataSet (i.e. {@code this}) that do *
not have a matching
 * element on the other side are joined with {@code null} and emitted *
to the resulting DataSet.
 *
 * @param other The other DataSet with which this DataSet is joined.
 * @return A JoinOperatorSet to continue the definition of the Join
 * transformation.
 *
 * @see
 * org.apache.flink.api.java.operators.join.JoinOperatorSetsBase
 * @see DataSet
 */
 public <R> JoinOperatorSetsBase<T, R> leftOuterJoin(DataSet<R> other) {
 return new JoinOperatorSetsBase<>(this, other, JoinHint.OPTIMIZER_CHOOSES,
JoinType.LEFT_OUTER);
 }
```

3. 右外连接

右外连接生成右表中的全部行，以及左、右表中的公共行（内连接）。据此，可得到右表中的所有行，以及左、右表中的公共行。如果左表中不存在对应内容，则填写 NULL。右外连接的性能类似于之前提到的左外连接。右外连接如图 8.27 所示。

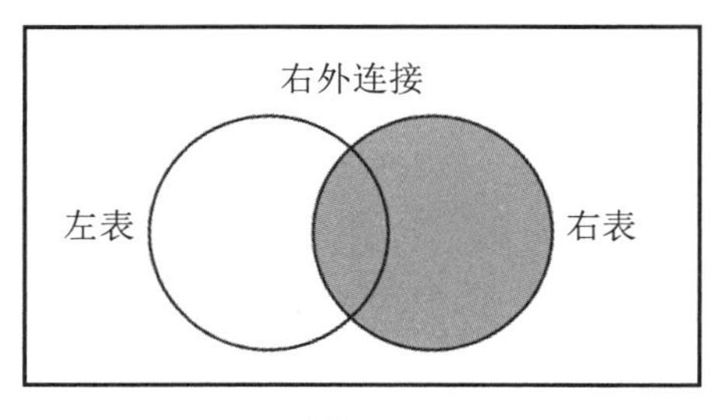

图 8.27

下面执行右外连接，进而连接两个元组 DataSet，如下所示：

```
cities2
    .rightOuterJoin(temp2)
```

```
    .where(0)
    .equalTo(0) {
        (x,y) => (if (x==null) (y._1,"unknown") else (y._1, x._2), y)
    }
    .map(x=> (x._1._2, x._2._2.toInt))
    .groupBy(0)
    .sum(1)
    .first(10).print()
```

该作业输出显示了源自两个 DataSet 的元组，其中，cityID 位于右表或者两个 DataSet 中，如下所示：

```
(Boston,111)
(Chicago,116)
(New York,119)
(Philadelphia,116)
(San Francisco,113)
(unknown,44) . // note that only right hand side temperatures DataSet //hasid
6 which is not in cities DataSet
```

当前作业可在 Flink UI 中进行查看，如图 8.28 所示。

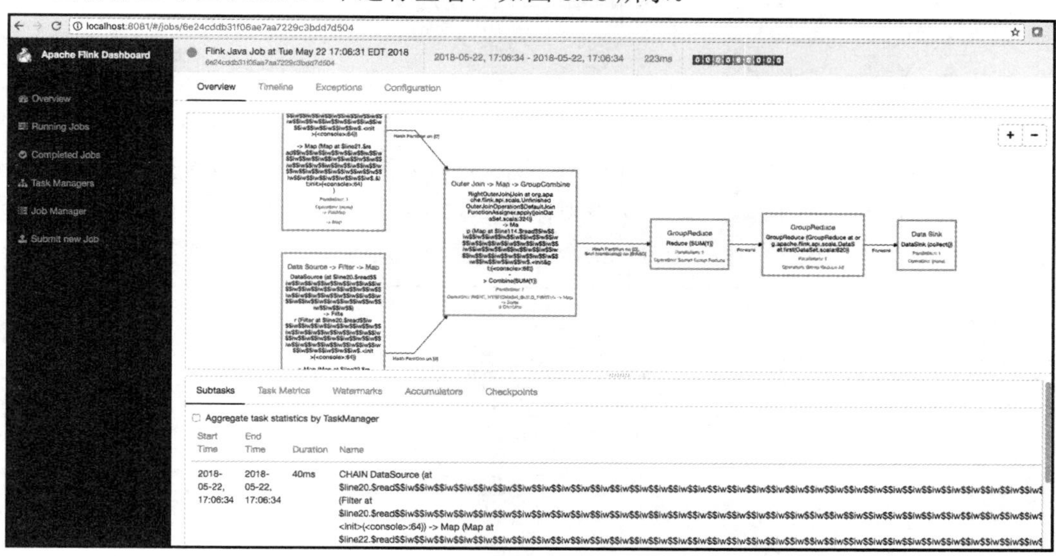

图 8.28

rightOuterJoin() API 的定义如下所示：

```
/**
* Initiates a Right Outer Join transformation.
```

```
 *
 * <p>An Outer Join transformation joins two elements of two
 * {@link DataSet DataSets} on key equality and provides multiple ways *
to combine
 * joining elements into one DataSet.
 *
 * <p>Elements of the <b>right</b> DataSet (i.e. {@code other}) that * do
not have a matching
 * element on {@code this} side are joined with {@code null} and emitted
 * to the resulting DataSet.
 *
 * @param other The other DataSet with which this DataSet is joined.
 * @return A JoinOperatorSet to continue the definition of the Join
 * transformation.
 *
 * @see
 * .apache.flink.api.java.operators.join.JoinOperatorSetsBase
 * @see DataSet
 */
 public <R> JoinOperatorSetsBase<T, R> rightOuterJoin(DataSet<R> other) {
 return new JoinOperatorSetsBase<>(this, other, JoinHint.OPTIMIZER_CHOOSES,
JoinType.RIGHT_OUTER);
 }
```

4. 全外连接

全外连接生成连接子句左、右表中的全部行（匹配和不匹配的）。当需要保留两个表中的所有行时，可使用全外连接。当一个表中存在匹配项时，全外连接返回所有行。如果两个表中的公共行较少，最终结果将十分庞大，因而性能也相对低下。全外连接如图 8.29 所示。

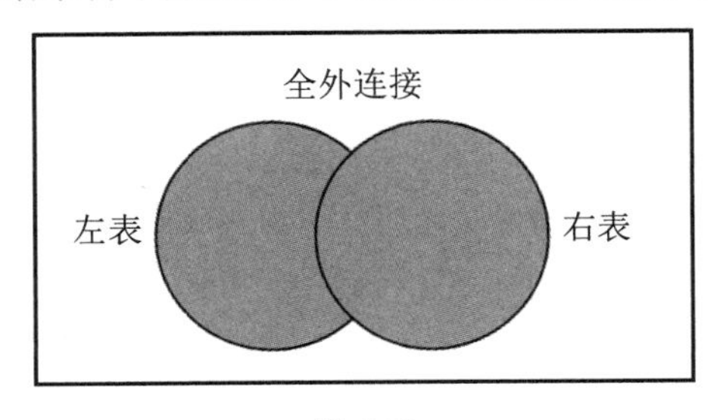

图 8.29

下面执行全外连接，并连接两个元组的 DataSet，如下所示：

```
cities2
    .fullOuterJoin(temp2)
    .where(0)
```

```
    .equalTo(0) {
        (x,y) => (if (x==null) (y._1,"unknown") else (x._1, x._2),
                if (y==null) (x._1,0) else (y._1, y._2.toInt))
    }
    .map(x=> (x._1._2, x._2._2.toInt))
    .groupBy(0)
    .sum(1)
    .first(10).print()
```

该作业的输出结果显示了源自两个 DataSet 的元组，其中，cityID 位于两个 DataSet（或之一）中，如下所示：

```
(Boston,111)
(Chicago,116)
(Las Vegas,0) // Las vegas has no records in temperatures DataSet so is
assigned 0
(New York,119)
(Philadelphia,116)
(San Francisco,113)
(unknown,44) // note that only right hand side temperatures DataSet has id
6 which is not in cities DataSet
```

当前作业可在 Flink UI 中进行查看，如图 8.30 所示。

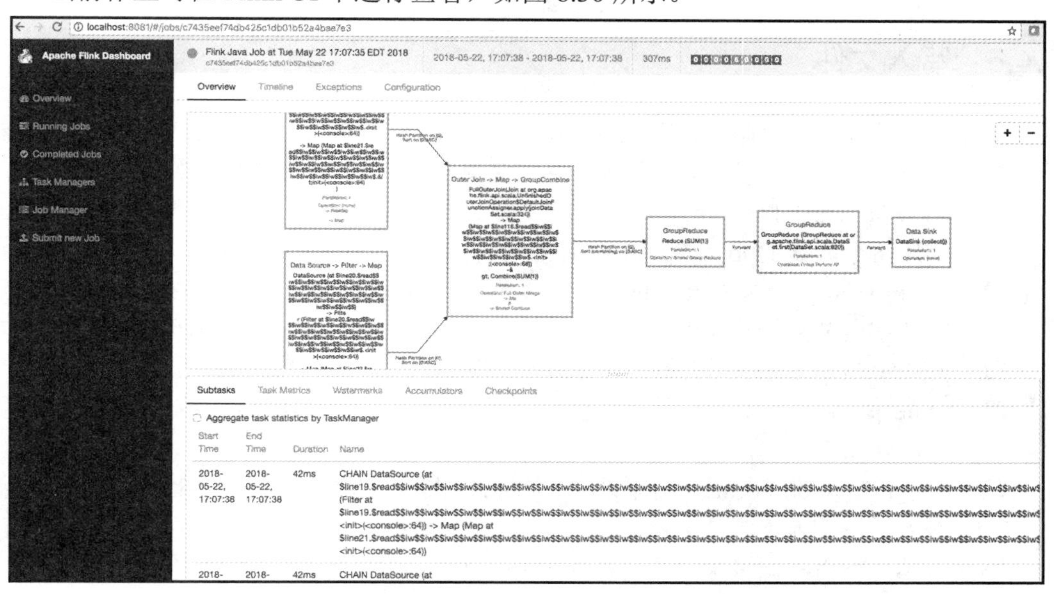

图 8.30

fullOuterJoin() API 的定义如下所示：

```
/**
 * Initiates a Full Outer Join transformation.
 *
 * <p>An Outer Join transformation joins two elements of two
 * {@link DataSet DataSets} on key equality and provides multiple ways *
to combine joining elements into one DataSet.
 *
 * <p>Elements of <b>both</b> DataSets that do not have a matching
 * element on the opposing side are joined with {@code null} and emitted
 * to the resulting DataSet.
 *
 * @param other The other DataSet with which this DataSet is joined.
 * @return A JoinOperatorSet to continue the definition of the Join
 * transformation.
 *
 * @see org.apache.flink.api.java.operators.join.JoinOperatorSetsBase
 * @see DataSet
 */
 public <R> JoinOperatorSetsBase<T, R> fullOuterJoin(DataSet<R> other) {
 return new JoinOperatorSetsBase<>(this, other, JoinHint.OPTIMIZER_CHOOSES,
JoinType.FULL_OUTER);
 }
```

8.4.6 写入文件

数据接收器使用、存储并返回 DataSet。相应地，数据接收器操作通过 OutputFormat 予以描述。Flink 中包含了各种内建输出格式，并封装在 DataSet 上的操作之后，如下所示。

- writeAsText()/TextOutputFormat：作为字符串逐行写入元素。另外，字符串通过调用每个元素的 toString()方法而得到。
- writeAsCsv(...)/CsvOutputFormat：作为以逗号分隔的数值文件写入元组。同时，行和字段分隔符均可配置。每个字段的数值源自对象的 toString()方法。
- print()/printToErr()：在标准输出/错误流中输出每个元素的 toString()值。
- write()/FileOutputFormat：自定义文件输出的方法和基类。支持自定义对象到字节的转换。
- output()/OutputFormat：对于非文件数据接收器（例如，将结果存储于数据库中），最为通用的输出方法。

下面尝试将城市和温度值的内连接结果通过 writeAsText()写入某个文件中。

提示：

在调用 benv.execute()之前不会看到任何输出结果。

首先针对城市和温度值的内连接创建一个 DataSet，如下所示：

```
val results = cities2
    .join(temp2)
    .where(0)
    .equalTo(0)
    .map(x=> (x._1._2, x._2._2.toInt))
    .groupBy(0)
    .sum(1)
```

随后，在结果 DataSet 上调用 writeAsText()，并调用 DataSink 上的 execute()，如下所示：

```
results.writeAsText("file:///Users/sridharalla/flink-1.4.2/results.txt").
setParallelism(1)
benv.execute()
```

若打开刚刚生成的文件，将会看到下列代码显示的连接操作结果：

```
(Boston,111)
(Chicago,116)
(New York,119)
(Philadelphia,116)
(San Francisco,113)
```

当前作业可在 Flink UI 中进行查看，如图 8.31 所示。

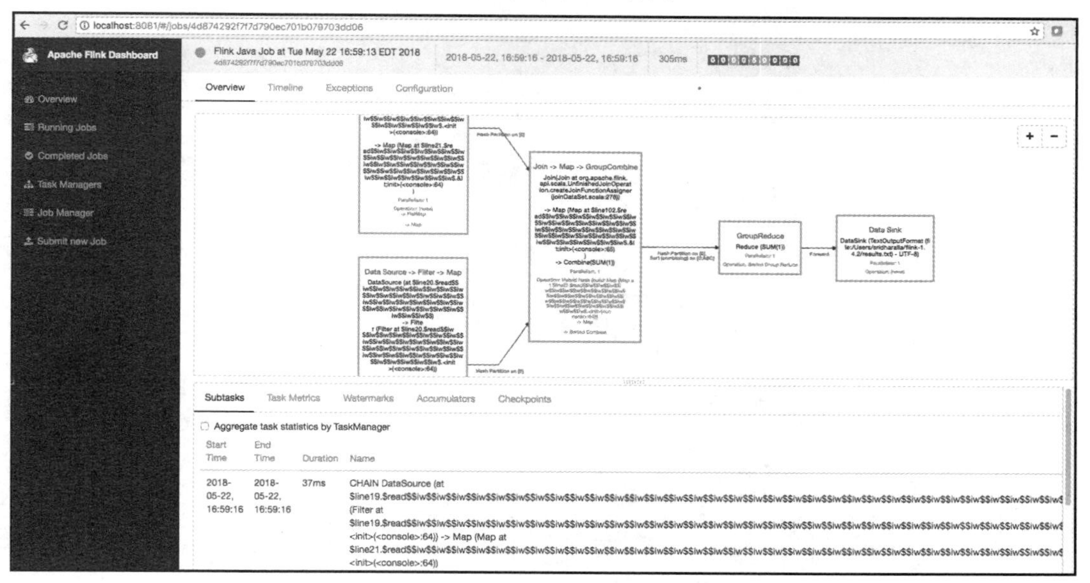

图 8.31

8.5 本章小结

本章讨论了 Apache Flink，以及如何使用 Flink 在大数据上执行批处理分析。除此之外，本章还学习了 Flink 及其内部工作机制。接下来，本章介绍了数据的加载和分析方法，进而执行转换和聚合操作。最后，我们还探讨了如何在大数据上执行 Join 操作。

第 9 章将利用 Apache Flink 进行实时数据分析。

第 9 章　Apache Flink 流式处理

本章介绍 Apache Flink 流式处理，以及如何利用该框架在构建实时应用程序时处理数据。本章首先讨论 DataStream API，并考查各类执行操作。

本章主要涉及以下主题：

- 基于 DataStream API 的数据处理机制。
- 转换。
- 聚合。
- 窗口操作。
- 物理分区。
- 调整数据尺度。
- 数据接收器。
- 事件时间和水印。
- Kafka 连接器。
- Twitter 连接器。
- Elasticsearch 连接器。
- Cassandra 连接器。

9.1　流式执行模型简介

Flink 是针对分布式流式处理的一个开源框架，其特征如下所示：

- 提供了准确的结果，即使对无序或后期到达的数据也是如此。
- 具有“有状态”和容错特性，当维护“仅一次”应用程序状态时，可无缝地从故障中恢复。
- 适合大规模场合，可运行于数千个节点上，且具有较好的吞吐量和延迟性。

图 9.1 显示了流式处理的整体示意图。

当在无界数据集上获取准确的结果时，Flink 的相关特性不可或缺，包括状态管理、处理无序数据、灵活的窗口机制，具体如下：

- 对于有状态计算来说，Flink 可保证“仅一次”语义。这里，“有状态”意味着应用程序可维护一段时间以来所处理数据的聚合或汇总结果；而对于故障事件中的应用程序状态，Flink 检查点机制保障了“仅一次”语义，如图 9.2 所示。

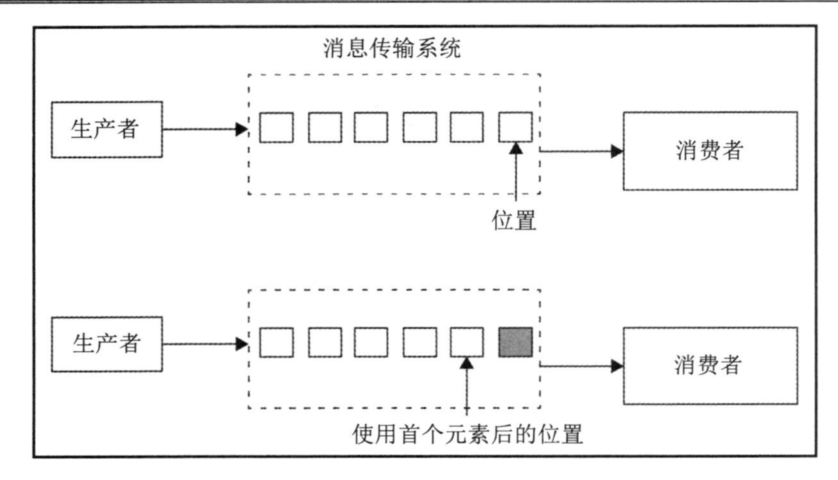

图 9.1

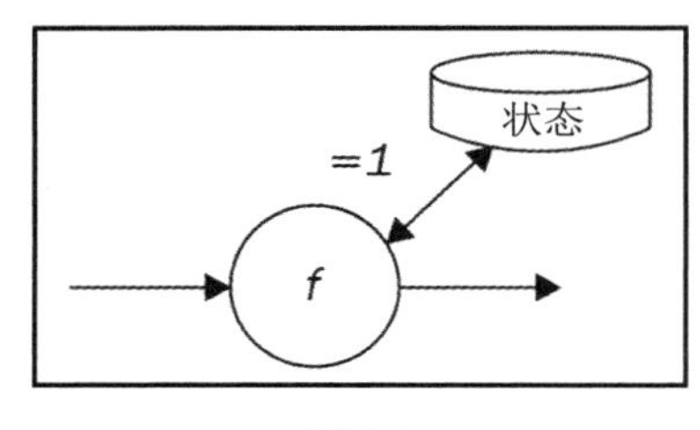

图 9.2

- Flink 支持流式处理和基于事件时间语义的窗口机制。事件时间可针对数据流计算准确的结果。其中，事件以无序状态到达且存在延迟现象，如图 9.3 所示。

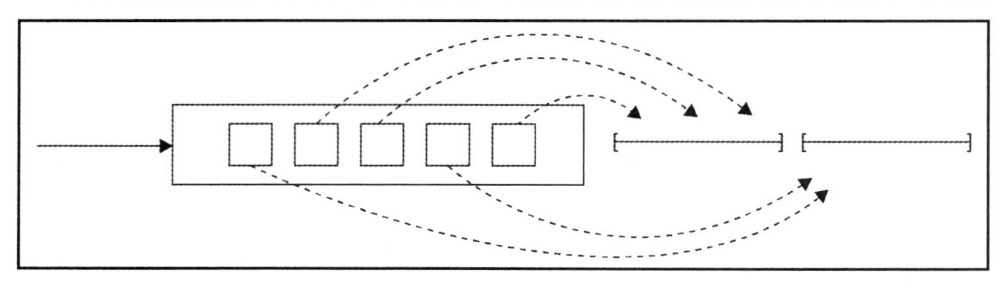

图 9.3

- 除了数据驱动窗口之外，Flink 还可根据时间、计数或会话支持灵活的窗口机制。利用灵活的触发条件，可对窗口进行自定义，进而支持更加高级的流模式。Flink 的窗口机制可对创建数据的真实环境进行建模，如图 9.4 所示。
- Flink 的容错机制是轻量级的，允许系统保持高吞吐量，并提供“仅一次”一致性保障。Flink 可在不损失数据的前提下从故障中恢复，而可靠性和延迟之间的权衡是可以忽略的，如图 9.5 所示。

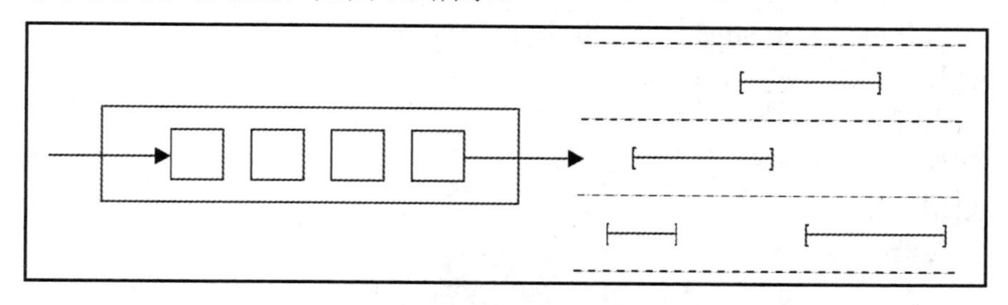

图 9.4

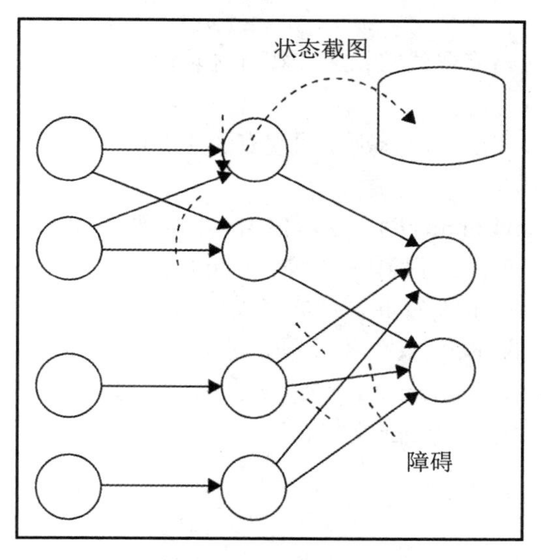

图 9.5

- ❑ Flink 支持高吞吐量和低延迟（也就是说，可快速处理大量的数据）。
- ❑ Flink 的保存点提供了状态版本机制，进而可更新应用程序，或者在不丢失状态以及最小停机时间的情况下重新处理历史数据。
- ❑ Flink 的设计目标是在包含数千个节点的大规模集群上运行，除了独立的集群节点之外，Flink 还提供了 YARN 和 Mesos 方面的支持。

9.2　利用 DataStream API 进行数据处理

健壮的分析机制对于实时数据的处理十分重要，对于数据驱动的领域来说尤其如此。

对此，Flink 可通过 DataStream API 执行实时分析。这种流式数据处理 API 适用于物联网应用程序，进而存储、处理和分析实时（或接近于实时）数据。

接下来将分析与 DataStream API 相关的每种元素，其中包括：

- 执行环境。
- 数据源。
- 转换。
- 数据接收。
- 连接器。

9.2.1 执行环境

当编写 Flink 程序时，需要建立相应的执行环境。对此，可使用现有的环境或者构建新环境。

根据具体需求条件，Flink 可分别使用现有的 Flink 环境、创建本地环境或者创建远程环境。

采用 getExecutionEnvironment()命令可根据具体需要实现不同任务，具体如下：

- 在 IDE 中，当在本地环境中执行时，Flink 将启动本地执行环境。
- 当执行 JAR 时，Flink 集群管理器将以分布方式运行程序。
- 当创建自己的本地或远程环境时，可采用 createLocalEnvironment() 和 createRemoteEnvironment 这一类方法（相关参数包含字符串形式的主机名、int 类型的端口、字符串和.jar 文件）。

9.2.2 数据源

Flink 可以从不同的数据源处获取数据，其中定义了许多内建源函数，从而可无缝地获取数据，Flink 中的一些预先实现的数据源函数可对这一过程予以简化。此外，当现有函数无法满足数据源的要求时，Flink 支持自定义数据源函数的编写。

读者可访问 https://ci.apache.org/projects/flink/flink-docs-release-1.4/dev/datastream_api.html，以查看 DataStream API 的相关文档。

Flink 中一些现有的数据源函数包括：

- 基于套接字的数据源。
- 基于文件的数据源。

1. 基于套接字的数据源

DataStream API 支持套接字的读取，下列代码片段展示了简单的流式 API 示例：

```
// Data type for words with count
case class WordWithCount(word: String, count: Long)
// get input data by connecting to the socket
val text = senv.socketTextStream("127.0.0.1", 9000, '\n')
// parse the data, group it, window it, and aggregate the counts
val windowCounts = text
 .flatMap { w => w.split("\\s") }
 .map { w => WordWithCount(w, 1) }
 .keyBy("word")
 .timeWindow(Time.seconds(5), Time.seconds(1))
 .sum("count")
// print the results with a single thread, rather than in parallel
windowCounts.print().setParallelism(1)
senv.execute("Socket Window WordCount")
```

上述代码连接至本地主机的 9000 端口上，检索并处理文本数据，将字符串划分为独立的单词（此处采用空格加以分隔）。随后，代码将在 5 秒的窗口中计算单词的出现频率，并对此予以输出。

当运行上述示例时，需要使用到 Flink 的 Scala Shell，如图 9.6 所示。

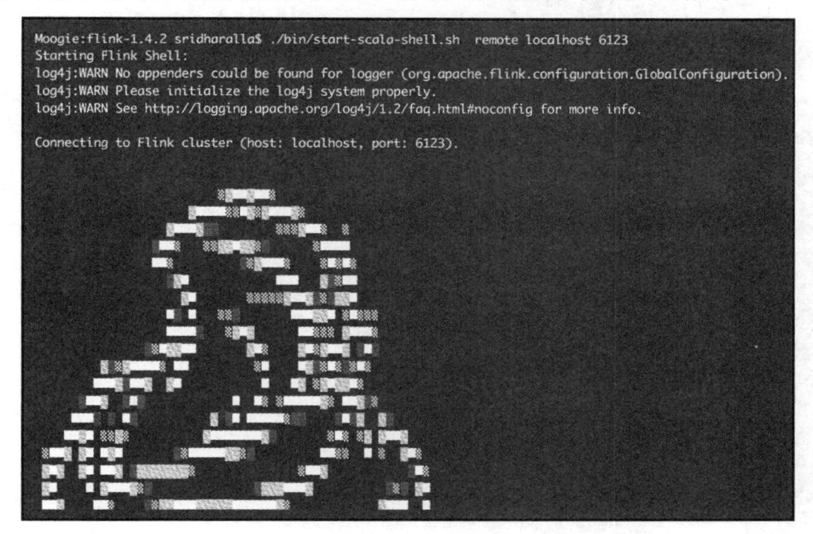

图 9.6

在任意 Linux 系统中运行 nc 命令，进而启动本地服务器，如图 9.7 所示。

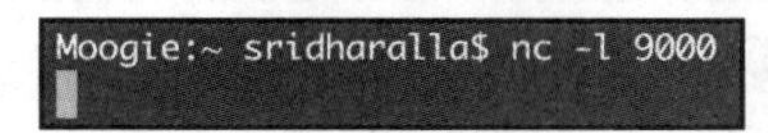

图 9.7

在 Shell 中运行代码，连接至端口 9000 并监听数据，如图 9.8 所示。

```
// Data type for words with count
case class WordWithCount(word: String, count: Long)

// get input data by connecting to the socket
val text = senv.socketTextStream("127.0.0.1", 9000, '\n')

// parse the data, group it, window it, and aggregate the counts
val windowCounts = text
    .flatMap { w => w.split("\\s") }
    .map { w => WordWithCount(w, 1) }
    .keyBy("word")
    .timeWindow(Time.seconds(5), Time.seconds(1))
    .sum("count")

// print the results with a single thread, rather than in parallel
windowCounts.print().setParallelism(1)

senv.execute("Socket Window WordCount")

// Exiting paste mode, now interpreting.

Submitting job with JobID: 557deeac32d01c1061ff4b558b97b7c8. Waiting for job completion.
Connected to JobManager at Actor[akka.tcp://flink@localhost:6123/user/jobmanager#-598835017] with leader session id 00000000-0000-0000-0000
0000000000.
04/29/2018 21:19:57     Job execution switched to status RUNNING.
04/29/2018 21:19:57     Source: Socket Stream(1/1) switched to SCHEDULED
04/29/2018 21:19:57     Flat Map -> Map(1/1) switched to SCHEDULED
04/29/2018 21:19:57     TriggerWindow(SlidingProcessingTimeWindows(5000, 1000), ReducingStateDescriptor{serializer=$line19.$read$$iw$$iw$$iw$
iw$$iw$$iw$$iw$$iw$$iw$$iw$$iw$$iw$$iw$$iw$$iw$$iw$$iw$$iw$$iw$$iw$$iw$$iw$$iw$$iw$$iw$$iw$$iw$$iw$$iw$$iw$$iw$$iw$$iw$$iw$$iw$$anon$2$$a
on$1@895bda3c, reduceFunction=org.apache.flink.streaming.api.functions.aggregation.SumAggregator@375febef}, ProcessingTimeTrigger(), Windowed
tream.reduce(WindowedStream.java:241))(1/1) switched to SCHEDULED
```

图 9.8

图 9.9 显示了运行于 Web 控制台中的当前作业。

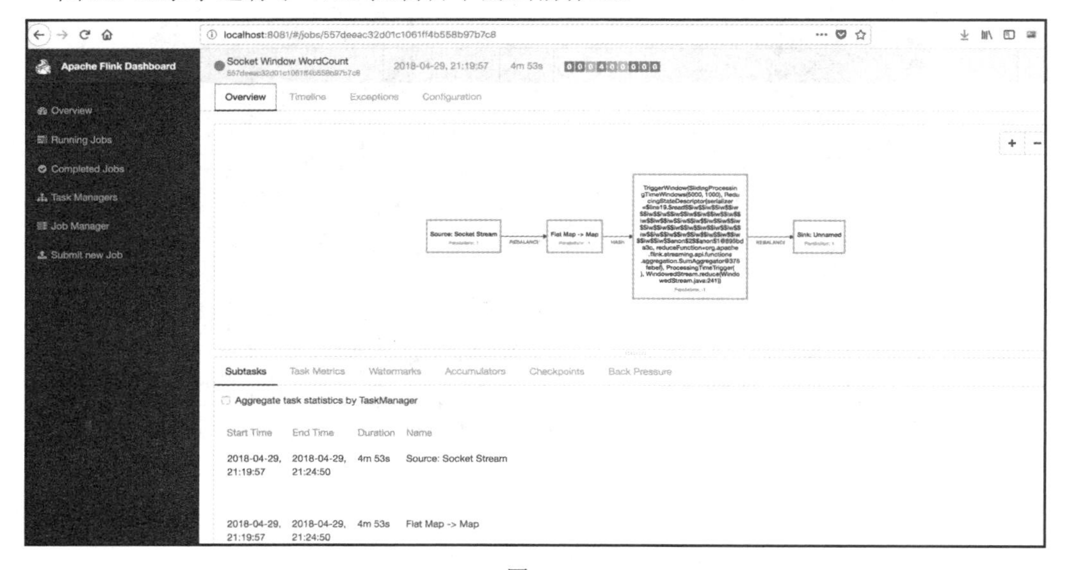

图 9.9

在图 9.10 中，可进一步理解当前所执行的任务。

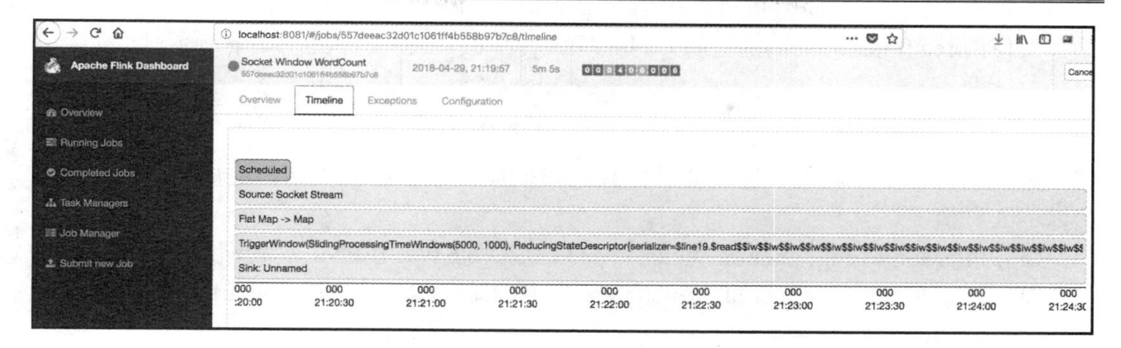

图 9.10

在将文本数据输入 nc 服务器控制台时，即可在 log 文件夹中看到输出结果。

在当前示例中，可以看到 taskmanager 的日志文件（log 文件），如下所示：

```
tail -f log/flink-sridharalla-taskmanager-1-Moogie.local.out
```

图 9.11 是跟踪日志文件时将看到的内容。

```
WordWithCount(hellow,1)
WordWithCount(hellow,1)
WordWithCount(world,1)
```

图 9.11

在介绍了上述示例代码后，接下来考查套接字数据流的 API。

API 中定义了主机名称和端口号，进而可从套接字中读取数据，如下所示：

```
socketTextStream(hostName, port);
```

除此之外，还可进一步定义分隔符，如下所示：

```
socketTextStream(hostName,port,delimiter)
```

同时，还可指定 API 从套接字中读取数据的最大次数，如下所示：

```
socketTextStream(hostName,port,delimiter, maxRetry)
```

2．基于文件的数据源

当在 Flink 中采用基于文件的源函数时，可使用 readTextFile(String path)。默认条件下，字符串路径涵盖了默认值 TextInputFormat，不是逐行读取文本和字符串。

如果文件格式与文本不同，可使用下列函数指定格式：

```
readFile(FileInputFormat<Out> inputFormat, String path)
```

当使用 readFileStream()函数时，Flink 可以读取产生的文件流，如下所示：

```
readFileStream(String filePath, long intervalMillis,
FileMonitoringFunction.WatchType watchType)
```

其中包含了文件路径、轮询文件路径的时间间隔以及监视类型。这里，监视类型包含以下 3 种情况。

- FileMonitoringFunction.WatchType.ONLY_NEW_FILES：仅处理新文件。
- FileMonitoringFunction.WatchType.PROCESS_ONLY_APPENDED：仅用于处理附加的文件内容。
- FileMonitoringFunction.WatchType.REPROCESS_WITH_APPENDED：不仅处理附加的文件内容，还包括文件中之前的内容。

如果文件并不是文本文件，那么，可使用该函数定义文件输入格式，如下所示：

```
readFile(fileInputFormat, path, watchType, interval, pathFilter,
typeInfo)
```

该命令将读取文件任务划分为两项子任务：

- 第一项子任务仅根据指定的 WatchType 监视文件路径。
- 第二项子任务以并行方式执行实际的文件读取操作。

9.2.3 转换

数据转换将数据从一种形式转换为另一种形式。其中，输入内容可以是一个或多个数据流；而输出内容则可以是 0 个、1 个或多个数据流。下面将考查不同的转换行为。

1. map

这是最简单的转换，其中，输入内容表示为一个数据流；输出内容同样为一个数据流。

在 Java 中，map 的定义如下所示：

```
inputStream.map(new MapFunction<Integer, Integer>() {
@Override
public Integer map(Integer value) throws Exception {
return 5 * value;
}
});
```

在 Scala 中，其定义方式如下所示：

```
inputStream.map { x => x * 5 }
```

2. flatMap

flatMap 接收一条记录作为输入，并生成 0 个、1 个或多条记录。

在 Java 中，flatMap 的定义如下所示：

```
inputStream.flatMap(new FlatMapFunction<String, String>() {
@Override
public void flatMap(String value, Collector<String> out)
throws Exception {
    for(String word: value.split(" ")){
        out.collect(word);
    }
});
```

在 Scala 中，其定义如下所示：

```
inputStream.flatMap { str => str.split(" ") }
```

3. filter

filter 函数计算相关条件，并根据所满足的条件生成输出记录。

提示：

除此之外，filter 也可生成 0 条记录。

在 Java 中，filter 的定义如下所示：

```
inputStream.filter(new FilterFunction<Integer>() {
@Override
    public boolean filter(Integer value) throws Exception {
        return value != 1;
    }
});
```

在 Scala 中，其定义如下所示：

```
inputStream.filter { _ != 1 }
```

4. keyBy

keyBy 根据对应键在逻辑上对数据流分区，并采用 hush 函数进行分区，同时返回 KeyedDataStream。

在 Java 中，keyBy 的定义如下所示：

```
inputStream.keyBy("someKey");
```

在 Scala 中，其定义如下所示：

```
inputStream.keyBy("someKey")
```

5. reduce

下列代码展示了 reduce 的求和计算。

在 Java 中，其定义方式如下所示：

```
keyedInputStream. reduce(new ReduceFunction<Integer>() {
@Override
    public Integer reduce(Integer value1, Integer value2)
        throws Exception {
            return value1 + value2;
        }
});
```

在 Scala 中，其定义方式如下所示：

```
keyedInputStream. reduce { _ + _ }
```

6. fold

利用当前记录整合最后一个 fold 的数据流，fold 将生成 KeyedDataStream，并返回数据流。

在 Java 中，fold 的实现方式如下所示：

```
keyedInputStream keyedStream.fold("Start", new
FoldFunction<Integer,String>() {
@Override
    public String fold(String current, Integer value) {
        return current + "=" + value;
    }
});
```

在 Scala 中，其实现如下所示：

```
keyedInputStream.fold("Start")((str, i) => { str + "=" + i })
```

在上述函数中，对于数据流(1,2,3,4,5)，将生成 Start=1=2=3=4=5。

7. 聚合

DataStream API 支持各种聚合操作，例如 min、max、sum 等。这一类函数可应用于 KeyedDataStream 上，在 Java 中，其实现方式如下所示：

```
keyedInputStream.sum(0)
keyedInputStream.sum("key")
keyedInputStream.min(0)
keyedInputStream.min("key")
keyedInputStream.max(0)
keyedInputStream.max("key")
keyedInputStream.minBy(0)
keyedInputStream.minBy("key")
keyedInputStream.maxBy(0)
keyedInputStream.maxBy("key")
```

在 Scala 中，其实现方式如下所示：

```
keyedInputStream.sum(0)
keyedInputStream.sum("key")
keyedInputStream.min(0)
keyedInputStream.min("key")
keyedInputStream.max(0)
keyedInputStream.max("key")
keyedInputStream.minBy(0)
keyedInputStream.minBy("key")
keyedInputStream.maxBy(0)
keyedInputStream.maxBy("key")
```

max 和 maxBy 之间的差别在于，max 返回数据流中的最大值；maxBy 则返回包含最大值的一个键。同样，这一差别也体现于 min 和 minBy 中。

8．window

window 函数可根据时间或其他条件对现有 KeyedDataStreams 进行分组。下列转换根据 10 秒的时间窗口生成记录分组。

在 Java 中，其实现方式如下所示：

```
inputStream.keyBy(0).window(TumblingEventTimeWindows.of(Time.seconds(10)));
```

在 Scala 中，其实现方式如下所示：

```
inputStream.keyBy(0).window(TumblingEventTimeWindows.of(Time.seconds(10)))
```

Flink 定义了称为窗口的数据片来处理潜在的无限数据流。

这有助于使用转换处理数据块。对于数据流上的窗口机制，可指定一个键，并在此键上进行分发；同时定义一个函数，该函数描述在窗口数据流上执行哪些转换。

当把数据流分片至窗口中时，可使用预先实现的 Flink 分配器，并选取滚动窗口、滑动窗口以及全局和会话窗口等选项。

通过扩展 WindowAssigner 类，Flink 还可编写自定义窗口分配器。

下面考查分配器的工作方式。

（1）全局窗口

除非由触发器指定，否则，全局窗口是一种不间断的窗口。通常情况下，每个元素被分配给一个全局窗口（每个键）。如果不指定任何触发器，则不会触发计算。

（2）滚动窗口

滚动窗口是一种固定长度的窗口，且不存在重叠。通过滚动窗口，可在特定时间点执行元素计算。例如，一个 10 分钟的滚动窗口可以用来计算 10 分钟内发生的一组事件。

（3）滑动窗口

滑动窗口与滚动窗口十分类似，唯一的差别在于，滑动窗口处于重叠状态。滑动窗口也是固定长度的窗口。通过用户给定的窗口滑动参数与上一个滑动窗口重叠。

通过该窗口机制，可从特定时间帧中出现的一组事件中执行相关计算。

（4）会话窗口

当需要根据输入数据确定窗口边界时，会话窗口十分有用。在窗口起始时间和窗口尺寸方面，会话窗口提供了较大的灵活性。

会话间隙配置参数是指，会话视为关闭之前的等待时间。

9. windowAll

windowAll 函数支持常规数据类的分组机制。鉴于运行于未分区的数据流上，因而该过程通常是一类非并行数据转换。在 Java 中，其定义方式如下所示：

```
inputStream.windowAll(TumblingEventTimeWindows.of(Time.seconds(10)));
```

在 Scala 中，其定义方式如下所示：

```
inputStream.windowAll(TumblingEventTimeWindows.of(Time.seconds(10)))
```

窗口数据流函数与常规的数据流函数较为类似，唯一的差别在于，前者工作于窗口数据流上。因此，窗口的 reduce 行为类似于 reduce 函数；窗口的 fold 行为类似于 fold 函数。除此之外，还存在相应的聚合操作。

10. union

union 函数执行两个或多个数据流的合并操作，并以并行方式整合数据流。如果数据流自身进行合并，将输出每条记录两次。在 Java 中，union 的定义方式如下所示：

```
inputStream. union(inputStream1, inputStream2, ...);
```

在 Scala 中，union 的定义如下所示：

```
inputStream. union(inputStream1, inputStream2, ...)
```

11．窗口的 join

通过公共窗口中的一些键连接两个数据流。下列示例显示了 5 秒窗口中两个数据流的连接操作，其中，第一个数据流的第一个属性的连接条件等于另一个数据流的第二个条件。在 Java 中，其实现方式如下所示：

```
inputStream. join(inputStream1)
.where(0).equalTo(1)
.window(TumblingEventTimeWindows.of(Time.seconds(5)))
.apply (new JoinFunction () {...});
```

在 Scala 中，其实现方式如下所示：

```
inputStream. join(inputStream1)
.where(0).equalTo(1)
.window(TumblingEventTimeWindows.of(Time.seconds(5)))
.apply { ... }
```

12．split

使用该函数可将数据流根据某种标准划分为两个或多个数据流。当获取混合数据流并需要单独处理数据时，这将十分有用。在 Java 中，其实现方式如下所示：

```
SplitStream<Integer> split = inputStream.split(new
OutputSelector<Integer>() {
@Override
public Iterable<String> select(Integer value) {
List<String> output = new ArrayList<String>();
if (value % 2 == 0) {
output.add("even");
}
else {
output.add("odd");
}
return output;
}
});
```

在 Scala 中，其实现方式如下所示：

```
val split = inputStream.split( (num: Int) =>(num % 2) match {
    case 0 => List("even")
    case 1 => List("odd")
})
```

13. select

使用该函数可以从划分的数据流中选取特定的数据流。在 Java 中，其实现方式如下所示：

```
SplitStream<Integer> split;
DataStream<Integer> even = split.select("even");
DataStream<Integer> odd = split.select("odd");
DataStream<Integer> all = split.select("even","odd");
```

在 Scala 中，其实现方式如下所示：

```
val even = split select "even"
val odd = split select "odd"
val all = split.select("even","odd")
```

14. project

使用 project 函数可以从事件流中选择属性的子集，并将所选的元素发送至下一个处理流中。在 Java 中，其实现方式如下所示：

```
DataStream<Tuple4<Integer, Double, String, String>> in = // [...]
DataStream<Tuple2<String, String>> out = in.project(3,2);
```

在 Scala 中，其实现方式如下所示：

```
val in : DataStream[(Int,Double,String)] = // [...]
val out = in.project(3,2)
```

上述代码从给定的记录中选择属性数值 2 和 3。下列内容显示了样本输入和输出记录：

```
(1,10.0, A, B )=> (B,A)
(2,20.0, C, D )=> (D,C)
```

15. 物理分区

当使用 Flink 时，可执行流式数据的物理分区。此外，还可通过相关选项实现自定义分区。下面逐一考查不同的分区类型。

（1）自定义分区

如前所述，在 Java 中，自定义分区器实现如下所示：

```
inputStream.partitionCustom(partitioner, "someKey");
inputStream.partitionCustom(partitioner, 0);
```

在 Scala 中，其实现如下所示：

```
inputStream.partitionCustom(partitioner, "someKey")
inputStream.partitionCustom(partitioner, 0)
```

当编写自定义分区器时，应确保实现了高效的 hash 函数。

（2）随机分区

随机分区以随机方式划分数据流。在 Java 中，其实现方式如下所示：

```
inputStream.shuffle();
```

在 Scala 中，其实现方式如下所示：

```
inputStream.shuffle()
```

（3）再平衡分区

这一分区类型有助于实现均匀的数据分布，并采用了循环分布方法。当数据呈现为倾斜状态时，这种分区类型十分有用。在 Java 中，其实现方式如下所示：

```
inputStream.rebalance();
```

在 Scala 中，其实现方式如下所示：

```
inputStream.rebalance();
```

16. 调整数据尺度

该方法用于在操作间分布数据，在数据的子集上执行转换并对其进行整合。注意，这一类再平衡方案仅出现于单一节点上，因而不存在网络间的数据传输行为。在 Java 中，其实现方式如下所示：

```
inputStream.rescale();
```

在 Scala 中，其实现方式如下所示：

```
inputStream.rescale();
```

17. 广播机制

广播机制将所有记录分布至每个分区中，这有助于将每个元素分布至全部分区上。在 Java 中，其实现方式如下所示：

```
inputStream.broadcast();
```

在 Scala 中，其实现方式如下所示：

```
inputStream.broadcast()Data Sinks
```

一旦数据转换完毕，则需要对结果予以保存。下列各项内容列出了 Flink 中的保存选项。

- writeAsText()：以字符串的形式一次写入一行。
- writeAsCsV()：将元组写入逗号分隔的值文件。另外，行和字段分隔符均可进行配置。
- print()/printErr()：将记录写入标准输出中；或者还可选择写入标准错误系统中。
- writeUsingOutputFormat()：可提供自定义输出格式。当确定自定义格式时，须扩展 OutputFormat——负责序列化和反序列化操作。
- writeToSocket()：Flink 还支持将数据写入特定的套接字中，并为正确的序列化和格式化定义 SerializationSchema。

18. 事件时间和水印

Flink Streaming API 的设计灵感源自 Google Dataflow 模型，该 API 支持不同的时间概念。下列内容列举了流式环境中用于捕捉时间的常见场景。

- 事件时间。事件时间是指事件在其生产设备上出现的时间。例如，在物联网项目中，事件时间可以是传感器捕捉读取操作时的时间。一般情况下，这一类事件时间应在进入 Flink 之前置入记录中。在时间处理过程中，此类时间戳将被析取并围绕窗口机制予以考查。事件时间处理可用于无序事件。
- 处理时间。处理时间则是执行数据处理流时的机器时间。处理时间的窗口机制仅查看事件被处理时的时间戳。相比之下，处理时间是最为简单的流处理方式——无须使用处理机器与生产机器之间的异步机制。时间不提供某种确定性，因为它取决于系统中记录的流动速度。
- 摄入时间。摄入时间是指特定事件进入 Flink 时的时间。所有基于时间的操作均会参考该时间戳。与处理时间相比，摄入时间是一类代价相对高昂的操作，但却会生成可预测的结果。摄入时间程序不能处理任何无序事件，因为它只在事件进入 Flink 系统后才分配时间戳。

下列示例展示了如何设置事件时间和水印。对于摄入时间和处理时间，仅分配了时间特征，水印的生成过程则被自动处理。

在 Java 中，其实现过程如下所示：

```
final StreamExecutionEnvironment env =
StreamExecutionEnvironment.getExecutionEnvironment();
env.setStreamTimeCharacteristic(TimeCharacteristic.ProcessingTime);
```

```
//or
env.setStreamTimeCharacteristic(TimeCharacteristic.IngestionTime);
```

在 Scala 中，其实现过程如下所示：

```
val env = StreamExecutionEnvironment.getExecutionEnvironment
env.setStreamTimeCharacteristic(TimeCharacteristic.ProcessingTime)
//or
env.setStreamTimeCharacteristic(TimeCharacteristic.IngestionTime)
```

在事件时间流程序中，可指定水印和时间戳的分配方式。对此，存在以下两种水印和时间戳的分配方式：

- 直接从数据源属性中分配。
- 使用时间戳分配器。

当与事件时间流协同工作时，在 Java 中，时间特征的分配实现过程如下所示：

```
final StreamExecutionEnvironment env =
StreamExecutionEnvironment.getExecutionEnvironment();
env.setStreamTimeCharacteristic(TimeCharacteristic.EventTime;
```

在 Scala 中，其实现过程如下所示：

```
val env = StreamExecutionEnvironment.getExecutionEnvironment
env.setStreamTimeCharacteristic(TimeCharacteristic.EventTime)
```

通常，在存储数据源中的记录时即存储事件时间是一种较好的方法。除此之外，Flink 还支持某些预定义时间戳析取器和水印生成器。

19. 连接器

Apache Flink 支持多种连接器，并通过各种技术实现数据的读取/写入操作。

（1）Kafka 连接器

Kafka 是一种发布-订阅分布式消息队列系统，用户可将消息发布至某一特定主题。随后，这些消息将被分布至主题的订阅器中。Flink 提供了多种选项，并可将 Kafka 使用者定义为 Flink 流中的数据源。当使用 Flink Kafka 连接器时，需要使用特定的 JAR 文件。

通过相应的 Maven 依赖关系即可使用连接器。例如，对于 Kafka 0.9 版本，可向 pom.xml 文件中添加下列依赖关系：

```
<dependency>
    <groupId>org.apache.flink</groupId>
    <artifactId>flink-connector-kafka-0.9_2.11/artifactId>
    <version>1.1.4</version>
</dependency>
```

下面考查如何将 Kafka 使用者用作 Kafka 源。在 Java 中，其实现方式如下所示：

```
Properties properties = new Properties();
properties.setProperty("bootstrap.servers", "localhost:9092");
properties.setProperty("group.id", "test");
DataStream<String> input = env.addSource(new
FlinkKafkaConsumer09<String>("mytopic", new SimpleStringSchema(),
properties));
```

在 Scala 中，其实现方式如下所示：

```
val properties = new Properties();
properties.setProperty("bootstrap.servers", "localhost:9092");
// only required for Kafka 0.8
properties.setProperty("zookeeper.connect", "localhost:2181");
properties.setProperty("group.id", "test");
stream = env
.addSource(new FlinkKafkaConsumer09[String]("mytopic", new
SimpleStringSchema(), properties))
.print
```

上述代码首先设置 Kafka 主机、Zookeeper 主机和端口的各项属性，并于随后指定了主题名称，在当前示例中表示为 mytopic。因此，如果消息发布至 mytopic 主题中，将被 Flink 流进行处理。

如果以不同的格式获取数据，还可以为反序列化指定自定义模式。默认情况下，Flink 支持字符串和 JSON 反序列化器。当启用容错机制时，可启用 Flink 中的检查点机制。Flink 周期性地获取状态快照。若出现故障，将恢复至最近一次检查点处并重新启动处理过程。除此之外，还可将 Kafka 生产者定义为一个接收器，这将向 Kafka 主题中写入数据。在 Java 中，其实现过程如下所示：

```
stream.addSink(new FlinkKafkaProducer09[String]("localhost:9092",
"mytopic", new SimpleStringSchema()))
```

在 Scala 中，其实现过程如下所示：

```
stream.addSink(new FlinkKafkaProducer09<String>("localhost:9092",
"mytopic", new SimpleStringSchema()));
```

（2）Twitter 连接器

随着社交媒体和网站功能性的不断提升，应可从 Twitter 中获取数据并对其进行处理。Twitter 数据可用于各种产品、服务和应用程序等的情感分析。

相应地，Flink 提供了 Twitter 连接器作为一种数据源。当使用这一类连接器时，须通过 Twitter 账号创建 Twitter 应用程序，并生成连接器所用的授权密钥。

Twitter 连接器可在 Java 或 Scala 中使用。一旦生成了令牌，即可编写程序并从 Twitter 中获取数据，具体步骤如下：

① 首先添加 Maven 依赖关系：

```
<dependency>
<groupId>org.apache.flink</groupId>
<artifactId>flink-connector-twitter_2.11/artifactId>
<version>1.1.4</version>
</dependency>
```

② 随后作为数据源添加 Twitter。在 Java 中，其实现方式如下所示：

```
Properties props = new Properties();
props.setProperty(TwitterSource.CONSUMER_KEY, "");
props.setProperty(TwitterSource.CONSUMER_SECRET, "");
props.setProperty(TwitterSource.TOKEN, "");
props.setProperty(TwitterSource.TOKEN_SECRET, "");
DataStream<String> streamSource = env.addSource(new
TwitterSource(props));
```

在 Scala 中，其实现方式如下所示：

```
val props = new Properties();
props.setProperty(TwitterSource.CONSUMER_KEY, "");
props.setProperty(TwitterSource.CONSUMER_SECRET, "");
props.setProperty(TwitterSource.TOKEN, "");
props.setProperty(TwitterSource.TOKEN_SECRET, "");
DataStream<String> streamSource = env.addSource(new
TwitterSource(props));
```

在上述代码中，首先设置了获得的令牌属性，并于随后加入 TwitterSource。如果给定的信息正确无误，即可开始从 Twitter 中获取数据。TwitterSource 将以 JSON 字符串形式生成数据。Twitter JSON 示例代码如下所示：

```
{
...
"text": ""Loyalty 3.0: How to Revolutionize Customer &amp; Employee
Engagement with Big Data &amp; #Gamification" can be ordered here:
http://t.co/1XhqyaNjuR",
"geo": null,
"retweeted": false,
"in_reply_to_screen_name": null,
```

```
"possibly_sensitive": false,
"truncated": false,
"lang": "en",
"hashtags": [{
"text": "Gamification",
"indices": [90,
103]
}],
},
"in_reply_to_status_id_str": null,
"id": 330094515484508160
...
}
```

TwitterSource 提供了各种 StatusesSampleEndpoint，并返回一组随机消息集。如果需要加入过滤器，且并不打算使用默认的端点，则可实现 TwitterSource.EndpointInitializer 接口。

一旦从 Twitter 中获取了数据，即可对数据进行处理、存储或分析。

（3）RabbitMQ 连接器

RabbitMQ 是一类广泛使用的高性能分布式消息队列系统。对于高吞吐量操作，RabbitMQ 一般用作消息传输系统。此外，还可生成分布式消息队列，并在队列中纳入发布者和订阅者。关于 RabbitMQ 的更多信息，读者可访问 https://www.rabbitmq.com/。

Flink 支持 RabbitMQ 间数据的获取和发布，并提供了一个连接器，可以充当数据流的数据源。

为了能够使 RabbitMQ 连接器正常工作，需要提供下列信息。

- RabbitMQ：诸如主机、端口、用户凭证等配置信息。
- 队列：希望订阅的 RabbitMQ 队列名称。
- 相关 ID：作为一个 RabbitMQ 特性，用于在分布式系统中通过唯一 ID 关联请求和响应。
- 反序列化模式：RabbitMQ 以序列化方式存储和传输数据，进而避免产生网络流量。因此，当接收消息时，订阅器知晓如何对数据执行反序列化操作。Flink 连接器提供了一些默认的反序列化器，例如字符串反序列化器。

RabbitMQ 源在数据传输方面提供了下列选项。

- “仅一次”方案：在 RabbitMQ 事务中使用 RabbitMQ 关联 ID 和 Flink 检查点机制。
- “至少一次”方案：当启用 Flink 检查点机制，但未设置 RabbitMQ 相关 ID 时。

RabbitMQ 自动提交模式并不具备可靠的传输保证。

下面尝试编写代码，以使连接器可正常工作。类似于其他连接器，需要向代码中添加 Maven 依赖关系，如下所示：

```
<dependency>
<groupId>org.apache.flink</groupId>
<artifactId>flink-connector-rabbitmq_2.11/artifactId>
<version>1.1.4</version>
</dependency>
```

下列代码片段展示了在 Java 中如何使用 RabbitMQ 连接器。

```
//Configurations
RMQConnectionConfig connectionConfig = new RMQConnectionConfig.Builder()
.setHost(<host>).setPort(<port>).setUserName(..)
.setPassword(..).setVirtualHost("/").build();

//Get Data Stream without correlation ids
DataStream<String> streamWO = env.addSource(new
RMQSource<String>(connectionConfig, "my-queue", new SimpleStringSchema()))
.print

//Get Data Stream with correlation ids
DataStream<String> streamW = env.addSource(new
RMQSource<String>(connectionConfig, "my-queue", true, new
SimpleStringSchema()))
.print
```

类似地，Scala 中的对应代码如下所示：

```
val connectionConfig = new RMQConnectionConfig.Builder()
.setHost(<host>).setPort(<port>).setUserName(..)
.setPassword(..).setVirtualHost("/").build()
streamsWOIds  =  env.addSource(new  RMQSource[String](connectionConfig,
" myqueue", new SimpleStringSchema))
.print
streamsWIds  =  env.addSource(new  RMQSource[String](connectionConfig,
"myqueue", true, new SimpleStringSchema))
.print
```

此外，还可将 RabbitMQ 连接器用作 Flink 接收器。

当把处理结果发送回不同的 RabbitMQ 队列中时，需要提供以下 3 项较为重要的配置：

- RabbitMQ 配置。
- 队列名称——所处理数据的返回位置。

- 序列化模式——用以将数据转换为字节的 RabbitMQ 模式。

下列 Java 示例代码显示了如何将该连接器用作 Flink 接收器。

```
RMQConnectionConfig connectionConfig = new RMQConnectionConfig.Builder()
.setHost(<host>).setPort(<port>).setUserName(..)
.setPassword(..).setVirtualHost("/").build();
stream.addSink(new RMQSink<String>(connectionConfig, "target-queue", new
StringToByteSerializer()));
```

Scala 中的实现方式如下所示：

```
val connectionConfig = new RMQConnectionConfig.Builder()
.setHost(<host>).setPort(<port>).setUserName(..)
.setPassword(..).setVirtualHost("/").build()
stream.addSink(new RMQSink[String](connectionConfig, "target-queue", new
StringToByteSerializer
```

（4）Elasticsearch 连接器

Elasticsearch 是一个分布式、低延迟、全文本搜索引擎，并可对所选取的文档进行索引，随后可在文档集合上执行全文本搜索。关于 Elasticsearch 的更多信息，读者可访问 https://www.elastic.co。

在某些场合下，可能需要通过 Flink 处理数据，随后将其存储于 Elasticsearch 中。对此，Flink 对 Elasticsearch 连接器提供了相应的支持。截至目前，Elasticsearch 包含了两个主要版本，Flink 对此均予支持。对于 Elasticsearch 1.x，需要添加下列 Maven 依赖关系：

```
<dependency>
\<groupId>org.apache.flink</groupId>
<artifactId>flink-connector-elasticsearch_2.11</artifactId>
<version>1.1.4</version>
</dependency>
```

Flink 连接器提供了一个接收器将数据写入 Elasticsearch 中。它使用了以下两种方法连接到 Elasticsearch：

- 嵌入式节点模式。在嵌入式节点模式中，接收器使用 BulkProcessor 将文档发送至 ElasticSearch 中。在将文档发送至 Elasticsearch 之前，可对缓冲的请求数量进行配置，如下所示：

```
DataStream<String> input = ...;
Map<String, String> config = Maps.newHashMap();
config.put("bulk.flush.max.actions", "1");
config.put("cluster.name", "cluster-name");
```

```
input.addSink(new ElasticsearchSink<>(config, new
IndexRequestBuilder<String>() {
@Override
public IndexRequest createIndexRequest(String element,
RuntimeContext ctx) {
    Map<String, Object> json = new HashMap<>();
    json.put("data", element);
    return Requests.indexRequest()
    .index("my-index")
    .type("my-type")
    .source(json);
}
}));
```

上述代码利用集群名称，以及发送请求前的缓冲文件数量创建了哈希映射。随后，我们向数据流中添加了接收器，并指定了索引、类型和存储的文档。类似地，Scala 中的代码实现如下所示：

```
val input: DataStream[String] = ...
val config = new util.HashMap[String, String]
config.put("bulk.flush.max.actions", "1")
config.put("cluster.name", "cluster-name")
text.addSink(new ElasticsearchSink(config, new
IndexRequestBuilder[String]
{
    override def createIndexRequest(element: String, ctx:
RuntimeContext):
    IndexRequest = {
        val json = new util.HashMap[String, AnyRef]
        json.put("data", element)
        Requests.indexRequest.index("my-index").`type`("mytype").
source(json)
    }
}))
```

- 传输客户端模式。Elasticsearch 允许通过 9300 端口上的传输客户端进行连接。Flink 支持基于其连接器的连接方式。随后，可在配置中指定集群中的所有 Elasticsearch 节点。下列代码显示了 Java 中的实现方式：

```
DataStream<String> input = ...;
Map<String, String> config = Maps.newHashMap();
config.put("bulk.flush.max.actions", "1");
```

```
config.put("cluster.name", "cluster-name");
List<TransportAddress> transports = new ArrayList<String>();
transports.add(new InetSocketTransportAddress("es-node-1", 9300));
transports.add(new InetSocketTransportAddress("es-node-2", 9300));
transports.add(new InetSocketTransportAddress("es-node-3", 9300));
input.addSink(new ElasticsearchSink<>(config, transports, new
IndexRequestBuilder<String>() {
@Override
public IndexRequest createIndexRequest(String element,
RuntimeContext ctx) {
Map<String, Object> json = new HashMap<>();
json.put("data", element);
return Requests.indexRequest()
.index("my-index")
.type("my-type")
.source(json);
}
}));
```

其中提供了与集群名称、节点、端口、批量发送的最大请求数量等相关的细节内容。类似地，Scala 中的实现代码如下所示：

```
val input: DataStream[String] = ...
val config = new util.HashMap[String, String]
config.put("bulk.flush.max.actions", "1")
config.put("cluster.name", "cluster-name")
val transports = new ArrayList[String]
transports.add(new InetSocketTransportAddress("es-node-1", 9300))
transports.add(new InetSocketTransportAddress("es-node-2", 9300))
transports.add(new InetSocketTransportAddress("es-node-3", 9300))
text.addSink(new ElasticsearchSink(config, transports, new
IndexRequestBuilder[String] {
override def createIndexRequest(element: String, ctx:
RuntimeContext):
IndexRequest = {
val json = new util.HashMap[String, AnyRef]
json.put("data", element)
Requests.indexRequest.index("my-index").`type`("mytype").
source(json)
}
}))
```

（5）Cassandra 连接器

Cassandra 是一种分布式、低延迟的 NoSQL 数据库，同时也是一个基于键-值的数据库。相应地，许多高吞吐量的应用程序采用 Cassandra 作为其主数据库。Cassandra 可与分布式集群节点协同工作，其中并不会涉及主从架构。任何节点均可执行相应的读、写操作。关于 Cassandra 的更多信息，读者可访问 http://cassandra.apache.org。

Apache Flink 提供了一个连接器，可将数据写入 Cassandra 中。在许多应用程序中，我们需要将 Flink 中的流数据存储至 Cassandra 中。

类似于其他连接器，此处也需要添加 Maven 依赖关系，如下所示：

```
<dependency>
<groupId>org.apache.flink</groupId>
<artifactId>flink-connector-cassandra_2.11</artifactId>
<version>1.1.4</version>
</dependency>
```

在加入了依赖关系后，可利用相关配置添加 Cassandra 接收器，在 Java 中，其实现方式如下所示：

```
CassandraSink.addSink(input)
.setQuery("INSERT INTO cep.events (id, message) values (?, ?);")
.setClusterBuilder(new ClusterBuilder() {
@Override
public Cluster buildCluster(Cluster.Builder builder) {
return builder.addContactPoint("127.0.0.1").build();
}
})
.build()
```

在 Scala 中，其实现方式如下所示：

```
CassandraSink.addSink(input)
.setQuery("INSERT INTO cep.events (id, message) values (?, ?);")
.setClusterBuilder(new ClusterBuilder() {
@Override
public Cluster buildCluster(Cluster.Builder builder) {
return builder.addContactPoint("127.0.0.1").build();
}
)
.build();
```

9.3 本章小结

本章学习了 Flink 中功能强大的 API，即 DataStream API；数据源、转换和接收器之间的协同工作方式，以及连接器技术（Elasticsearch、Cassandra、Kafka、RabbitMQ 等）。除此之外，本章还讨论了基于 Apache Flink 的流式处理技术。

在第 10 章中，我们将改变一下话题，讨论令人兴奋的可视化数据领域方面的内容。

第 10 章　大数据可视化技术

本章讨论大数据和分析处理过程中最为重要的内容之一，即数据的可视化技术。俗话说，一图胜千言。在数据处理期间，我们需要持续地对数据予以理解、应用和交互，与表、列或文本文件的读取操作相比，数据的可视化行为将更加简单明了。当通过数据分析方法（例如 Python、R、Spark、Flink、Hive、MapReduce 等）得到某种结论后，我们往往需要在数据背景下对其加以进一步的理解。针对于此，需要使用到图形显示技术。

本章主要涉及以下主题：

- 数据可视化简介。
- Tableau。
- 图表类型。
- 使用 Python。
- 使用 R。
- 数据可视化工具。

10.1　数据可视化简介

数据可视化是理解大数据进而体现数据价值的有效方法之一。数据可视化依赖于相关用例，图形和图表则是数据的可视化表达结果，同时也提供了一种强大的汇总和显示数据的方法，而这种方法往往更容易理解。我们可通过图表和图形观察某些数据的主要特征，这不仅体现了学习过程中的数值结果，同时还展示了数据的形状和模式，这对于数据分析和决策的制定来说十分重要。对于数据可视化，需要注意以下几点内容：

- 数据类型所对应的图形表达类型。
- 如何设计支持交互特征的可视化方案。
- 如何采用图形方式搜索和调整数据集。
- 区分数据和结论间的差异。
- 随着大数据规模的增长，如何选择可视化技术。

- ❑ 如何处理延迟问题，以便消除可视化数据中的延迟现象。
- ❑ 对于高速数据或流式数据，如何优化设计方案，进而展示实时可视化效果。
- ❑ 如何实现数据库的数据可视化技术。
- ❑ 如何实现内存中的数据可视化技术。

数据可视化包含了多种不同的实现方式，图 10.1 展示了一些例子，描述了图表类型的选择如何改变可视化方案的使用和效果。

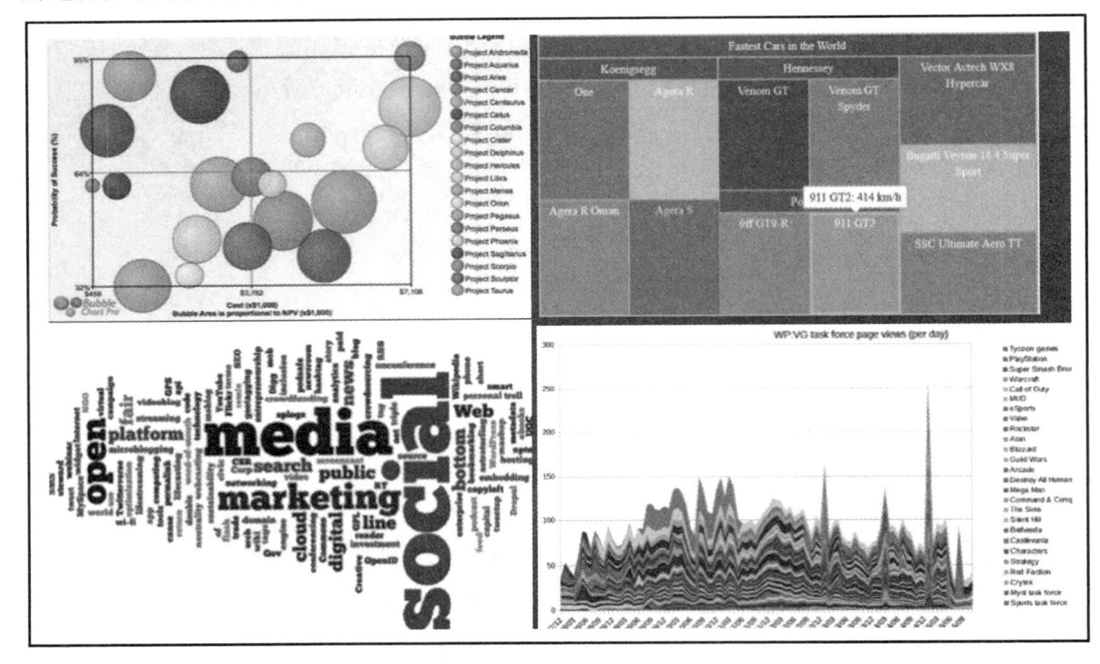

图 10.1

图 10.2 展示了一些可视化示例。

10.2　Tableau

本节将介绍 Tableau 这一十分流行的可视化工具。对此，可尝试下载 Tableau 试用版，并在本地机器上进行安装。读者可访问 https://www.tableau.com/下载 Tableau。

图 10.3 显示了 Tableau 的下载链接页面。

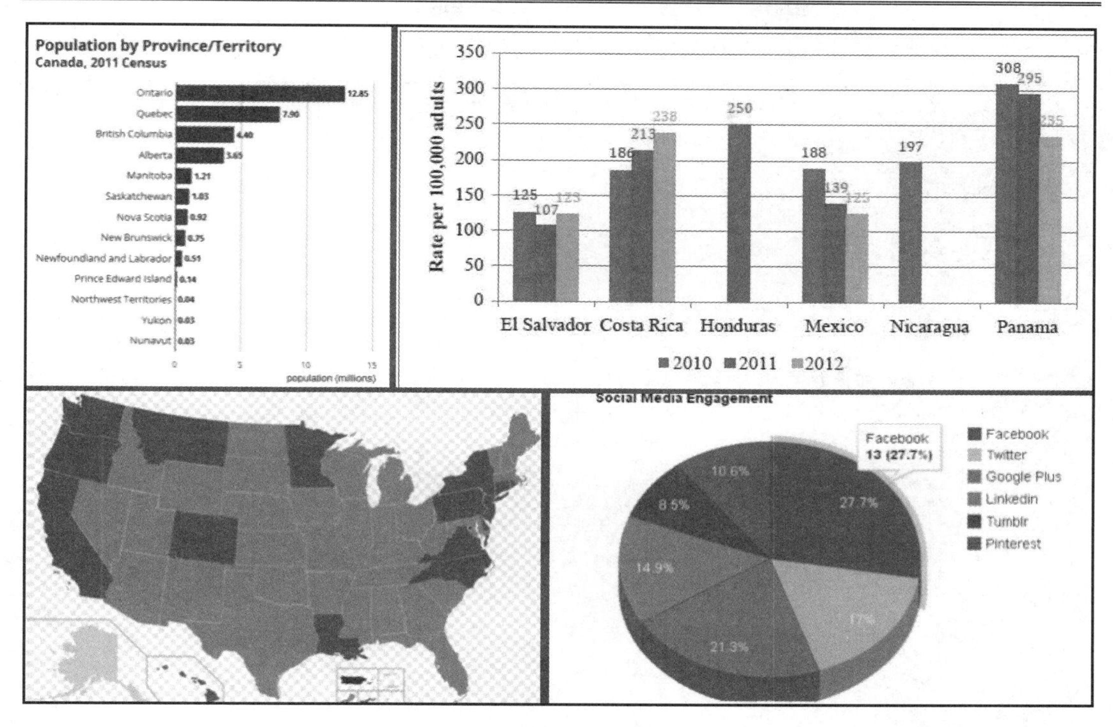

图 10.2

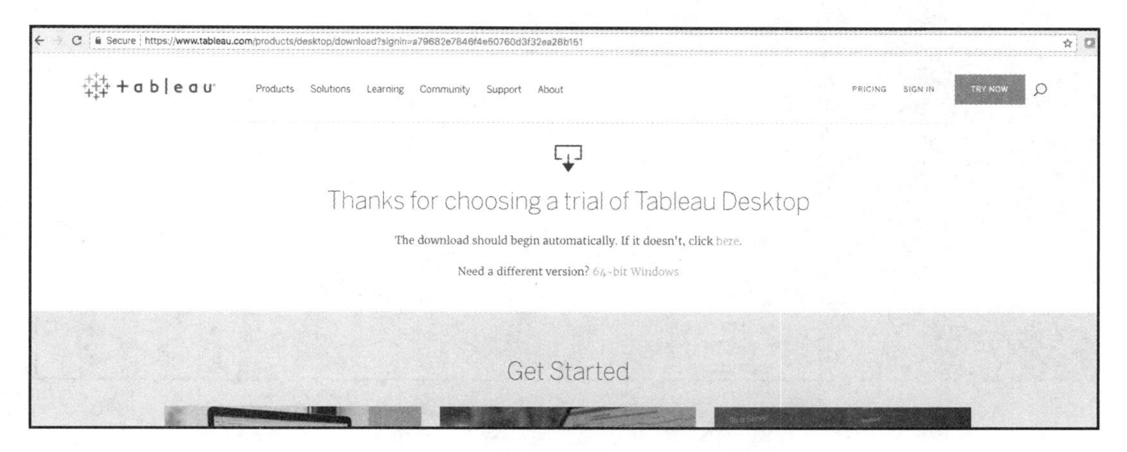

图 10.3

当 Tableau 试用版安装完毕后（或者用户拥有经许可使用的软件副本），即可尝试进行某些较为基本的可视化操作。

图 10.4 显示了 Tableau 的启动画面，其中包含了可供选择的各种数据源。

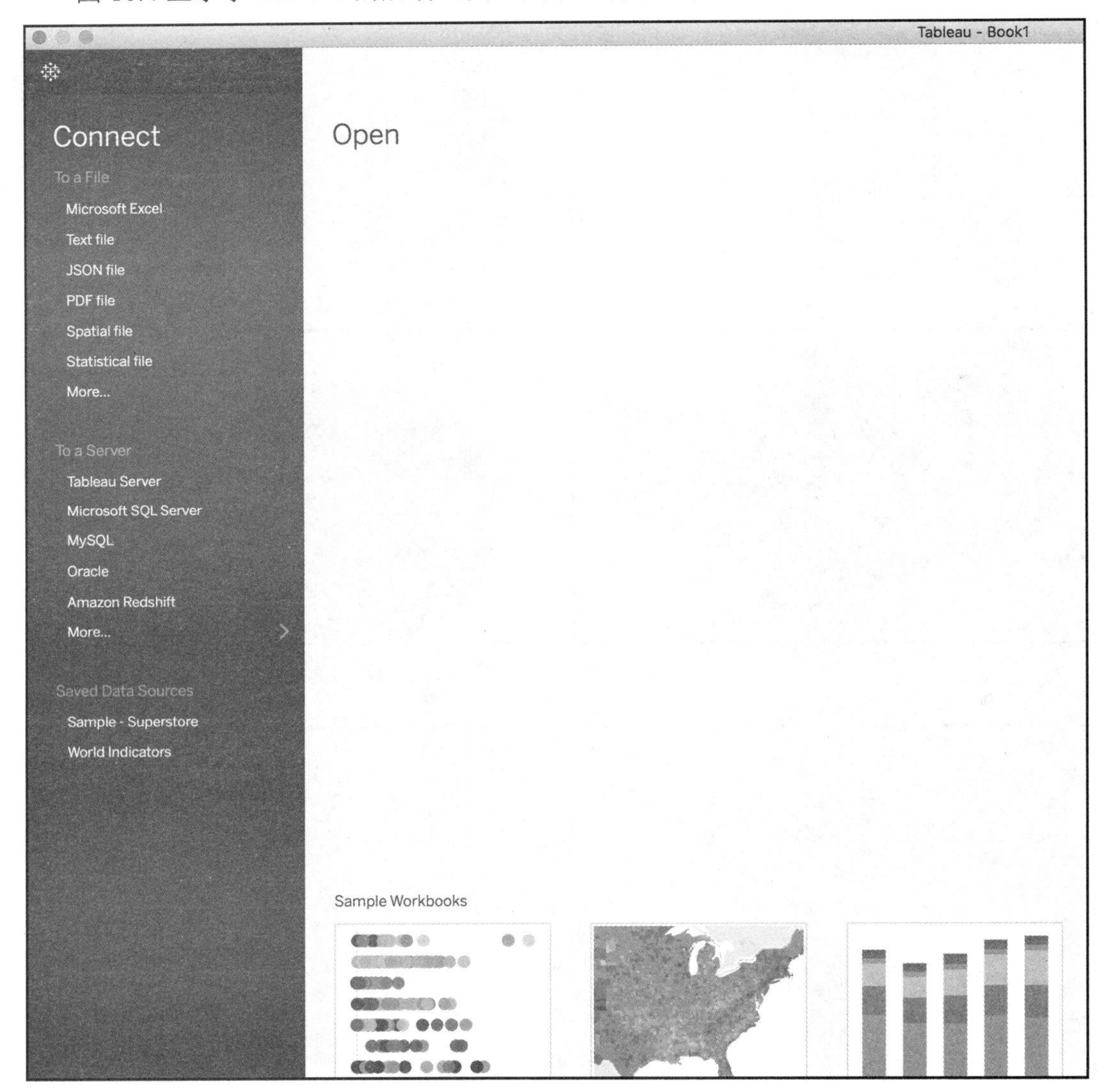

图 10.4

下面打开 OnlineRetail.csv 文件，图 10.5 显示了一个空白工作区。
其中，选取 Quantity 作为列，并查看条状图，如图 10.6 所示。
选取 Description 作为行，并查看各项数量结果，如图 10.7 所示。

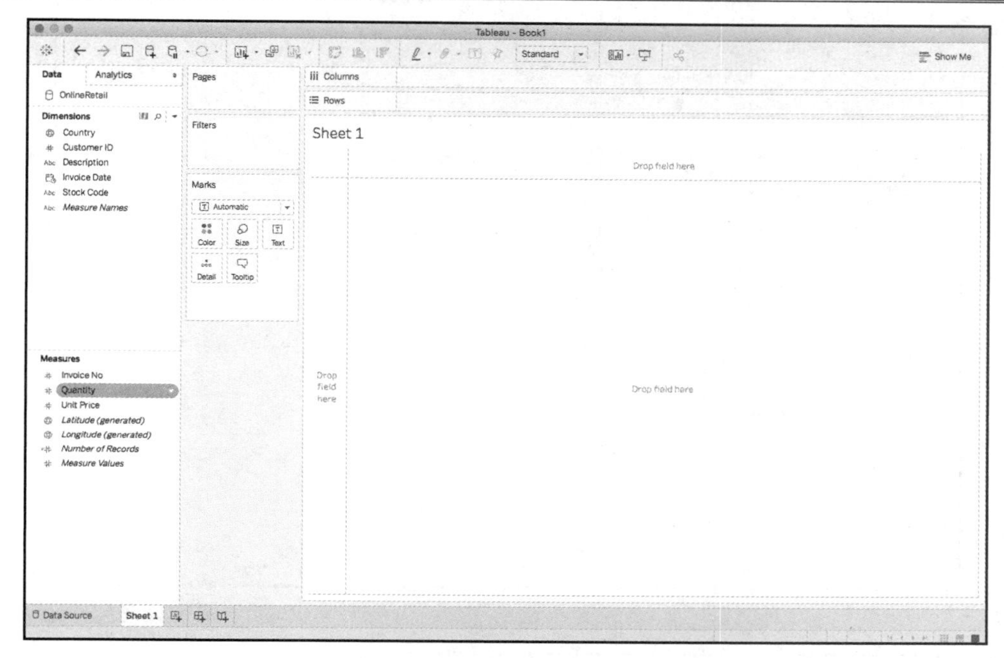

图 10.5

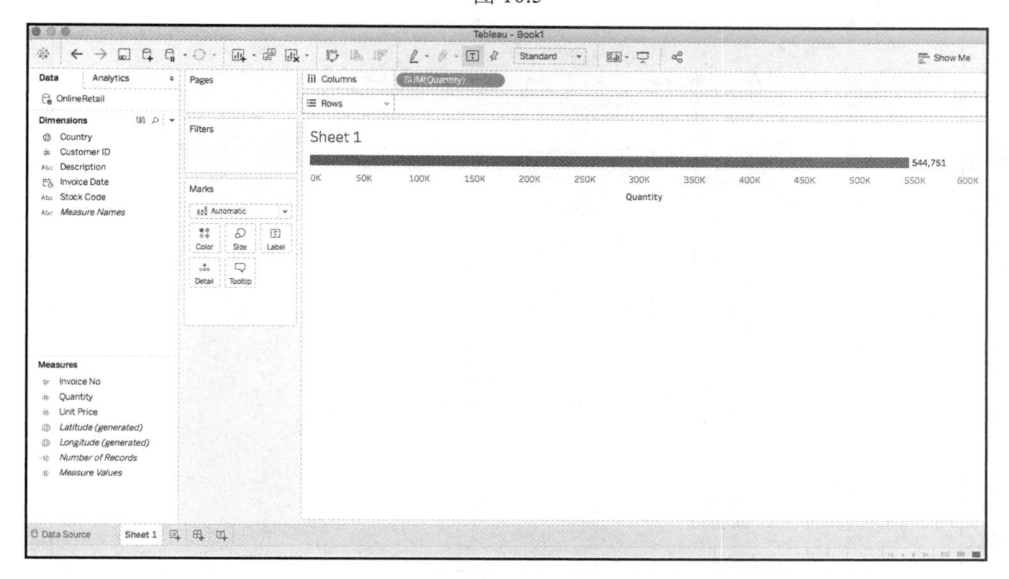

图 10.6

相应地，还可使用过滤器消除负值，如图 10.8 所示。

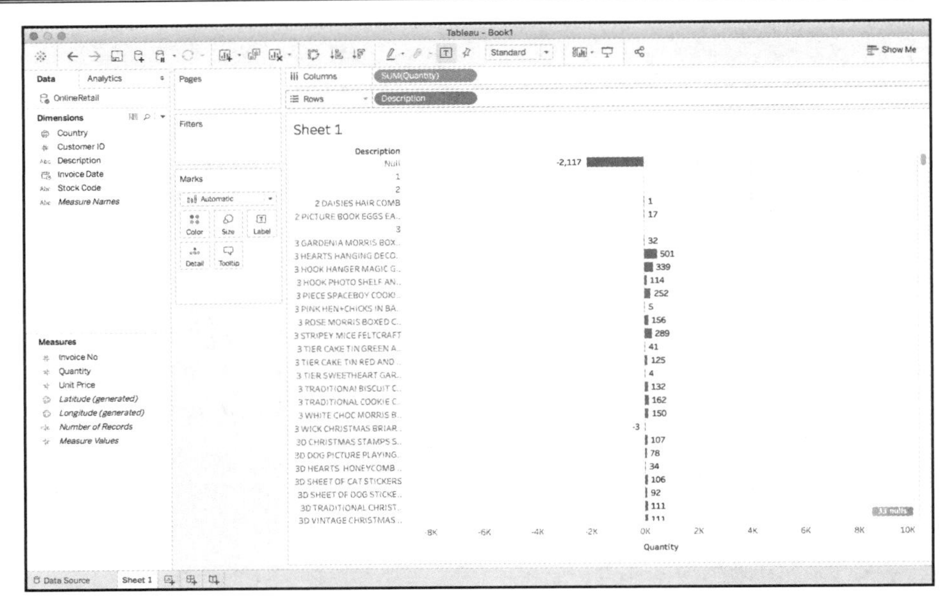

图 10.7

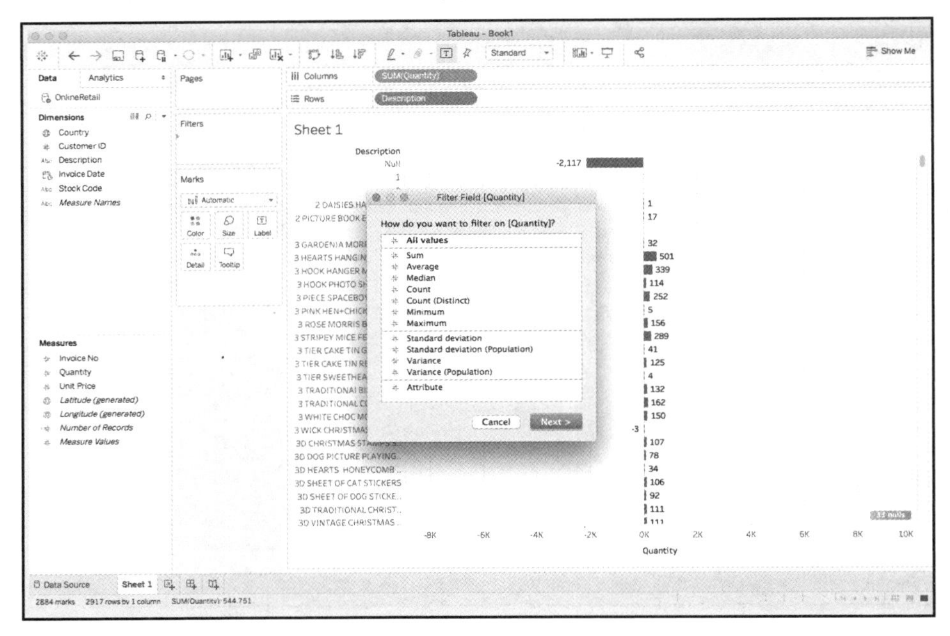

图 10.8

图 10.9 显示了任意数字列的数值范围，例如 Quantity。

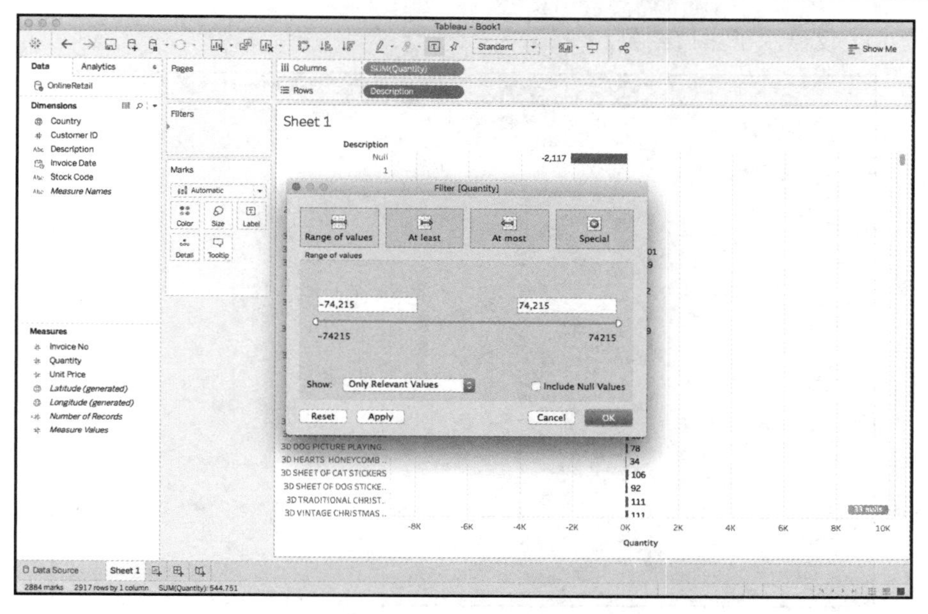

图 10.9

接下来，可针对 Quantity 选择有效的数值范围，如图 10.10 所示。

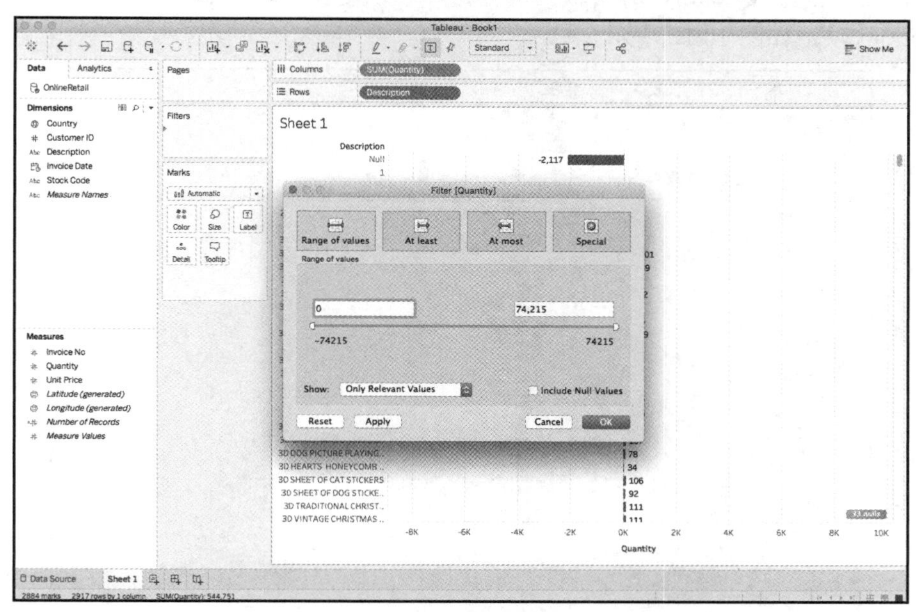

图 10.10

图 10.11 仅显示了正值结果。

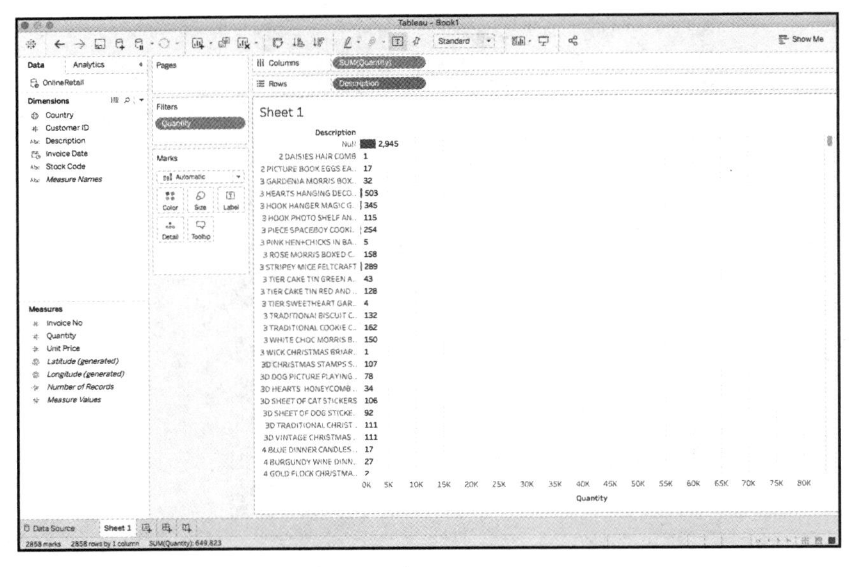

图 10.11

我们还可根据 Quantity 对条状图进行排序，以便在上方查看包含最大 Quantity 的 Descriptions 项，如图 10.12 所示。

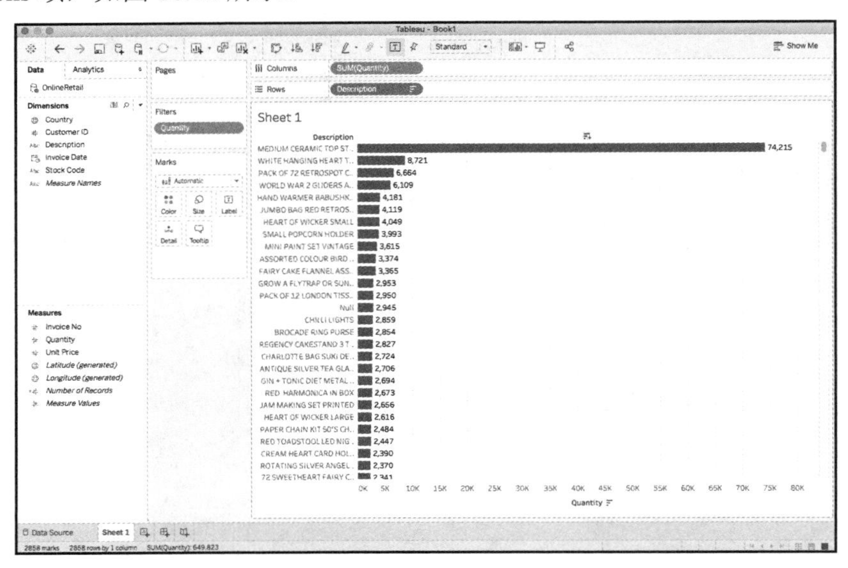

图 10.12

图 10.13 显示了如何创建新的工作表。

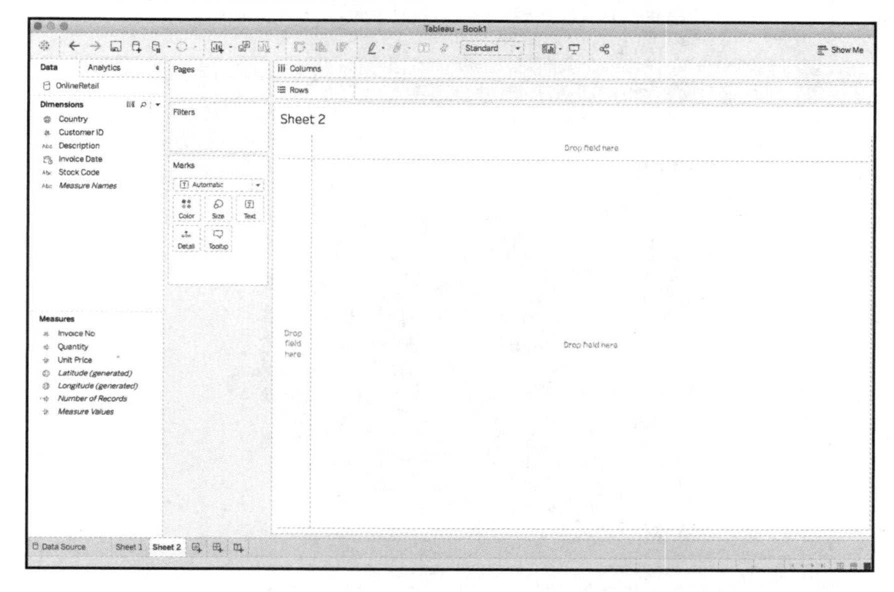

图 10.13

与上一个工作表类似，此处选择了 Description and Quantity，如图 10.14 所示。

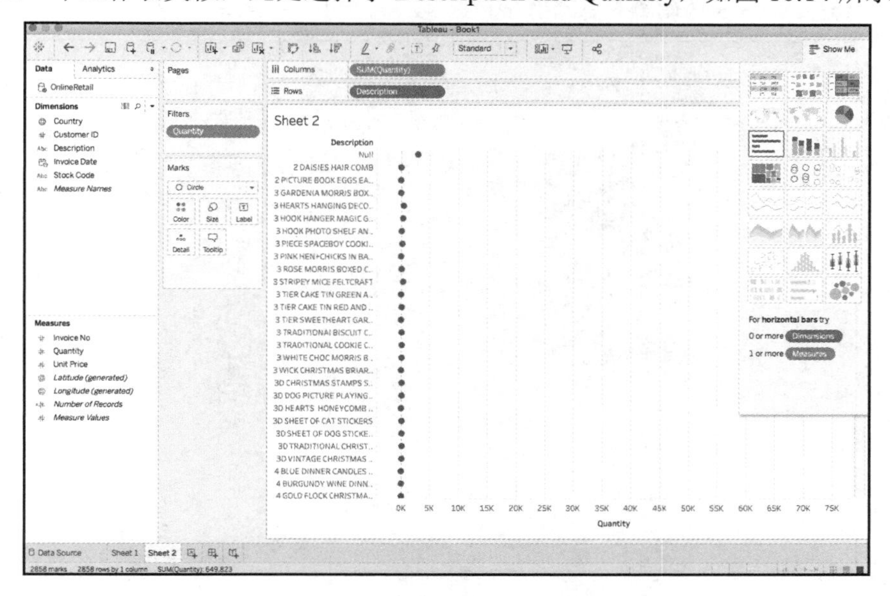

图 10.14

当然，我们也可以从右侧面板中选择不同的图表类型。此处选择了气泡图，如图 10.15 所示。

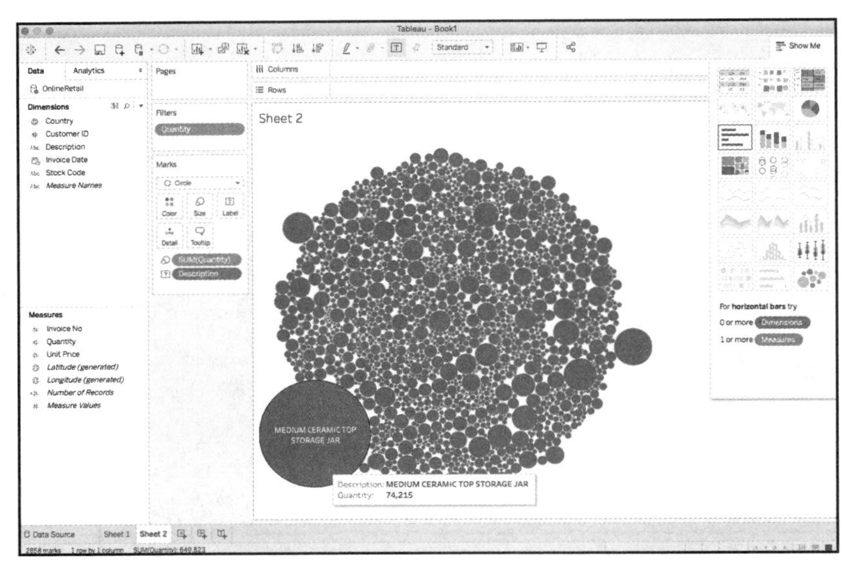

图 10.15

图 10.16 显示了选择矩阵树图作为图表类型后的可视化效果。

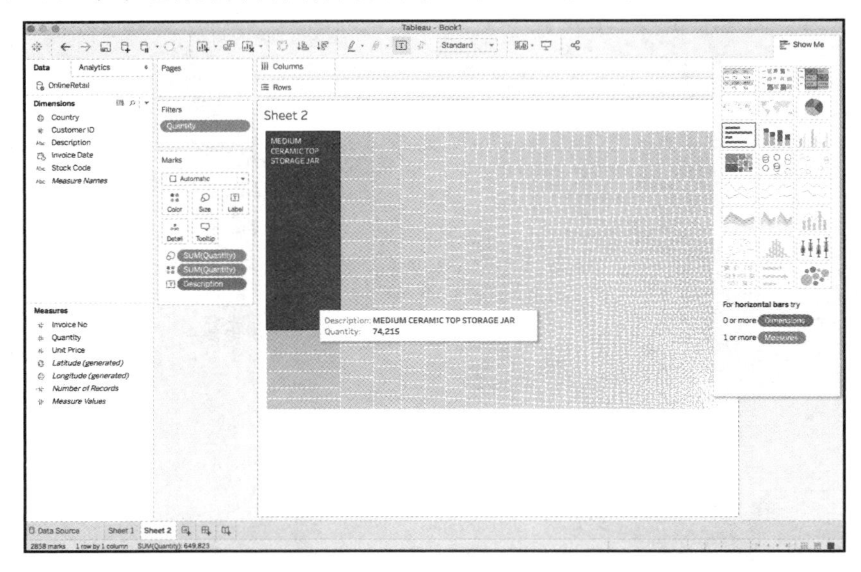

图 10.16

除此之外，还可进一步选取图表的颜色和其他属性，如图 10.17 所示。

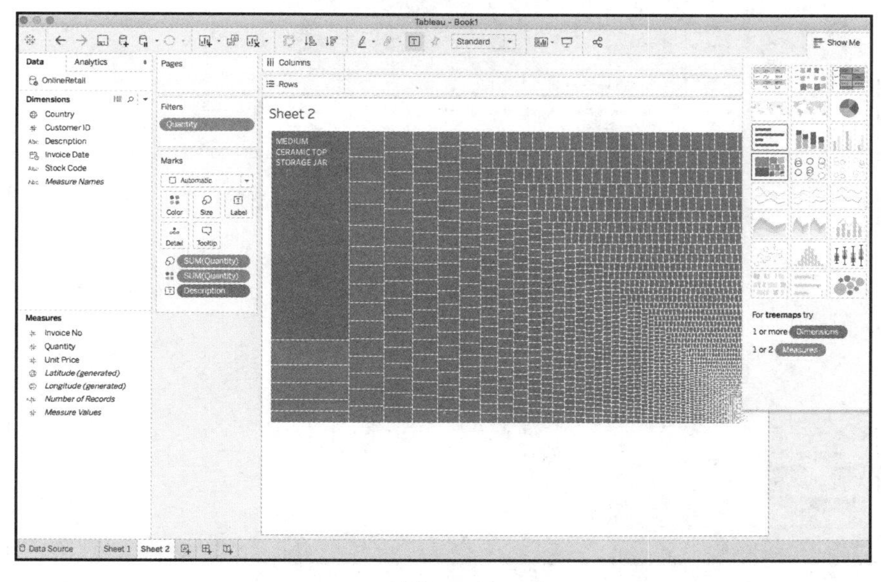

图 10.17

同时，排除某些行/列或值/数据的操作也十分简单，如图 10.18 所示。

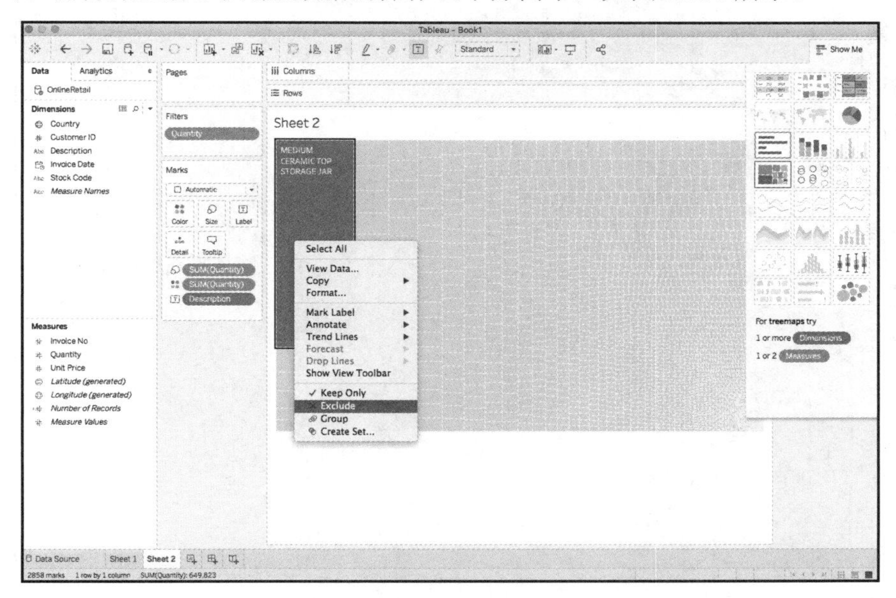

图 10.18

另外，还可生成包含多个工作表的 Dashboard，如图 10.19 所示。

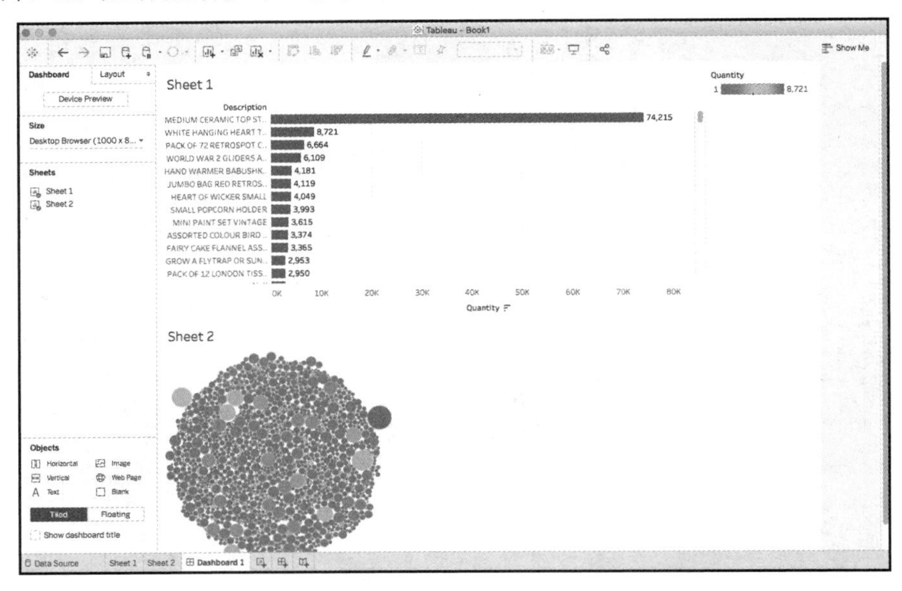

图 10.19

图 10.20 创建了其他类型的图标（线状图）。

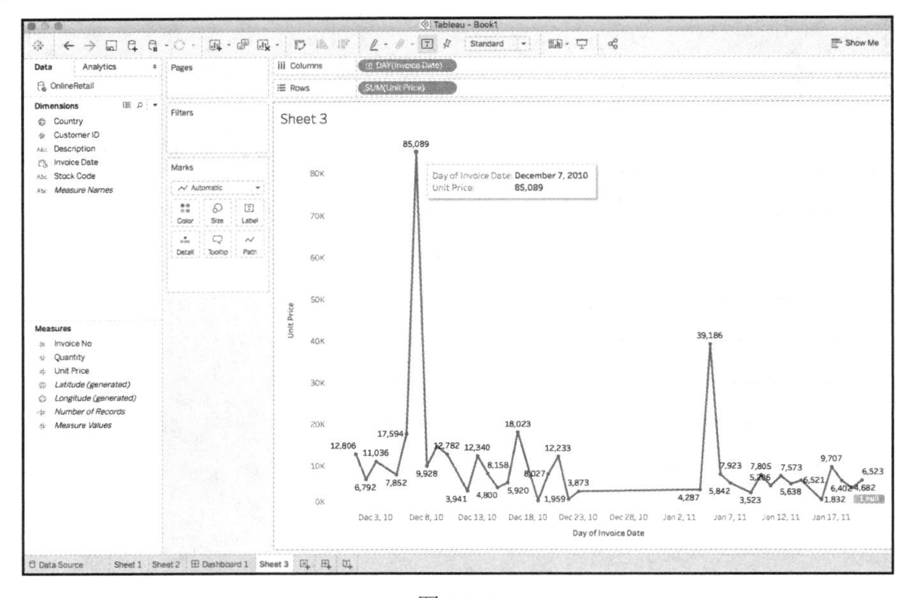

图 10.20

图 10.21 显示了向 Dashboard 中添加的新工作表。

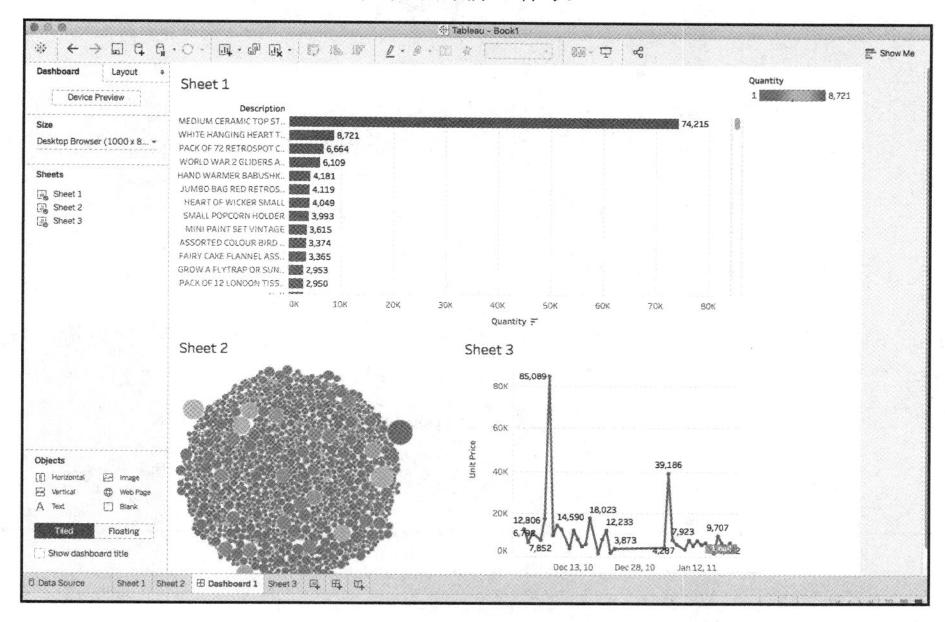

图 10.21

10.3　图　表　类　型

图表可采用多种形式予以呈现。然而。一些常见的特性为图表提供了从数据中提取相关含义的能力。通常情况下，图表中的数据以图形方式加以显示，与文本内容相比，我们可迅速领悟其中的意义。相应地，文本一般仅用于对数据进行注解。

在图形中，文本较为重要的应用之一是标题。图形中的标题通常在其上方显示，同时提供了图形中所用数据的扼要描述。另外，数据的维度一般显示于各个轴向上。当采用水平和垂直轴时，则分别表示为 x 轴和 y 轴。每个轴向上具有对应的量度，并采用周期性的刻度表示，通常伴有数字或分类指示。另外，每个轴向上还包含了一个显示于一侧的标记，用于简要描述所表示的维度。如果量度表示为数值，该标记一般采用括号中的量度单位作为后缀。在图形表达中，直线网格可用于辅助显示数据的视觉对齐效果。例如，可在视觉上利用某些突出的线条体现变化过程中网格的增强效果。相应地，突显的线条称作主要网格线；其余线条则称作次要网格线。

图表中的数据可通过各种格式予以呈现，并可包含独立的文本标记，用以描述与图表中所示位置相关联的数据。相应地，数据可通过点或各种形状、连接或非连接，以及任意颜色和模式的组合结果加以展示。而一些推断结果或者关注点则可直接叠加于图形上，以进一步帮助信息提取。

当显示于图表中的数据包含了多个变量时，图表中可以设置对应的图例。这里，图例包含了图表中显示的变量列表，以及显示示例。此类信息有助于在图表中识别源自变量的数据。

10.3.1　线状图

线状图可查看一段时间内一个或多个变量的行为，并对发展趋势进行判断。在传统的 BI 中，线状图可用于显示过去 12 个月以来的规模、利润、收入方面的发展状况。当与大数据协同工作时，公司可采用这一类可视化技术跟踪产品的购买量（以星期计算）、销售部门的平均月订单数量等。

图 10.22 显示了线状图示例。

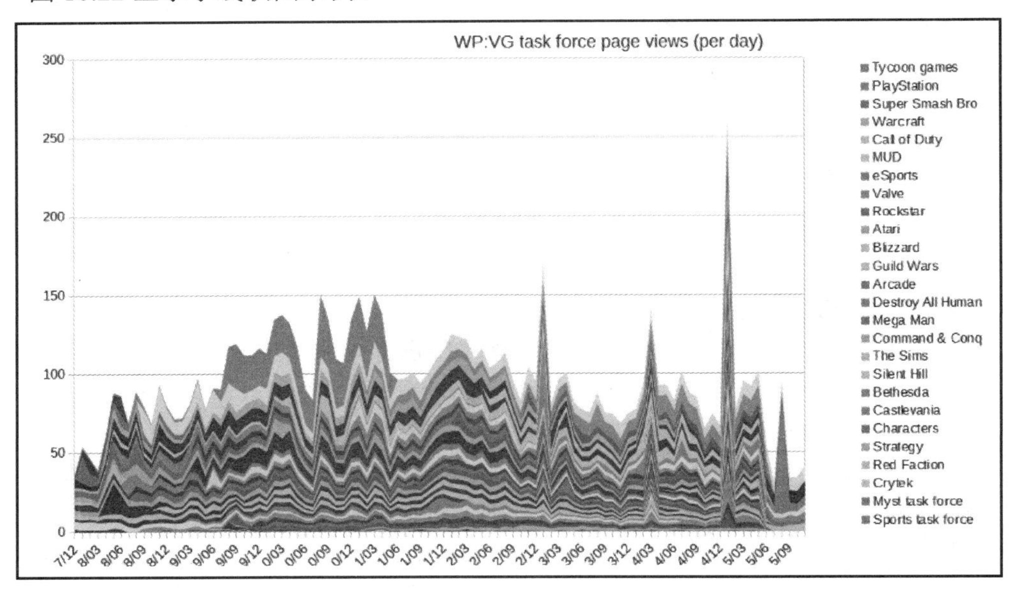

图 10.22

10.3.2　饼图

饼图展示了整体内容中的各个组成部分。同时使用传统数据和大数据的公司可通过这种技术观察客户的细分状态和市场份额。不同之处在于，这一些公司采用原始数据作为分析的来源。

图 10.23 显示了饼图示例。

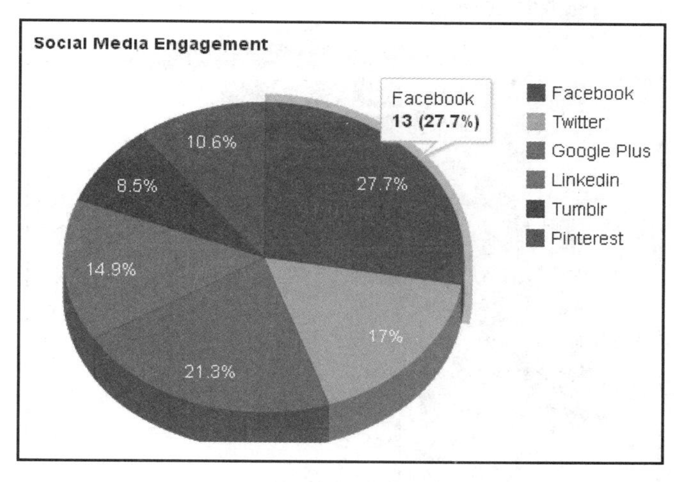

图 10.23

10.3.3　柱状图

饼状图可对不同的变量值进行比较。在传统的 BI 中，公司可通过类别分析销售额，并通过渠道查看市场的营销成本等。当对大数据进行分析时，公司可以按小时查看客户的参与度、销售数据等内容。

图 10.24 显示了垂直柱状图示例。

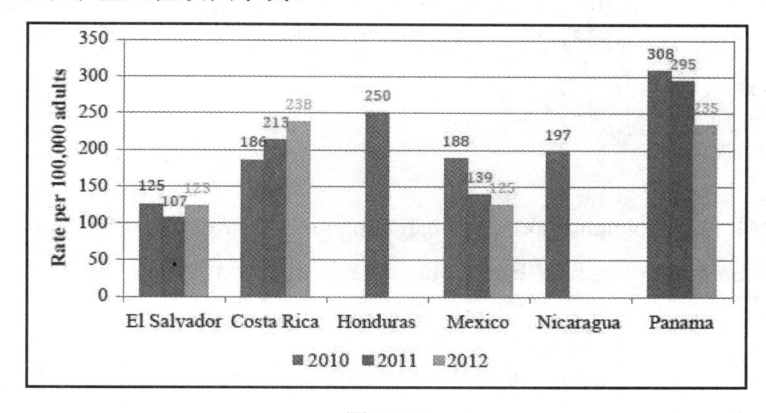

图 10.24

图 10.25 显示了水平柱状图示例。

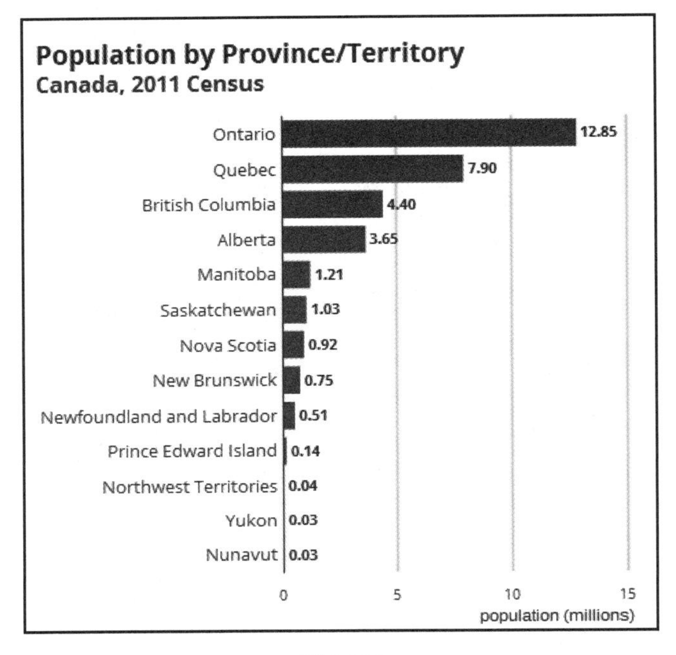

图 10.25

10.3.4　热图

热图通过颜色展示数据。读者可能对 Excel 中的热图有所耳闻。其中，采用绿色突出显示销售业绩最好的分公司，而以红色展示销售业绩最差的分公司。如果零售商打算了解店内顾客最常光顾的购物通道，则可使用销售楼层的热图加以展示。例如来自视频监控系统中的数据。

图 10.26 显示了热图示例。

注意：

读者可访问 https://blog.hubspot.com/marketing/great-data-visualization-examples 和 http://www.mastersindatascience.org/blog/10-cool-big-datavisualizations/，以了解其他一些较为有趣的可视化技术。

可视化本身就是一门艺术，每个用例都需要关注可视化的内容，包括图表类型、数据点的数量、元素的颜色等。

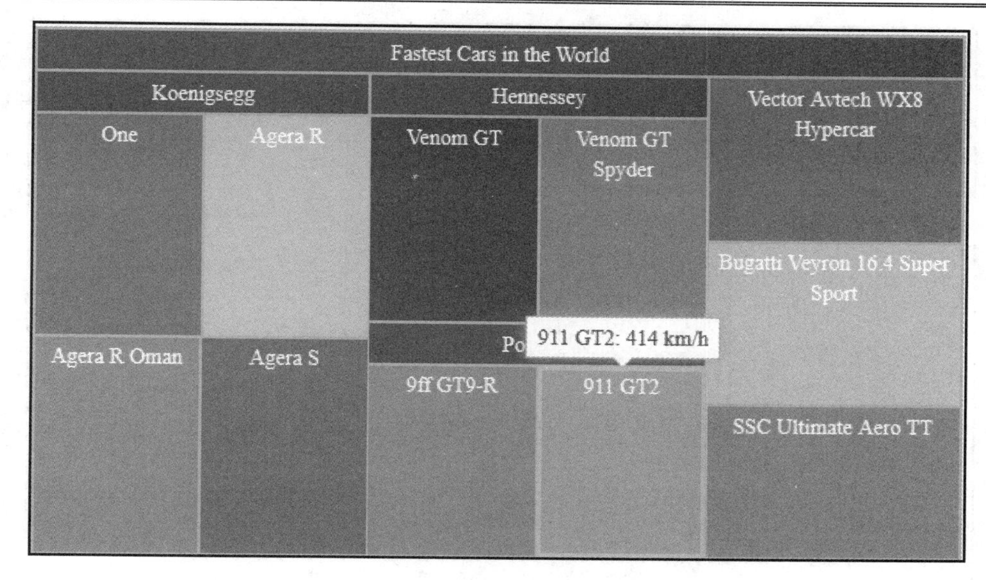

图 10.26

10.4　基于 Python 的数据可视化

Python 提供了多项大数据分析的扩展功能，其中也涉及数据的图形表示和可视化技术。

提示：

第 4 章曾对基于 Python 的大数据分析和可视化技术有所讨论。

下列代码显示了一个 Python 示例（仅涉及单列）。

```
d8 = pd.DataFrame(df, columns=['Quantity'])[0:100]
d8.plot()
```

此处仅选取了前 100 个元素，以避免图形数据过于拥挤，进而较好地展示当前示例。对应结果如图 10.27 所示。

下列代码显示了多个列的可视化结果：

```
d8 = pd.DataFrame(df, columns=['Quantity', 'UnitPrice'])[0:100]
d8.plot()
```

对应结果如图 10.28 所示。

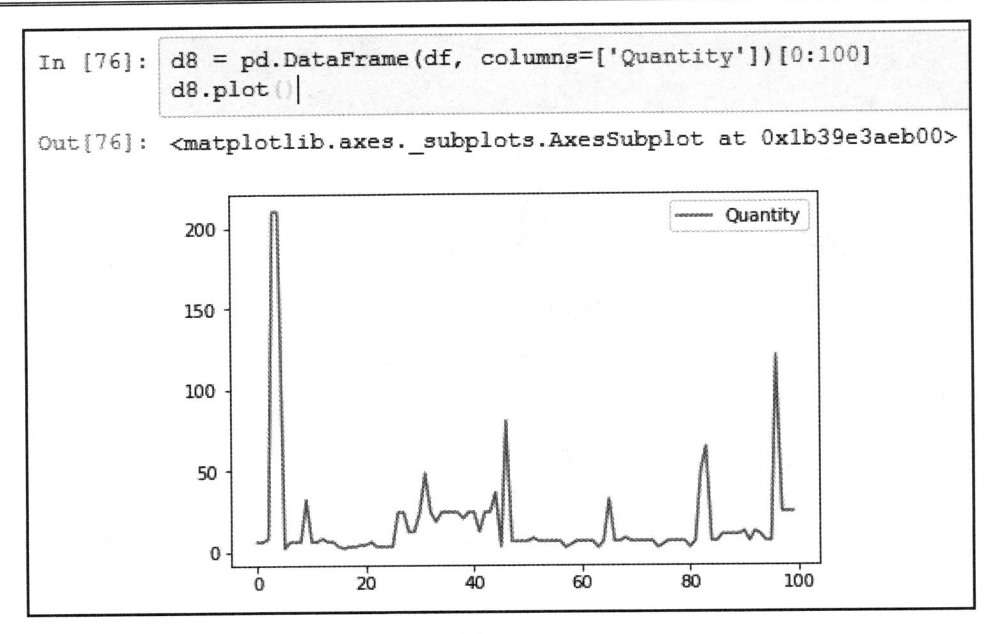

图 10.27

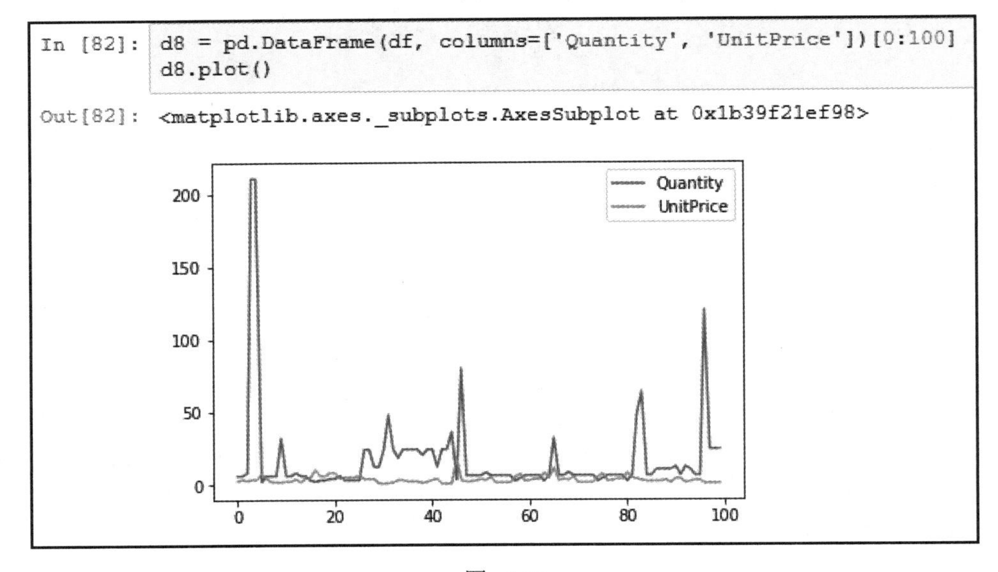

图 10.28

需要注意的是，Python 并不会绘制定性的数据列（例如 Description），而是仅显示图形化的数据。

10.5　基于 R 的数据可视化

R 提供了多项大数据分析的扩展功能，其中也涉及数据的图形表示和可视化技术。

提示：

第 5 章曾对基于 R 的大数据分析和可视化技术有所讨论。

当采用 R 时，我们也可对所选列进行绘制，如下所示：

```
plot(df$UnitPrice)
```

对应结果如图 10.29 所示。

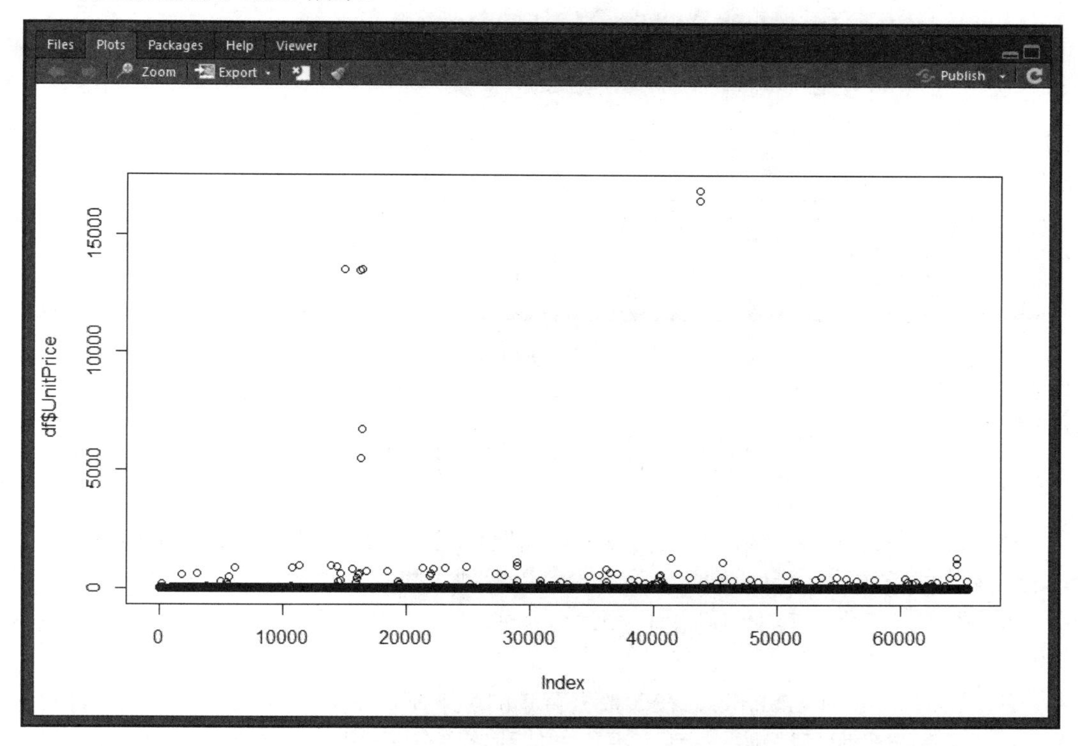

图 10.29

又如：

```
plot(d1, type="b")
```

对应结果如图 10.30 所示。

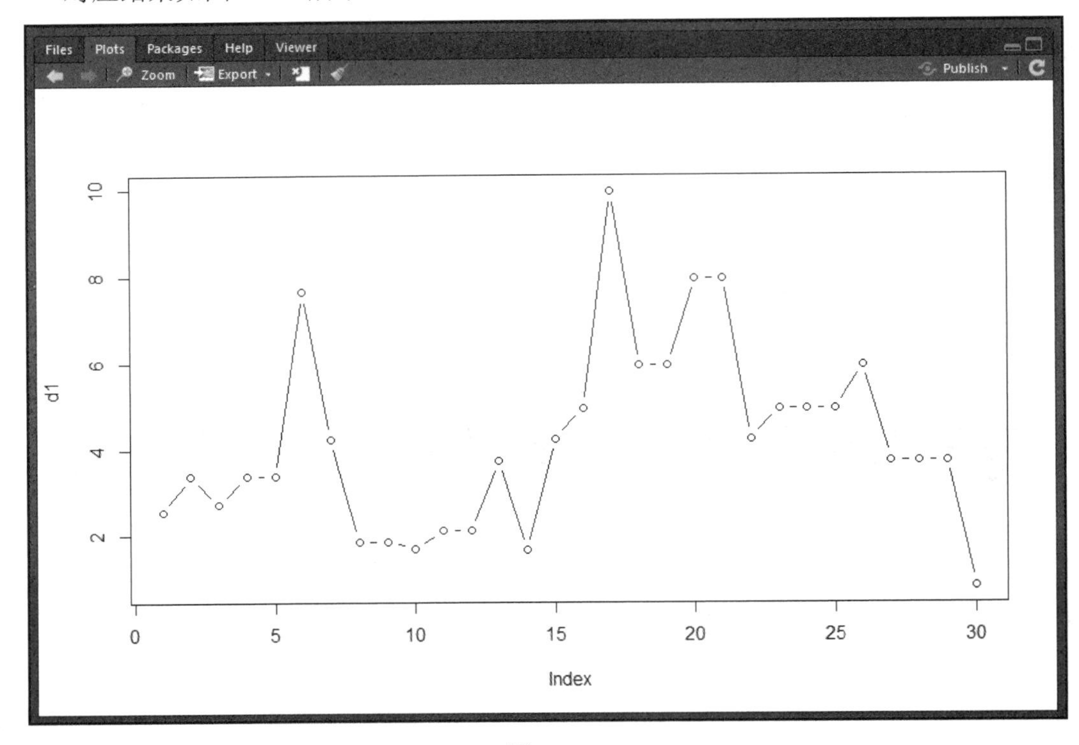

图 10.30

10.6　大数据可视化工具

通过对大数据工具市场的快速调查，可以发现包括微软（Microsoft）SAP、IBM 和 SAS 在内的大公司的身影。除此之外，也可以看到一些专业软件供应商发布的技术领先的大数据可视化工具，其中包括 Tableau、Qlik 和 TIBCO。下面列出了一些领先的数据可视化产品。

- IBM Cognos Analytics：在大数据背景的驱动下，IBM 发布的数据分析包提供了各种自助服务选项，从而可更加方便地获取分析结果。读者可访问 https://www.ibm.com/analytics/us/en/technology/products/cognosanalytics/以了解更多信息。
- QlikSense and QlikView：Qlik 解决方案能够执行更复杂的数据分析，进而获取更加隐藏的分析结果。

- Microsoft PowerBI：Power BI 工具可关联数百个数据源，并于随后在 Web 和移动设备上发布分析报告。读者可访问 https://powerbi.microsoft.com/en-us/以了解更多信息。
- Oracle Visual Analyzer：Oracle Visual Analyzer 是一个基于 Web 的可视化工具，允许创建可管理的视图，以帮助发现数据中的相关性和模式。读者可访问 https://docs.oracle.com/cloud/latest/reportingcs_use/BILUG/GUID-7DC34CA8-3F7C-45CF-8350-441D8D9898EA.htm#BILUG-GUID-7DC34CA8-3F7C-45CF-8350-441D8D9898EA 以了解更多信息。
- SAP Lumira：Lumira 号称是面向所有用户的、具有自助服务的数据可视化工具，并可将可视内容整合至故事板（storyboard）中。
- SAS Visual Analytics：SAS 解决方案提升了可伸缩性和管理能力，同时包含了动态可视化和灵活的部署选项。读者可访问 https://www.sas.com/en_us/software/business-intelligence/visual-analytics.html 以了解更多内容。
- Tableau Desktop：Tableau 的交互式视图可动态地发现隐藏的分析结论，高级用户还可以对元数据进行管理，以充分利用不同的数据源。读者可访问 https://www.tableau.com/products/desktop 以了解更多信息。
- TIBCO Spotfire：TIBCO Spotfire 将分析软件作为一项服务予以提供，并可为小型开发团队甚至是整个机构提供相应的解决方案。读者可访问 http://spotfire.tibco.com/以了解更多信息。

10.7　本 章 小 结

本章介绍了可视化功能及其背后的各种概念。第 11 章将讨论云计算的力量，以及它如何改变大数据和大数据分析的格局。

第 11 章　云计算简介

本章讨论云计算、基础设施即服务（IaaS）、平台即服务、软件即服务等概念。除此之外，本章还将简要介绍一些云供应商。

本章主要涉及以下主题：

- 云计算基础知识。
- 概念和术语。
- 目标和收益。
- 风险与挑战。
- 角色和界线。
- 云计算特征。
- 云交付模型。
- 云部署模型。

无论是在数百万名移动用户间分享图片，抑或支撑较为重要的业务操作，云服务平台均针对这一类灵活、低成本的 IT 资源提供了快速访问方式。当采用云计算时，一般不需要在管理硬件上投入大量资金。相反，我们可提供适当的计算资源以实现相关理念，或者对 IT 部门的工作进行管理。

对于服务器、存储、数据库和互联网上应用程序服务集，云计算可提供简单的访问方式。云服务平台，例如亚马逊 Web 服务（AWS）即持有并负责维护应用程序所需的网络连接硬件，并可通过 Web 应用程序使用相关内容。

11.1　概念和术语

本节主要介绍云计算及其构件的基本概念。

11.1.1　云

云是指，为远程提供可伸缩和可度量的 IT 资源而设计的独特的 IT 环境。这一术语最初是对互联网的一种隐喻，用来描述一组网络，这些网络提供了对一组分散的 IT 资源的远程访问。在云计算成为正式的 IT 术语之前，在基于 Web 体系结构的各种规范和核心文档中，云通常用于表示互联网。

11.1.2　IT 资源

IT 资源是指与 IT 相关的物理或虚拟组件，可以是虚拟服务器或自定义软件程序这一类软件，抑或物理服务器或网络设备这一类硬件设施。

11.1.3　本地环境

作为一种独特的、可远程访问的环境，云代表了部署 IT 资源的一种选择。在组织边界内（organizational boundary，不具体体现某种云），托管于传统企业中的 IT 资源视为位于 IT 企业中，或者是受控制的 IT 环境中（并非基于云）。该术语用于将 IT 资源限定为基于云的替代资源。本地的 IT 资源不能基于云，反之亦然。

11.1.4　云使用者和云供应商

云供应商是提供了基于云 IT 资源的实体，而云使用者则是使用云 IT 资源的实体。

11.1.5　扩展

扩展表示 IT 资源处理应用需求的能力。

1. 扩展类型

扩展一般涉及以下几种类型。

- 横向扩展：内外扩展。
- 纵向扩展：上下扩展。

（1）横向扩展

分配或释放相同类型的 IT 资源称为水平扩展。其中，资源的横向分配称作向外扩展，而资源的横向释放则称作向内扩展。在云环境中，横向扩展则是一种较为常见的扩展形式。

（2）纵向扩展

当现有资源被另一个更高或更低的容量所替代时，即会产生纵向扩展。其中，利用另一个较高容量资源替换 IT 资源称作向上扩展，而利用较低容量资源替换 IT 资源则称作向下扩展。考虑到替换下载时间问题，纵向扩展在云环境中一般较少出现。

2. 云服务

虽然云是一类远程访问环境，但云内的所有资源并不是均可采用远程方式被访问。例如，部署于云中的数据库或物理服务器仅可通过位于同一云内的其他 IT 资源予以访问。对此，可专门部署包含发布 API 的软件程序，并启用远程客户端的访问功能。

云服务表示为任意 IT 资源，并可通过云实现远程访问。与涉及服务技术的其他 IT 领域不同，例如面向服务架构（SOA），云计算环境中的服务其范围则更加广泛。云服务可作为一个简单的 Web 软件程序而存在，其中包含了通过消息传输协议调用的技术接口；或者作为管理工具或大型环境的远程访问点。

3. 云服务使用者

云服务使用者是软件程序在运行期间访问云服务时假定的临时角色。

常见的云服务使用者类型包括：通过发布的服务契约远程访问云服务的软件程序和服务，以及工作站、笔记本电脑、移动设备（作为服务，运行远程访问其他 IT 资源的软件）。

11.2 目标和收益

与批发商一样，公共云提供商的商业模式也是基于对 IT 资源的大规模收购，并向云使用者提供诱人的价格，这有助于机构在不需要投入任何成本的情况下访问强大的基础设施。

对于云 IT 资源来说，减少初始 IT 投资是最为常见的经济因素，例如硬件、软件的购买以及运营成本。云的可度量应用特性体现为一个特征集，该特征集允许运营支出（与业务性能直接相关）取代预期的资本支出，这也称作比例成本。

降低成本可以让企业从小规模做起，并根据需要增加相应的 IT 资源配置。此外，较低的初始费用允许资本重新定向到核心业务投资中去。成本的降低主要来自于云供应商对大型数据中心的部署和运营。这一类数据中心通常坐落于成本相对较低的区域，包括地价、IT 专业人员、网络带宽等。

同样的道理也适用于操作系统、中间件、平台软件以及应用软件。汇集的 IT 资源可以由多个云使用者所共享，从而提高或优化利用率。通过优化云架构、管理和治理方案，可以进一步降低运营成本并提高生产效率。

相应地，云使用者的收益主要体现在以下几个方面：

- 可在短期内，按需访问即付即用的计算资源（例如按小时收费的处理器），并在不需要时释放这些计算资源。

- 可以访问无限的计算资源，这些资源可以根据需要使用，且无须事先准备。
- 在较为基础的层次上添加或删除 IT 资源，例如以千兆字节增量的方式调整可用的磁盘存储空间。
- 支持基础结构的抽象，以使应用程序不会被锁定至相关设备或位置处；必要时，还可方便地对其加以移动。

例如，如果某家公司需要执行大量的批处理任务，那么，可以像应用程序软件扩展那样快速地完成任务。使用 100 台服务器/1 小时的成本与 1 台服务器/100 小时的成本基本相同。在不需要注入大量的初始投资的情况下即可构建大规模的计算基础设施，这种 IT 资源的弹性颇具吸引力。

尽管云计算的优点十分明显，但实际的经济学计算和评估过程可能十分复杂。云计算策略的制定所涉及的问题不再仅仅是租赁成本和购买成本之间的简单比较。

11.2.1　可扩展性的提升

通过提供 IT 资源池，并设计综合利用此类资源的工具和技术方案，云计算可根据需要或者使用云消费者的配置内容，以即时、动态方式将 IT 资源分配至云消费者。这使得云消费者可适当地调整基于云的 IT 资源，并通过自动或手动方式适应于波动或峰值的处理过程。类似地，随着处理需求的减少，还可释放基于云的 IT 资源（通过手动或自动方式）。

在 IT 资源可扩展性的灵活程度方面，云平台固有的内在特征与之前提到的比例成本收益直接相关。除了规模减少所带来的财政收益之外，IT 资源满足未知需求的能力还可避免出现使用阈值时可能发生的业务损失。

11.2.2　可用性和可靠性的提升

IT 资源的可用性和可靠性与业务利益直接相关。对于客户来说，运行中断限制了业务的开放时间，因而在使用和收益方面也将大打折扣。在大容量应用期间，运行故障则会带来更大的影响。其间，IT 资源不仅不能响应客户的请求，而且意外产生的故障还会降低客户的整体信心。

对于提升云环境中 IT 资源的可用性，进而最小化甚至是消除中断问题，典型的云环境通常可给予广泛的支持，这也是其体现的一种内在能力。另外，可靠性的提升还可使运行期故障条件所带来的影响降至最低。

特别地：

- 在提升可用性后，IT 资源将可在更长的时间内予以访问（例如 1 天中的 22 个小时）。云供应商通常会提供具有弹性的 IT 资源，进而可保证高水平的可用性。
- 具有高可靠性的 IT 资源可有效地消除一些异常条件，或者从这些异常情况中予以恢复。另外，云环境的模块化体系结构还支持故障转移机制，从而进一步提高了可靠性。

需要注意的是，在考虑租用基于云的服务和 IT 资源时，机构必须仔细检查云供应商提供的 SLA。尽管许多云环境均可提供较高的可用性和可靠性，但 SLA 提供了最终的保证，即实际的合同义务。

11.3　风险和挑战

本节讨论云计算所面临的几项关键性挑战，这些挑战主要与公共云中使用 IT 资源的消费者有关。

11.3.1　安全漏洞

将业务数据迁移至云意味着数据安全这一类责任将与云供应商所共享。IT 资源的远程应用需要云使用者扩展信任边界，进而将外部云也纳入进来。考虑到第三方云供应商一般会在地价相对便宜，或者交通方便的地理位置处建立数据中心，因而难以解决多区域协调性以及法律方面的问题。当采用公共云进行托管时，云使用者通常并不了解 IT 资源和数据的物理位置。对于某些机构来说，这可能会导致与行业或政府法规相关的法律问题，相关法律规定了相应的数据隐私和存储策略。

考虑到多重边界的存在，因而很难建立切实可行的安全体系结构，同时不会引入任何漏洞，除非云使用者和云供应商恰好支持相同或兼容的安全框架。然而，公共云的这种兼容机制通常难以实现。重叠的信任边界所引发的另一个问题则与云供应商和云使用者的权限访问相关。当前，数据的安全程度仅限于云使用者和云供应商所用的安全控制策略。除此之外，鉴于基于云的 IT 资源一般处于共享状态，因此，不同的云使用者之间可能存在重叠的信任边界。

信任边界的重叠以及数据暴露问题将会为恶意的云使用者（采用人工方式或自动化方式）提供更大的机会攻击 IT 资源，或者窃取、破坏相关业务数据。假设存在某种场景，需要两个访问相同云服务的机构将各自的信任边界扩展到云，这将导致重叠的信任边界。

对于云服务供应商来说，挑战来自于如何提供能够满足两个云服务使用者需求的安全机制。

11.3.2　减少运营治理控制

通常情况下，云使用者所分配的控制级别一般低于本地 IT 资源。这可能会引发下列问题：云供应商如何对云环境进行操控，以及云与其使用者之间通信所需的外部连接。

11.3.3　云提供商之间有限的可移植性

由于在云计算行业中缺乏既定的行业标准，公共云在不同程度上通常是专有的。当用户依赖于专有环境的定制解决方案时，挑战来自于云供应商之间的迁移行为。

11.4　角色和边界

根据云与 IT 资源的关系和交互方式，机构和人员可承担不同类型的预定义角色。其中，每个角色都会参与、执行与云操作相关的各项职责。下面将对这一类角色加以定义，并确定其主要的交互行为。

11.4.1　云供应商

云供应商是提供基于云的 IT 资源的机构。在扮演云供应商这一角色时，该机构负责按照协议的 SLA 条款向云使用者提供云服务。同时，云供应商还将进一步承担必要的管理措施和管理职责，以便确保整个云基础设施的正常运行。

云供应商一般持有云使用者租用的 IT 资源；相应地，一些云供应商也会转卖其他云供应商租用的 IT 资源。

11.4.2　云使用者

云使用者一般是一个机构（或个人），并与云供应商签署正式的合同，或者与云供应商进行协调，进而使用云供应商提供的 IT 资源。

11.4.3　云服务持有者

合法拥有云服务的个人或机构称作云服务持有者。云服务持有者可以是云使用者，或者是拥有云服务所在云环境的提供者。

11.4.4　云资源管理员

云资源管理员是负责管理基于云的 IT 资源（包括云服务）的个人或组织。云资源管理员可以是（或隶属于）云使用者或云服务的提供者；或者，它可以是（或隶属于）一个第三方组织，负责管理基于云的 IT 资源。

1．附加角色

NIST 的云计算参考体系结构定义了以下补充角色。

- 云审计者：对云环境进行独立评估的第三方（通常是经过认证的），负责承担云审计者的角色。与这一角色相关的一般职责包括评估安全控制、隐私影响和性能问题。云审计者这一角色的主要目的是为云环境提供公正的评估（以及相关认证），以增强云消费者和云供应商之间的信任关系。
- 云代理：该角色由管理、协商（云使用者和云供应商间的）云服务应用的一方承担。云代理提供的中介服务包括服务中介、聚合和套利。
- 云运营商：云运营商负责在云使用者和云供应商之间提供连接级的一方，一般是指网络和电信供应商。

2．组织边界

组织边界表示围绕由组织拥有和管理的一组 IT 资源的物理边界。当组织承担云使用者这一角色访问基于云的 IT 资源时，它需要将信任扩展到组织的物理边界之外，以包括部分云环境。

3．信任边界

当某个组织作为云使用者角色访问基于云的 IT 资源时，需要将信任扩展到组织的物理边界之外，以包含部分云环境。

11.5　云　特　征

IT 环境需要一组特定的特征集，以便以有效的方式远程提供可伸缩和度量的 IT 资源。大多数云环境均具备以下 6 个共同特性：

- 按需使用。
- 无处不在的访问。

- 多租户机制（和资源池机制）。
- 弹性。
- 监测应用状态。
- 弹性计算。

11.5.1　按需使用

云使用者可以单方面访问基于云的 IT 资源，并可自行提供这些 IT 资源。待配置完毕后，即可自动使用自给的 IT 资源，从而降低与云使用者或云供应商间的人员合作。这将产生一种按需使用的环境。这种特性也称为按需自助服务使用，它支持主流云中基于服务和应用驱动的特性。

11.5.2　无处不在的访问

这体现了云服务广泛的访问能力。为云服务建立此类访问机制需要获得设备、传输协议、接口和安全技术方面的支持。该访问级别通常需要云服务体系结构适应不同云服务使用者的特定需求。

11.5.3　多租户机制（和资源池机制）

软件程序的特性之一是程序的一个实例能够服务于不同的使用者（租户），因此它们彼此间是孤立的，这被称为多租户机制。云供应商将其 IT 资源置入池中，以便对多个云服务使用者提供服务。也就是说，使用多租户模型，该模型依赖于虚拟化技术。通过多租户技术，IT 资源可根据云服务使用者需求动态地分配和重新分配。

11.5.4　弹性

弹性是指云的自动化能力，可根据运行期条件或者云使用者、供应商预先决定的需求透明地扩展 IT 资源。弹性通常被认为是云计算应用的一个关键因素，主要是因为它与所减少的投资和比例成本收益密切相关。拥有大量 IT 资源的云提供商可以提供最大范围的弹性。

11.5.5　监测应用状态

这表示云平台能够跟踪其 IT 资源（主要由云使用者使用）的使用情况。基于所监测的内容，云供应商只能根据实际使用的 IT 资源和/或对 IT 资源授权访问的时间帧向云使用者收费。在这种情况下，测量的用法与按需特性密切相关。

11.5.6　弹性计算

弹性计算可视为一种故障转移形式，它在物理位置之间分布冗余的 IT 资源实现。相应地，可以预先配置 IT 资源，以便在出现不足时，自动将处理工作移交给另一个冗余实现。在云计算中，弹性特性可以指同一云（但在不同物理位置）或跨多个云的冗余 IT 资源。

11.6　云交付模型

对于云供应商提供的 IT 资源，云交付模型表示其特定的、预先打包的组合。对此，存在 3 种常见的云交付模型，已被广泛建立并予以形式化，如下所示：

- IaaS。
- PaaS。
- SaaS。

11.6.1　基础设施即服务

IaaS 交付模型表示一个独立的 IT 环境，它包含以基础设施为中心的 IT 资源，可以使用基于云服务的接口、工具访问和管理这些资源。该环境可包含硬件、网络、连接、操作系统和其他原始 IT 资源。与传统的托管或外包环境不同，通过 IaaS，IT 资源通常被虚拟化并被打包成束，从而简化了基础设施的运行期扩展和定制行为。

IaaS 的主要目标是，在配置和应用方面，向云使用者提供较高的控制能力和相关职责。IaaS 提供的 IT 资源通常未经预先配置，而是直接将管理责任交付至云使用者。该模型将被云消费者使用，进而对所创建的云环境予以控制。

某些时候，云供应商会从其他云供应商那里获得 IaaS 产品，以扩展自己的云环境。相应地，对于不同云供应商提供的 IaaS 产品，其 IT 资源的类型和品牌也可能有所不同。基于 IaaS 环境的 IT 资源一般作为初始化的虚拟实例予以提供。在典型 IaaS 环境中，主要的核心 IT 资源是虚拟服务器。

11.6.2　平台即服务

PaaS 交付模型表示一个预定义的、随时可用的环境，通常由已经部署和配置的 IT 资源组成。具体来说，PaaS 依赖于现有环境的使用（并以此加以定义），该环境建立了一组预先打包的产品和工具，用于支持定制应用程序的整个交付生命周期。

在 PaaS 中，云使用者的应用和投资主要涉及以下几方面内容：

- 出于可伸缩性和经济目的，云使用者希望将内部环境扩展到云。
- 云使用者通过现成的环境完全替代本地环境。
- 云使用者也希望成为云提供者，并部署自己的云服务以供其他外部云使用者使用。

当在一个现成的平台中工作时，对于使用 IaaS 模型建立和维护 IT 资源的基础设施，云消费者可以消除其间的管理负担。

11.6.3　软件即服务

一个定位于共享云服务，并作为产品或通用工具的软件程序体现了 SaaS 的典型特征。SaaS 交付模型通常用于将可重用的云服务（通常是商业上的服务）广泛提供给一系列的云消费者。围绕 SaaS 产品存在着一个完整的市场，并可针对不同的目的对其加以租用、使用。一般情况下，云使用者仅对 SaaS 包含有限的管理控制权限，此类权限通常由云供应商提供，但是它可以由担任云服务持有者角色的任何实体合法拥有。例如，如果某个组织机构作为云使用者并与 PasA 环境协同工作，即可构建一个云服务，同时将其部署到与 SaaS 产品相同的环境中。随后，同一个组织可有效地承担云供应商这一角色，因为在使用该云服务时，基于 SaaS 的云服务可充当云使用者的其他组织。

11.6.4　整合云交付模型

上述 3 种基本的云交付模型构成了一个自然的层次结构，并可对模型应用加以整合。下面将简要介绍两种常见的组合模式。

1．IaaS+PaaS

PaaS 环境将建立在与物理和虚拟服务器，以及 IaaS 环境中提供的其他 IT 资源类似的底层基础设施之上。

2．IaaS+PaaS+SaaS

这 3 种云交付模型可进行适当组合，进而构建彼此间的 IT 资源层。例如，通过添加到前述分层体系结构中，PaaS 环境提供的现有环境可以被云消费者组织机构用于开发和部署自己的 SaaS 云服务，然后将其作为商业产品予以提供。

图 11.1 显示了 IaaS、PaaS 和 SaaS 中的各层结构。

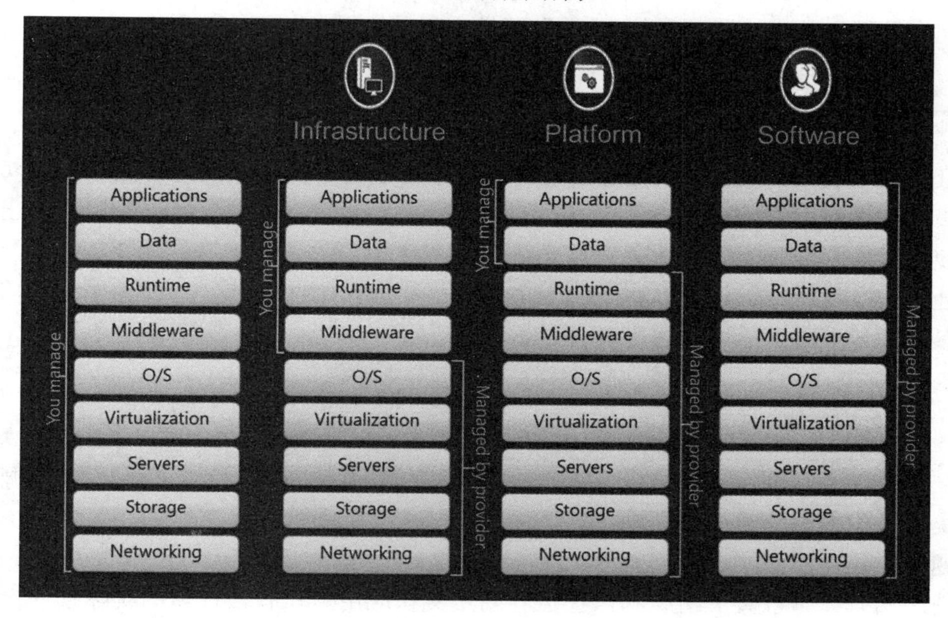

图 11.1

11.7　云部署模型

云部署模型表示特定类型的云环境，主要根据所有权、大小和访问方式加以区分。下面将讨论以下 4 种常见的云部署模型：

- 公共云。
- 社区云。
- 私有云。
- 混合云。

11.7.1　公共云

公共云是第三方云供应商拥有的可公开访问的云环境。公共云上的 IT 资源通常使用前述云交付模型提供，通常以一定的成本提供给云使用者，或者通过其他途径（如广告）实现商业化操作。

云供应商负责公共云及其 IT 资源的创建和持续维护。后续章节中的一些场景和体系结构均会涉及公共云，以及公共云 IT 资源供应商和使用者之间的关系。

11.7.2　社区云

社区云与公共云类似，只是它的访问仅限于特定的云使用者社区。社区云可以由社区成员共同拥有，也可以由提供有限访问权限的公共云的第三方云供应商拥有。社区中的云使用者通常会共同分担社区云的定义和改进职责。

社区成员身份并不一定能够保证访问或控制云的所有 IT 资源。除非社区允许，否则社区以外的各方通常不被允许访问。

11.7.3　私有云

私有云由单个组织机构拥有。私有云使机构能够使用云计算技术，并对 IT 资源的访问汇集至机构的不同部分、位置或部门。当私有云作为受控环境存在时，第 3 章中描述的问题将不复存在。

使用私有云可以改变组织和信任边界的定义和应用。私有云环境的实际管理行为可能由内部或外包人员执行。

对于私有云，同一组织在技术上既是云使用者又是云供应商。为了区分这些角色，应注意以下几点内容：

- 一个独立的组织部门通常负责提供云（因此承担云供应商这一角色）。
- 需要访问私有云的部门承担云使用者这一角色。

注意：

组织内部环境中的云服务使用者通过虚拟专用网络访问位于同一组织的私有云上的云服务。

在私有云环境中，正确地使用术语“本地”和“云”是十分重要的。尽管私有云可能实际驻留在组织机构的空间中，但只要云使用者能够对其进行远程访问，它所承载的 IT 资源仍然被认为是基于云的。因此，相对于基于私有云的 IT 资源而言，托管在私有云之外的 IT 资源仍被认为是本地资源。

11.7.4　混合云

混合云是由两个或多个不同的云部署模型组成的云环境。例如，云使用者可以选择将处理敏感数据的云服务部署到私有云中，而将其他不太敏感的云服务部署到公共云中，这种组合结果称作混合部署模型。

图 11.2 显示了一个使用混合云架构的组织机构，其中同时使用了私有云和公共云。

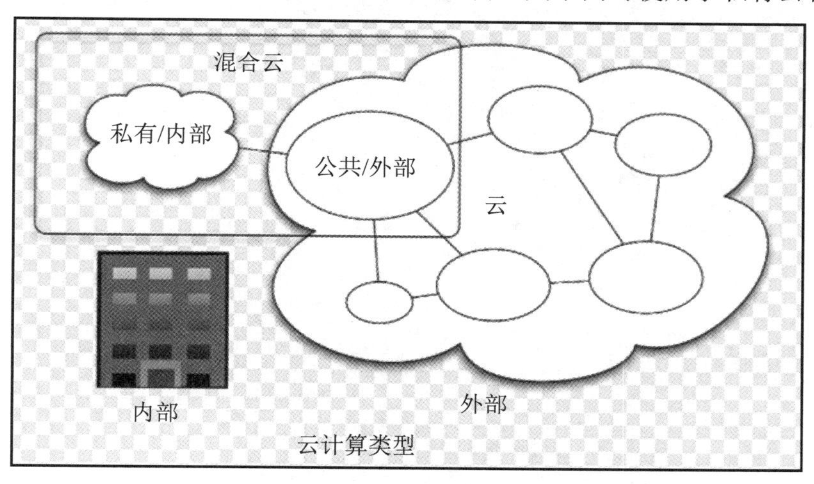

图 11.2

考虑到云环境中的潜在差异，以及管理职责通常在私有云供应商和公共云供应商之间进行划分，因此，混合部署体系结构的创建和维护可能非常复杂且颇具挑战性。

11.8　本章小结

本章讨论了云计算以及理解、实现该技术的关键技术。

第 12 章将介绍一家较为流行的云供应商，即 Amazon 的 AWS。

第 12 章　使用亚马逊 Web 服务

本章讨论 AWS 及其服务等概念。当在 AWS 云中设置 Hadoop 集群时，针对基于 Elastic MapReduce（EMR）的大数据分析过程，这些概念将会十分有用。除此之外，我们还将学习 AWS 提供的一些关键组件和服务，进而了解 AWS 组件和服务的各项功能。

本章主要涉及以下主题：

- Amazon Elastic Compute Cloud。
- 从 AMI 中启用多个实例。
- AWS Lambda。
- Amazon S3 简介。
- Amazon DynamoDB。
- Amazon Kinesis Data Stream。
- AWS Glue。
- Amazon EMR。

12.1　Amazon Elastic Compute Cloud

Amazon Elastic Compute Cloud（Amazon EC2）是一项 Web 服务，并可在云中提供安全、可调整大小的计算能力，其设计目的是让开发人员更容易地进行 Web 级别的云计算。

Amazon EC2 的简单 Web 服务接口使我们可轻松地获取和配置容量，并提供了对计算资源的完全控制，进而可使用 Amazon 的计算环境。Amazon EC2 将获取和启动新服务器实例所需的时间缩短为几分钟，并可在计算需求发生变化时快速扩展容量（上、下方向）。Amazon EC2 还可节省计算成本，且只需要为实际使用的容量付费。另外，Amazon EC2 为开发人员提供了构建具有故障恢复能力的应用程序工具，并将它们与常见的故障场景隔离开来。

12.1.1　弹性 Web 计算

Amazon EC2 可在几分钟内增加或减少容量，还可以同时委托一个或多个服务器实例。此外，还可使用 Amazon EC2 Auto Scaling 维护 EC2 Fleet 的可用性，并根据需要在上、下方向上自动扩展 Fleet，进而实现性能的最大化和成本的最小化。当对多项服务进行扩展时，还可使用 AWS Auto Scaling。

12.1.2　对操作的完整控制

我们可以对实例实施整体控制，包括根访问，进而像其他机器那样与其进行交互，并在启动分区上保留数据的同时终止任何实例的运行，然后使用 Web 服务 API 重新启动相同的实例。除此之外，还可以使用 Web 服务 API 远程重启实例，并访问它们的控制台输出结果。

12.1.3　灵活的云托管服务

用户可以选择多个实例类型、操作系统以及软件包。相应地，Amazon EC2 还可针对内存、CPU、实例存储和启动分区等配置进行选择，进而优化操作系统和应用程序选项。例如，在操作系统方面，可选取各种 Linux 和 Microsoft Windows Server 版本。

12.1.4　集成

Amazon EC2 集成了大多数 AWS 服务，例如 Amazon Simple Storage Service（Amazon S3）、Amazon 关系数据库服务（Amazon RDS）和 Amazon 虚拟私有云（Amazon VPC），进而为各种应用程序的计算、查询处理和云存储提供完整、安全的解决方案。

12.1.5　高可靠性

Amazon EC2 提供了一个高度可靠的环境，并可于其中快速、可预测地委托替换实例。这一项服务在亚马逊经过验证的网络基础设施和数据中心内运行。同时，Amazon EC2 服务级别协议（SLA）为每个 Amazon EC2 区域提供了 99.99%的可用性。

12.1.6　安全性

AWS 的云安全是重中之重。对于满足安全敏感组织的需求而构建的数据中心和网络体系结构，AWS 客户将受益良多。Amazon EC2 可与 Amazon VPC 一起协同工作，从而为计算资源提供安全和健壮的网络功能。

12.1.7　经济性

通过 Amazon EC2，用户可从 Amazon 规模效应中获取经济受益，同时仅需为实际消耗的计算能力支付很低的费用。关于 Amazon EC2 的更多实例购买选项，读者可访问 https://aws.amazon.com/ec2/pricing/以了解更多内容。

12.1.8　易于启动

存在多种方式可启动 Amazon EC2，具体包括 AWS Management Console、AWS 命令行工具（通过 CLI 获得，即命令行接口），或者使用 AWS SDK（软件开发工具包）。AWS 的启动和操作过程十分简单，读者可访问相关教程以了解更多内容，对应网址为 https://aws.amazon.com/getting-started/tutorials/。

12.1.9　亚马云及其镜像

亚马逊机器镜像（AMI）表示为一个模板，并包含了相应的软件配置内容（例如操作系统、应用程序服务器和应用程序）。我们可从 AMI 中启用一个实例，即作为云虚拟服务器运行的 AMI 副本。此外，还可启用多个 AMI 实例，如图 12.1 所示。

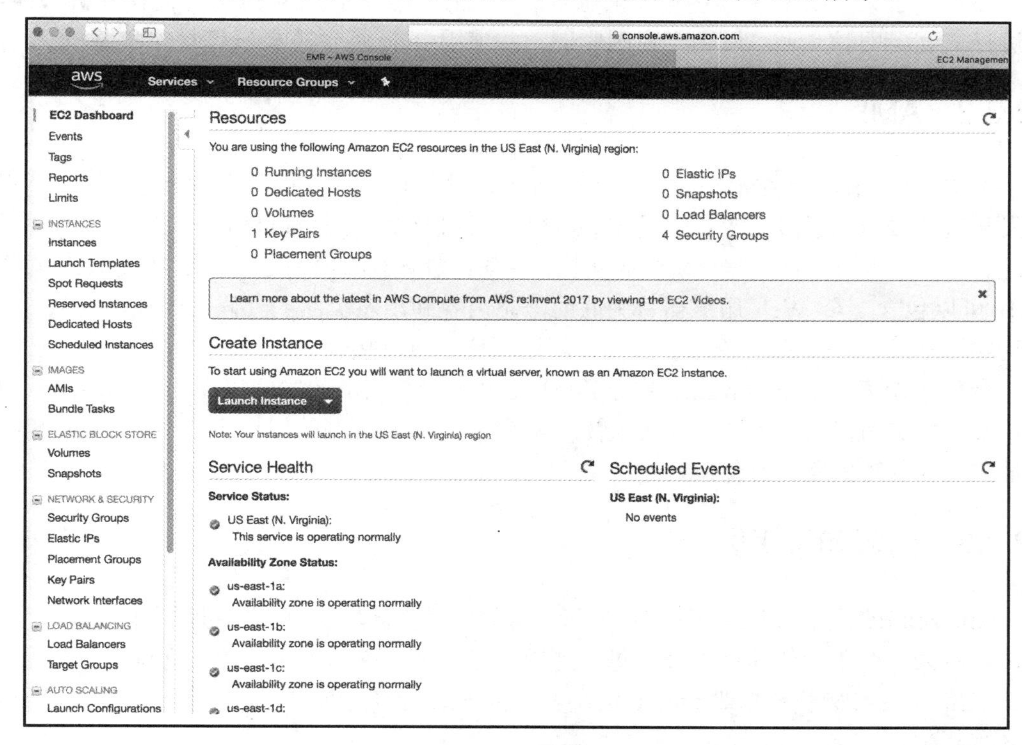

图 12.1

12.2　启用多个 AMI 实例

除非终止实例的运行状态，或者实例出现某种故障，否则，实例将一直处于运行状态。如果实例出现故障，则可从 AMI 中启动一个新的实例。

12.2.1　实例

我们可从单一 AMI 中启动不同的实例类型。本质上，实例类型决定了实例所用的主机硬件，每个实例类型提供了不同的计算机和内存功能。因此，可根据应用程序所需的内存量和计算能力，或者计划运行于实例上的软件选择相应的实例类型。针对每种 Amazon EC2，读者可访问 https://aws.amazon.com/ec2/instance-types/以了解更多硬件规范方面的信息。

在启用了某个实例后，它看起来像是一个传统的主机，并可像与任何计算机一样与其进行交互。同时，还可以完全控制实例，并使用 sudo 运行需要根权限的命令。

12.2.2　AMI

AWS 发布了多个 AMI，这些 AMI 包含了针对公共应用的通用软件配置。此外，AWS 开发社区的成员已经发布了自己的定制 AMI。除此之外，还可以创建自定义 AMI，并可快速、轻松地启动拥有所需一切的新实例。例如，如果应用程序是一个网站或 Web 服务，AMI 可以包含一个 Web 服务器、关联的静态内容和动态页面的代码。因此，在从 AMI 启动实例之后，Web 服务器即会启动，应用程序就可以接受请求了。

全部 AMI 基于 Amazon EBS 被分类，这表示从 AMI 启动的实例根设备是 Amazon EBS 卷；或者由实例存储所支持，这意味着，从 AMI 启动的实例的根设备是由 Amazon S3 中存储的模板创建的实例存储卷。

12.2.3　区域和可用区

Amazon EC2 在全球多个地点被托管，这些位置由区域和可用区组成。其中，每个区域都表示为一个单独的地理区域，每个区域都包含多个称为可用区的独立位置。Amazon EC2 提供了将实例和数据等资源放置在多个位置的能力。除非特别需要，否则资源不会跨区域复制。

亚马逊运营着最先进的、具有高可用性数据中心。注意，故障仍会出现，并对位于同一位置的实例的可用性带来影响，尽管这种情况较少发生。如果将所有实例均托管在一个受此故障影响的位置，则所有实例均处于不可用状态。

12.2.4　区域和可用区概念

每个区域完全独立；每个可用区处于隔离状态，但是某个区域中的可用区通过低延迟的链接进行连接。

图 12.2 显示了区域和可用区之间的关系。

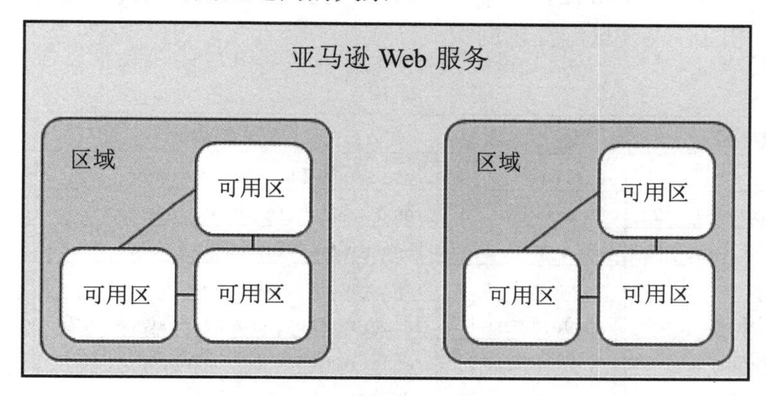

图 12.2

12.2.5　区域

每个 Amazon EC2 区域都被设计成与其他 Amazon EC2 区域完全隔离，这可以实现最大的容错性和稳定性。

当查看资源时，将只会看到与指定区域相关联的资源，其原因在于，区域之间是相互隔离的，因而不会自动地跨区域复制资源。

12.2.6　可用区

当启动一个实例时，可以选择一个可用区或分配一个可用区。如果将实例分布在多个可用区域中，且某个实例出现故障，则可以设计相应的应用程序，使另一个可用区中的实例能够处理请求。

此外，还可使用弹性 IP 地址，通过快速将地址重新映射到另一个可用区中的实例，来掩盖某个可用区中的实例故障。读者可访问 https://docs.aws.amazon.com/AWSEC2/

latest/UserGuide/elastic-ip-addresses-eip.html 以了解更多信息。

12.2.7 可用区域

用户的账号决定了可以使用的区域。例如，AWS 提供了多个区域，以便在满足相关位置条件的基础上启动 Amazon EC2 实例。又如，用户可能希望在欧洲启动实例，以便更加接近欧洲客户，抑或满足相关的法律条文。

AWS GovCloud（US）账号仅提供了 AWS GovCloud （US）区域的访问权限。对此，读者可参考 AWS GovCloud (US) Region 以查阅更多信息。

表 12.1 列出了 AWS 账号提供的相关区域。

表 12.1

区 域 名	区 域	端 点	协 议
US East (Ohio)	us-east-2	rds.us-east-2.amazonaws.com	HTTPS
US East (N. Virginia)	us-east-1	rds.us-east-1.amazonaws.com	HTTPS
US West (N. California)	us-west-1	rds.us-west-1.amazonaws.com	HTTPS
US West (Oregon)	us-west-2	rds.us-west-2.amazonaws.com	HTTPS
Asia Pacific (Tokyo)	ap-northeast-1	rds.ap-northeast-1.amazonaws.com	HTTPS
Asia Pacific (Seoul)	ap-northeast-2	rds.ap-northeast-2.amazonaws.com	HTTPS
Asia Pacific (Osaka-Local)	ap-northeast-3	rds.ap-northeast-3.amazonaws.com	HTTPS
Asia Pacific (Mumbai)	ap-south-1	rds.ap-south-1.amazonaws.com	HTTPS
Asia Pacific (Singapore)	ap-southeast-1	rds.ap-southeast-1.amazonaws.com	HTTPS
Asia Pacific (Sydney)	ap-southeast-2	rds.ap-southeast-2.amazonaws.com	HTTPS
Canada (Central)	ca-central-1	rds.ca-central-1.amazonaws.com	HTTPS
China (Beijing)	cn-north-1	rds.cn-north-1.amazonaws.com.cn	HTTPS
China (Ningxia)	cn-northwest-1	rds.cn-northwest-1.amazonaws.com.cn	HTTPS
EU (Frankfurt)	eu-central-1	rds.eu-central-1.amazonaws.com	HTTPS
EU (Ireland)	eu-west-1	rds.eu-west-1.amazonaws.com	HTTPS
EU (London)	eu-west-2	rds.eu-west-2.amazonaws.com	HTTPS
EU (Paris)	eu-west-3	rds.eu-west-3.amazonaws.com	HTTPS
South America (SaoPaulo)	sa-east-1	rds.sa-east-1.amazonaws.com	HTTPS

12.2.8 区域和端点

当与使用命令行接口或 API 操作的实例协同工作时，需要指定其区域端点。关于

Amazon EC2 的区域和端点，读者可访问 https://docs.aws.amazon.com/general/latest/gr/rande.html 以了解更多信息。

12.2.9　实例类型

当启动一个实例时，所指定的实例类型确定了用于该实例的主机硬件。每个实例类型提供了不同的计算机、内存和存储功能，并根据实际功能被分组至实例族中。相应地，可根据计划运行于实例上的应用程序或软件的需求条件选取某个实例类型。

1．标签

标签可通过不同方式对 AWS 资源进行分类，例如通过用途、持有者或环境。当拥有同一类型的大量资源时，这将十分有用——可根据所分配的标签快速识别特定的资源。每个标签由一个键和可选的数值构成，且需要对此加以定义。例如，可针对账号的 Amazon EC2 实例定义一组标签，进而跟踪每个实例的持有者和栈深度。

2．Amazon EC2 密钥对

Amazon EC2 使用公钥加密对登录信息进行加密和解密。公钥密码技术使用公钥加密数据片段（如密码），然后接收方使用私钥解密数据。公钥和私钥称作密钥对。

3．针对 Linux 实例的 Amazon EC2 安全组

安全组可充当虚拟防火墙，并控制一个或多个实例的流量。在启动实例时，需要将一个或多个安全组与实例相关联。用户可向支持流量的安全组（或者从所关联的实例）中添加各项规则，并可在任意时刻修改安全组的规则。一段时间后，新的规则将自动应用于与安全组关联的所有实例上。当确定是否允许流量到达某个实例时，可对实例关联的所有安全组中的全部规则进行评估。

4．弹性 IP 地址

弹性 IP 地址是一种针对动态云计算而设计的静态 IPv4 地址，并与 AWS 账号进行关联。利用弹性 IP 地址，可通过迅速地将该地址重新映射至另一个账号中的实例，进而掩盖实例故障。

12.2.10　Amazon EC2 和亚马逊虚拟私有云

Amazon VPC 可在 AWS 云中的逻辑隔离区域（称为 VPC）中定义虚拟网络。具体来说，可在 VPC 中启用 AWS 资源（例如实例）。这里，VPC 与传统的网络非常类似，用户可以在自己的数据中心中进行操作，这得益于使用 AWS 的可伸缩基础设施。用户可以

配置 VPC、选择其 IP 地址范围、创建子网，并配置路由表、网络网关和安全设置。另外，还可将 VPC 中的实例连接到互联网，或者将 VPC 连接到公司数据中心，以使 AWS 云成为数据中心的扩展。为了保护每个子网中的资源，此处可以使用多个安全层，包括安全组和网络访问控制列表。对此，读者可访问 https://aws.amazon.com/documentation/vpc/以了解更多信息。

1. Amazon Elastic Block Store

Amazon Elastic Block Store（Amazon EBS）提供与 EC2 实例协同使用的块级存储卷。EBS 卷是高度可用和可靠的存储卷，可以连接到同一可用区中的任何运行实例。连接到 EC2 实例的 EBS 卷作为存储卷被公开，且独立于实例的生命周期而存在。在 Amazon EBS 中，仅需为所用内容支付费用。关于 Amazon EBS 价格机制，读者可访问 https://aws.amazon.com/ebs/以了解更多内容（参见 Amazon EBS 页面中的 Projecting costs 部分）。

当数据需要快速访问且长期持久存储时，建议使用 Amazon EBS。EBS 适用于以下应用程序的主存储：作为文件系统、数据库，或者需要细粒度更新和访问原始、未格式化的块级存储。Amazon EBS 非常适合于依赖于随机读写的数据库应用程序，以及通过密集型应用程序执行长期、连续的读写操作。

2. Amazon EC2 实例存储

实例存储可为实例提供临时块级存储。该存储位于物理连接至主机的磁盘上。实例存储适用于数据的临时存储，例如频繁更改的信息（如缓冲区、缓存、暂存数据和其他临时内容），或实例（如负载平衡的 Web 服务器池）间的复制数据。

实例存储由作为块设备而公开的一个或多个实例存储卷组成。实例存储的大小以及可用设备的数量因实例类型而异。虽然实例存储用于特定的实例，但是磁盘子系统可在主机上的实例之间实现共享。

12.3　AWS Lambda

AWS Lambda 是一种计算服务，并可在不提供服务器或缺少服务器管理机制的情况下运行代码。仅在需要时，AWS Lambda 才会执行代码并且会自动扩展——从每天几个请求扩展到每秒数千个请求。用户只需要支付所消耗的计算时间即可。当代码未处于运行状态时则是不收费的。当使用 AWS Lambda 时，几乎可以为任何类型的应用程序或后端服务运行代码，且无须对此进行管理。AWS Lambda 在高可用性计算基础设施上运行代码，并执行所有计算资源的管理，包括服务器和操作系统维护、容量供应、自动伸缩、代码监视和日志记录。用户的工作则是采用 AWS Lambda 支持的语言之一编写代码（例如 Node.js、Java、C#、Go 和 Python）。

我们可以使用 AWS Lambda 运行代码以响应事件，例如 Amazon S3 存储桶或 Amazon DynamoDB 表中的数据更改；使用 Amazon API 网关运行响应 HTTP 请求的代码；或者使用 AWS SDK 和 API 调用代码。根据这些功能，我们可使用 Lambda 针对 AWS 服务（例如 Amazon S3 和 Amazon DynamoDB）轻松地构建数据处理触发器，进而处理 Kinesis 中的数据流；或者创建 AWS 级别的后端操作程序，以围绕当前系统提供较好的性能和安全措施。

此外，还可以构建由事件触发的函数组成的无服务器应用程序，并使用 AWS CodePipeline 和 AWS CodeBuild 对其进行自动部署。

此外，还可以构建由事件触发的函数组成的无服务器应用程序，并使用 AWS CodePipeline 和 AWS CodeBuild 自动部署它们。对此，读者可访问 https://docs.aws.amazon.com/lambda/latest/dg/deploying-lambda-apps.html 以了解更多内容。

AWS Lambda 是许多应用程序方案的理想计算平台，前提是可以使用 AWS Lambda 所支持的语言编写应用程序代码，例如 Node.js、Java、Go、C#和 Python，并可在 AWS Lambda 标准运行期环境和 Lambda 提供的资源中运行。

12.4　Amazon S3 简介

Amazon S3 运行于世界上最大的全球云基础设施上，其耐久性高达 99.999999999%。其中，数据自动分布在至少 3 个物理设施之间，这些物理设施在地理上散布于 AWS 区域内，Amazon S3 还可以自动将数据复制到任何其他 AWS 区域。

关于 AWS 全球云基础设施，读者可访问 https://aws.amazon.com/以了解更多内容。

12.4.1　Amazon S3 功能

Amazon S3 可视为一类互联网存储。用户可以使用 Amazon S3 在任何时候从 Web 上的任何地方存储和检索任意数量的数据。此外，还可以使用 AWS Management Console 来完成这些任务，这是一个简单、直观的 Web 界面。本节主要介绍 Amazon S3 以及如何使用 AWS Management Console 管理 Amazon S3 提供的存储空间。

时至今日，公司应具备能够轻松安全地收集、存储和大规模分析数据的能力。Amazon S3 是一种对象存储，用于存储和检索来自任何地方的数据——网站和移动应用程序、企业应用程序，以及来自物联网传感器或设备的数据，并存储各个行业使用的、数以百万计的应用程序数据。Amazon S3 提供了全面的安全性和协从能力，甚至可以满足最严格的法规要求。它使客户能够灵活地管理数据，以实现成本优化、访问控制和协从性等功能。同时，Amazon S3 还提供了就地查询功能，并可直接在 Amazon S3 中对数据进行强大的分析。Amazon S3 是目前受到广泛支持的存储平台，拥有 ISV 解决方案和系统集成商合作伙伴的最大生态系统。

12.4.2 全面的安全和协从能力

Amazon S3 是唯一支持 3 种不同加密形式的云存储平台，同时 S3 提供了与 AWS CloudTrail 之间的集成，并针对审计机制记录、监视和保留存储 API 的调用活动。Amazon S3 是 Amazon Macie 的唯一云存储平台，并通过机器学习自动发现、分类和保护 AWS 中的敏感数据。Amazon S3 支持安全标准和合格认证，包括 PCI-DSS、HIPAA/HITECH、FedRAMP、EU Data Protection Directive 和 FISMA，以帮助全球监管机构符合协从要求。

关于安全性和协从性问题，读者可访问 https://aws.amazon.com/security/和 https://aws.amazon.com/compliance/以了解更多信息。

12.4.3 就地查询

Amazon S3 可在数据上运行复杂的大数据分析，而无须将数据移动到单独的分析系统中。对于大量非结构化数据，Amazon Athena 向 SQL 人员提供了按需查询的访问能力。Amazon Redshift Spectrum 可运行跨越数据仓库和 S3 的查询操作。目前，仅 AWS 提供了 Amazon S3 Select 功能（当前仅用于测试预览），这是一种仅从 S3 对象中检索所需数据子集的方法，进而提升从 S3 访问数据的应用程序的性能，最高可达 400%。

关于就地查询，读者可访问 https://aws.amazon.com/blogs/aws/amazonredshift-spectrum-exabyte-scale-in-place-queries-of-s3-data/以了解更多内容。

12.4.4 灵活的管理机制

Amazon S3 提供了最灵活的存储管理和管理功能。存储管理员可以对数据的应用趋势执行分类、报告和可视化操作，以降低成本并提高服务水平。对象可以用唯一的、可定制的元数据进行标记，以便客户可以分别查看和控制每项工作负载的存储消耗、成本和安全性等特征。针对维护、协从性和分析操作方面的问题，S3 可提供与对象和元数据相关的进度报告。除此之外，S3 还可以分析对象访问模式，以构建与自动化分层、删除和保留相关的生命周期策略。由于 Amazon S3 使用了 AWS Lambda，因而客户可以记录活动、定义警报和调用工作流，而无须任何额外基础设施。

关于 S3 存储的管理机制，读者可访问 https://aws.amazon.com/s3/以了解更多内容。

12.4.5 最受支持的平台以及最大的生态系统

除了与大多数 AWS 服务集成之外，Amazon S3 生态系统还包括一些咨询系统集成商和独立软件供应商合作伙伴，且每月都有更多的机构在不断地加入其中。AWS Marketplace 提供了 35 个类别和 3500 多个软件清单，分别来自 1100 多个 ISV，这些软件都是预先配置完毕并部署在 AWS 云上的。AWS Partner Network（APN）合作伙伴已经调整了他们的服务和软件策略，并朝向与 S3 协同工作的方向迈进，进而实现备份和恢复、归档和灾难恢复等解决方案。

关于 AWS 存储合作商，读者可访问 https://aws.amazon.com/backup-recovery/partner-solutions/以了解更多信息。

12.4.6 简单、方便的数据传输机制

用户可以从多种选项中选择将数据传输到 Amazon S3 中（或反向）。S3 简单可靠的 API 使得互联网数据传输变得很容易。对于较大地理范围的数据上传行为，Amazon S3 Transfer Acceleration 可视为一种较为理想的解决方案。对于采用专用的网络连接时 AWS 间大量的数据移动行为，AWS 直连供应商提供了一致的高带宽、低延迟数据传输方案。对于 Pb 级的数据传输，可使用 AWS Snowball 和 AWS Snowball Edge 应用程序；对于更大的数据集，则可尝试使用 AWS Snowmobile。AWS Storage Gateway 通过物理或虚拟数据传输设备，并采用本地方式轻松地将数据卷或文件移动至 AWS 云中。

关于云计算迁移，读者可访问 https://aws.amazon.com/cloud-migration/以了解更多内容。

12.4.7 备份和恢复

Amazon S3 为关键数据的备份和归档提供了持久、可扩展和安全的场所。用户可以使用 S3 的版本控制功能来保护存储的数据。另外，还可定义生命周期规则，并将较少使用的数据迁移至 S3 Standard-Infrequent Access 中，同时将对象集归档至 Amazon Glacier 中。

关于备份和归档，读者可访问 https://aws.amazon.com/backup-restore/以了解更多内容。

12.4.8 数据存档

Amazon S3 和 Amazon Glacier 提供了一系列的存储类，以满足受监管行业遵从性方面的存档需求；或者为需要快速、不频繁访问存档数据的组织提供活动存档。Amazon

Glacier Vault Lock 提供了“一次写入多次读取（WORM）”的存储方案，以满足记录的保存需求。生命周期策略简化了从 Amazon S3 到 Amazon Glacier 间的数据转换，并帮助解决了基于客户定义策略的转换自动化问题。

关于数据存档，读者可访问 https://aws.amazon.com/archive/以了解更多内容。

12.4.9 数据湖和数据分析

无论存储的是药品还是金融数据，抑或是照片和视频等多媒体文件，Amazon S3 都可以用作大数据分析的数据湖。AWS 提供了全面的服务组合方案，并通过降低成本、扩大规模、提升创新速度等方式管理的数据。

关于数据湖和数据分析，读者可访问 https://aws.amazon.com/blogs/big-data/introducing-the-data-lake-solution-on-aws/以了解更多内容。

12.4.10 混合云存储

AWS 存储网关可帮助用户构建混合云存储，并通过 Amazon S3 的持久性和规模性，增强现有的本地存储环境。据此，可将一个工作负载从站点释放到云中进行处理，然后返回最终结果。另外，将主存储之外的较低价值的数据分层置入云计算中，可降低成本并扩大本地投资。或者，也可简单地将数据增量地移至 S3 中，作为备份或迁移项目的一部分内容。

关于混合云存储，读者可访问 https://aws.amazon.com/enterprise/hybrid/以了解更多内容。

12.4.11 原生云应用程序数据

Amazon S3 提供了高性能、高可用性的存储方案，使其易于扩展和维护具有成本效益的、运行较快的移动和物联网应用程序。当使用 S3 时，可以添加任何数量的内容，并从任何位置对其进行访问，这样就可以更快地部署应用程序并接触到更多客户。

12.4.12 灾难恢复

Amazon S3 的全球安全基础设施提供了健壮的灾难恢复解决方案，旨在提供更好的数据保护机制。跨区域复制（CRR）将每个 S3 对象自动复制到位于不同 AWS 区域的目标桶中。

关于灾难恢复，读者可访问 https://aws.amazon.com/disaster-recovery/以了解更多内容。

12.5　Amazon DynamoDB

Amazon DynamoDB 是一个经全面管理的 NoSQL 数据库服务，提供了快速、可预测的性能和无缝的可伸缩性。DynamoDB 减轻了操作和扩展分布式数据库的管理负担，这样，用户就不必担心硬件供应、设置和配置、复制、软件补丁或集群扩展等方面的问题。此外，DynamoDB 还提供了静态加密，从而消除了保护敏感数据所涉及的操作负担和复杂性。读者可访问 https://docs.aws.amazon.com/amazondynamodb/latest/developerguide/EncryptionAtRest. html 以了解更多内容。

当使用 DynamoDB 时，可以创建存储、检索任意数量数据的数据库表，并服务于任何级别的请求流量。用户可调整表的吞吐量，且不会导致停机或性能下降等问题。另外，还可使用 AWS Management Console 来监视资源利用率和性能指标。

Amazon DynamoDB 提供了按需备份的功能，并为表创建完整的备份，以便长期保存和存档，进而满足法规遵从性需求。

DynamoDB 可从表中自动删除过期项，以减少存储应用以及数据的存储成本。

DynamoDB 自动将表数据和流量分散到足够多的服务器上，以处理吞吐量和存储需求，同时保持一致和快速的性能。其间，所有数据都存储在固态磁盘（ssd）中，并在 AWS 区域的多个可用区中自动复制，同时提供内置的高可用性和数据持久性。最后，还可使用全局表以保持 DynamoDB 表在 AWS 区域之间的同步状态。

12.6　Amazon Kinesis Data Streams

我们可以使用 Amazon Kinesis Data Streams 并以实时方式收集和处理较大的数据流。对此，将创建数据处理应用程序称作 Amazon Kinesis Data Streams 应用程序。典型的 Amazon Kinesis Data Streams 将作为数据记录从 Kinesis 数据流中读取数据。这一类应用程序可使用 Kinesis Client Library 并可运行于 Amazon EC2 实例上。处理后的记录将被发送至显示输出中，用于生成警告信息，或者动态地调整价格和广告策略；抑或将数据发送至其他 AWS 服务中。

Kinesis Data Streams 连同 Amazon Kinesis Data Firehose 一起表示为 Kinesis 流数据平台中的部分内容。

用户可以使用 Kinesis Data Streams 快速、连续地接收和聚合数据。其间，使用的数据类型包括 IT 基础设施日志数据、应用程序日志、社交媒体、市场数据提要和 Web 点击流数据。由于数据接收和处理的响应时间是实时的，因此处理过程通常是轻量级的。

下面考查 Kinesis Data Streams 典型的应用场景。

12.6.1　加速日志和数据提要的输入和处理

生产者可以将数据直接推送到数据流中。例如，推送系统和应用程序日志可以在几秒钟内进行处理。这可以防止前端或应用服务器出现故障时日志数据丢失。Kinesis 数据流提供了加速的数据输入，因为在提交数据之前，一般不会在服务器上批量处理数据。

12.6.2　实时度量和报告机制

对于简单的数据分析和实时报告，可将所采集的数据应用于 Kinesis Data Streams 中。例如，数据处理应用程序可以在数据流进入时处理系统和应用程序日志的度量和报告，而不是等待接收批量数据。

12.6.3　实时数据分析

实时数据分析将并行处理和实时数据结合在一起。例如，可通过实时方式处理网站的点击流，随后利用多个并行运行的 Kinesis Data Streams 应用程序分析该网站的可用性。

12.6.4　复杂的数据流处理

对此，可创建 Amazon Kinesis Data Streams 应用程序和数据流的有限无环图（DAG）。这通常会将来自多个 Amazon Kinesis Data Streams 应用程序的数据置入另一个流中，以便由不同的 Amazon Kinesis Data Streams 应用程序进行下游处理。

12.6.5　Kinesis Data Streams 的优点

当使用 Kinesis Data Streams 解决各类流数据问题时，一种常见的应用是数据的实时聚合，随后将该数据加载至数据仓库或 MapReduce 集群中。

考虑到持久性和弹性等问题，数据将被置于 Kinesis 数据流中。对于记录置于数据流及其检索（即置入-获取）时间，二者间相差小于 1 秒；在数据被加入后，Amazon Kinesis Data Streams 应用程序几乎可即刻使用数据流中的数据。Kinesis Data Streams 在管理服务方面可减轻创建、运行数据接收管道过程中的一些操作负担。相应地，用户可创建流式 MapReduce 类型的应用程序，而 Kinesis Data Streams 的弹性特征也使得我们可以在上、下两个方向上“缩放”数据流。据此，数据记录在过期之前将不会再丢失。

多个 Amazon Kinesis Data Streams 应用程序可以使用来自流的数据。因此，多个操作（如归档和处理）可以并发且独立地进行。例如，两个应用程序可以从同一个数据流读取数据。其中，第一个应用程序计算处于运行状态的聚合操作，并更新 DynamoDB 表；第二个应用程序将数据压缩和归档至数据存储，例如 Amazon S3。随后，针对最新的报告内容，包含处于运行状态下的聚合结果的 DynamoDB 表将被显示输出所读取。

12.7　AWS Glue

AWS Glue 是一个完全托管的析取、转换和加载（ETL）服务，并采用相对简单、经济的方式对数据进行分类、清洗和充实，进而可在各种数据存储之间安全地移动数据。AWS Glue 由一个名为 AWS Glue Data Catalog 的中央数据存储库、一个自动生成 Python 代码的 ETL 引擎，以及一个灵活的调度器（用于处理依赖项解析、作业监视和故障时作业的重试）组成。AWS Glue 是无服务器的，所以不需要设置或管理基础架构。

用户可以使用 AWS Glue 控制台获取数据、转换数据，并使其可用于搜索和查询操作。控制台调用底层服务来编排转换数据所需的各项工作。除此之外，还可以使用 AWS Glue API 操作与 AWS Glue 服务接口，并利用熟悉的开发环境编辑、调试和测试 Python 或 Scala Apache Spark ETL 代码。

我们可以使用 AWS Glue 构建数据仓库，进而组织、清理、验证和格式化数据。同时，还可将 AWS 云数据转换并移至数据存储中。另外，还可将源自不同数据源的数据加载到数据仓库中，以便进行定期报告和分析。通过将数据存储在数据仓库中，可以集成来自业务不同部分的信息，并提供用于决策的公共数据源。

当构建数据仓库时，AWS Glue 可简化诸多操作任务，具体如下：

- 获取并将与数据存储相关的元数据编目到中心目录中。
- 可以处理半结构化数据，例如点击流或处理日志。
- 使用调度爬虫程序的表定义填充 AWS Glue Data Catalog。
- 爬虫程序调用分类器逻辑推断数据的模式、格式和数据类型。这些元数据存储为 AWS Glue Data Catalog 中的表，并在 ETL 作业的编写过程中使用。
- 生成 ETL 脚本，并在源和目标之间转换、扁平化、丰富数据内容。
- 检测模式变化，并根据首选项进行调整。
- 根据调度或事件触发 ETL 作业，并可自动初始化作业，将数据转移到数据仓库中。触发器可用于在作业之间创建依赖流。
- 收集运行期度量结果以监视数据仓库的活动。
- 处理错误并自动执行重试操作。

- 根据需要扩展资源并运行各项作业。

在对 Amazon S3 数据湖执行无服务器查询时，可以使用 AWS Glue。AWS Glue 可以对 Amazon S3 数据进行编目，以便利用 Amazon Athena 和 Amazon Redshift Spectrum 进行查询。通过爬虫，元数据与底层数据保持同步。通过 AWS Glue Data Catalog、Athena 和 Redshift Spectrum 可直接查询 Amazon S3 数据湖。当采用 AWS Glue 时，可通过一个统一的接口访问和分析数据，而无须将其加载至多个数据存储中。

利用 AWS Glue，可创建事件驱动的 ETL 管道。在 Amazon S3 中，只要有新的数据可用，即可从 AWS Lambda 函数中调用 AWS Glue ETL 作业来运行 ETL 作业。此外，还可将这个新数据集注册到 AWS Glue Data Catalog 中，并作为 ETL 作业的一部分内容。

用户可使用 AWS Glue 进一步理解数据集。除此之外，还可以使用各种 AWS 服务存储数据，同时仍然使用 AWS Glue Data Catalog 维护数据的统一视图。查看 Data Catalog 可快速搜索和发现所持有的数据集，并在一个中央存储库中维护相关的元数据。另外，Data Catalog 还可以作为外部 Apache Hive Metastore 的替代方案。

12.8　Amazon EMR

Amazon EMR 是一个托管集群平台，它简化了 AWS 上的 Apache Hadoop 和 Apache Spark 等大数据框架的运行过程，进而对大数据进行处理和分析。通过使用这些框架和相关的开源项目（例如 Apache Hive 和 Apache Pig），可针对分析功能和业务智能工作负载处理数据。此外，还可以使用 Amazon EMR 在其他 AWS 数据存储和数据库（如 Amazon S3 和 Amazon DynamoDB）间转换和移动大量数据。

Amazon EMR 提供了一个托管的 Hadoop 框架，该框架简单、快速且具有成本效益，以便在动态可伸缩的 Amazon EC2 实例间处理大量的数据。此外，还可以在 Amazon EMR 中运行其他流行的分布式框架，如 Apache Spark、HBase、Presto 和 Flink，并与 Amazon S3 和 Amazon DynamoDB 等其他 AWS 数据存储中的数据进行交互。

Amazon EMR 可安全、可靠地处理大量的大数据用例，包括日志分析、Web 索引、数据转换（ETL）、机器学习、财务分析、科学模拟和生物信息学。

当读者尝试对此进行操作时，需要访问 aws.amazon.com 并创建一个 AWS 账号。

注意：

创建和使用 EMR 集群需要支付一定的费用（通常是每天 10 美元）。当任务结束时，应终止集群的工作状态。

登录后，将会看到如图 12.3 所示的画面。

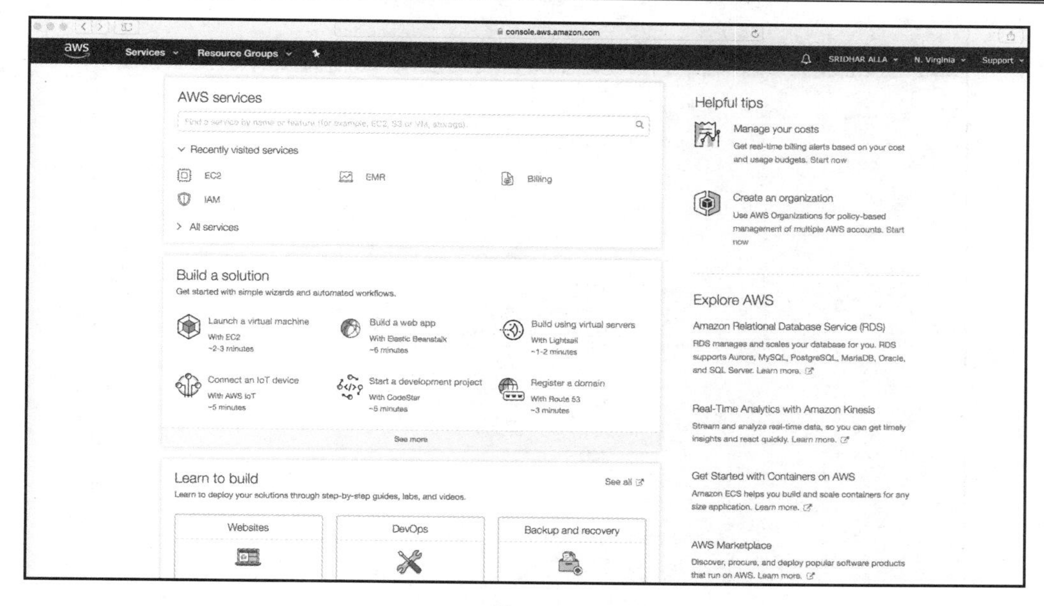

图 12.3

当选择 EMR 作为服务时，对应结果如图 12.4 所示。

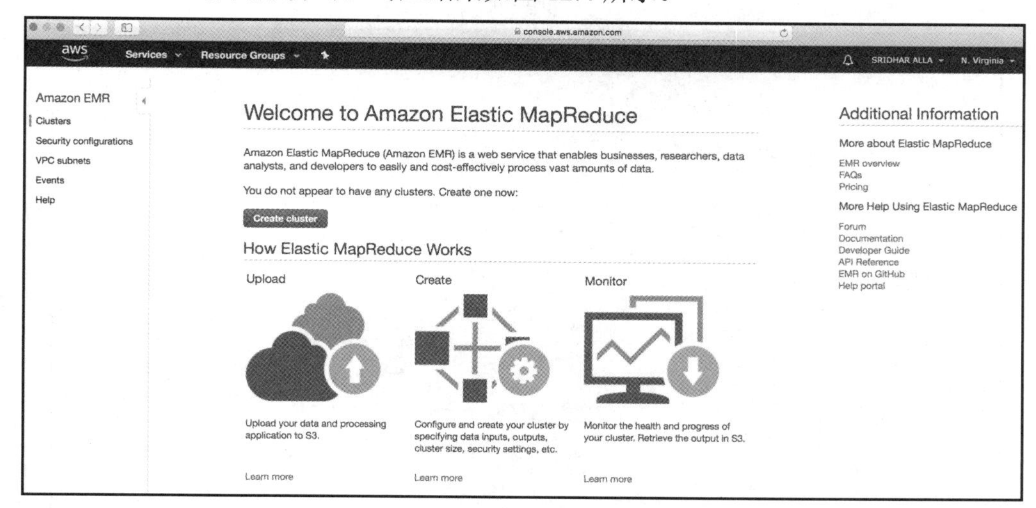

图 12.4

通过如图 12.5 所示的各种选项，即可创建一个 EMR 集群。

一组键对对于 EMR 来说不可或缺。对此，可打开一个新的选项卡，并访问 AWS 控制台中的 EC2 服务，如图 12.6 所示。

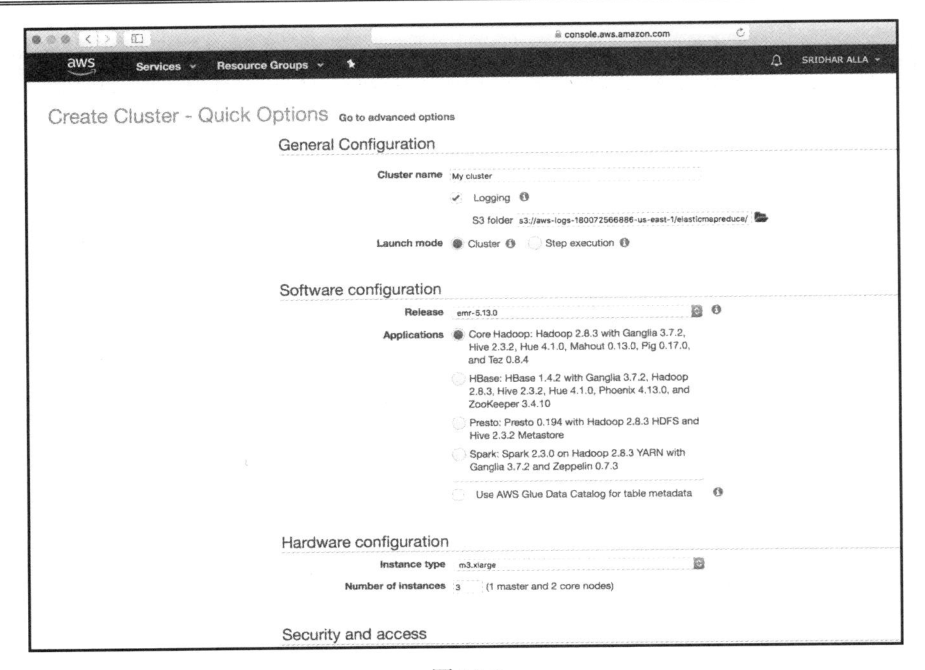

图 12.5

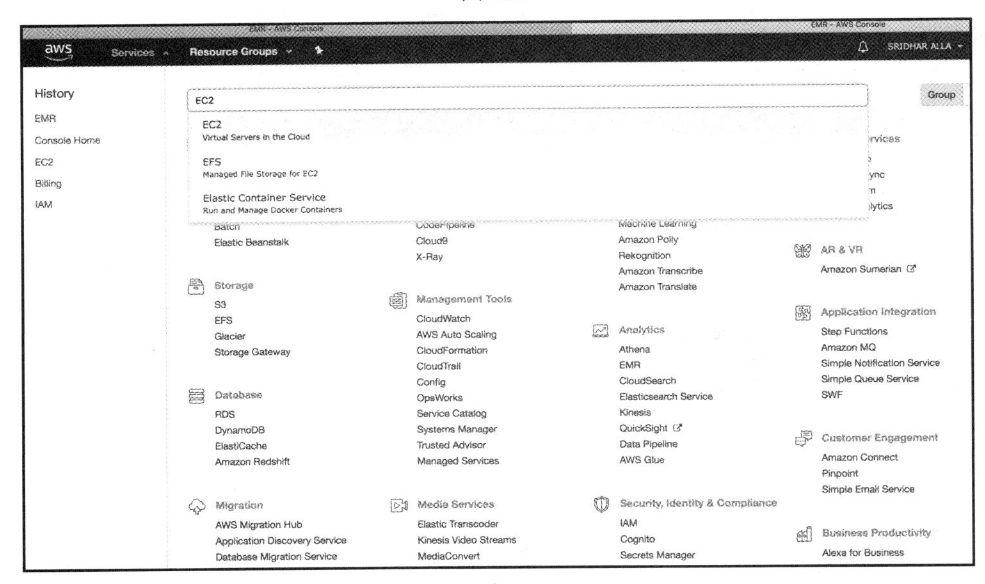

图 12.6

图 12.7 显示了当前的 EC2。

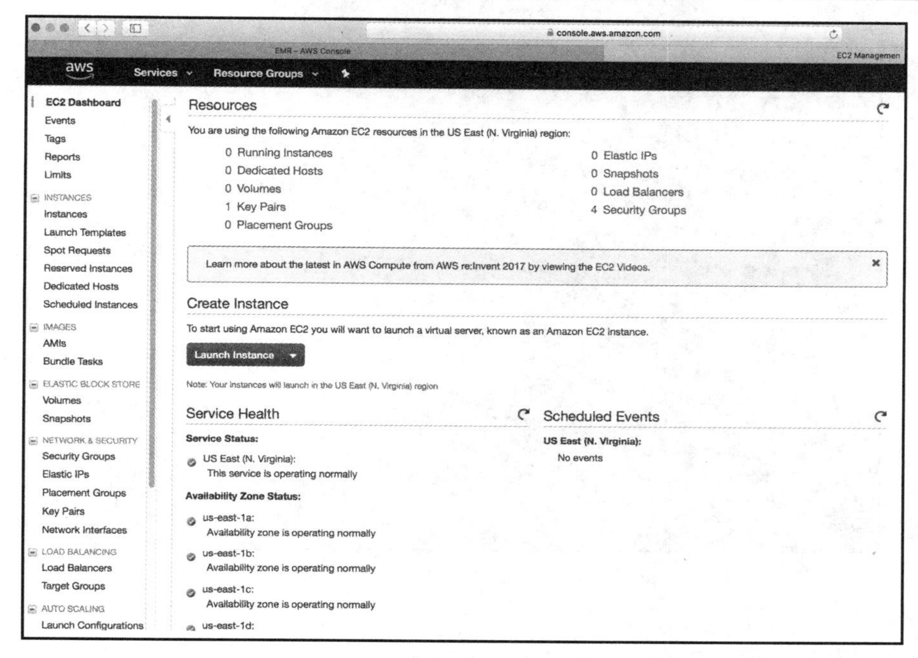

图 12.7

通过选取左侧面板中的 Key Pairs，即可在 EC2 中生成新的键对，如图 12.8 所示。

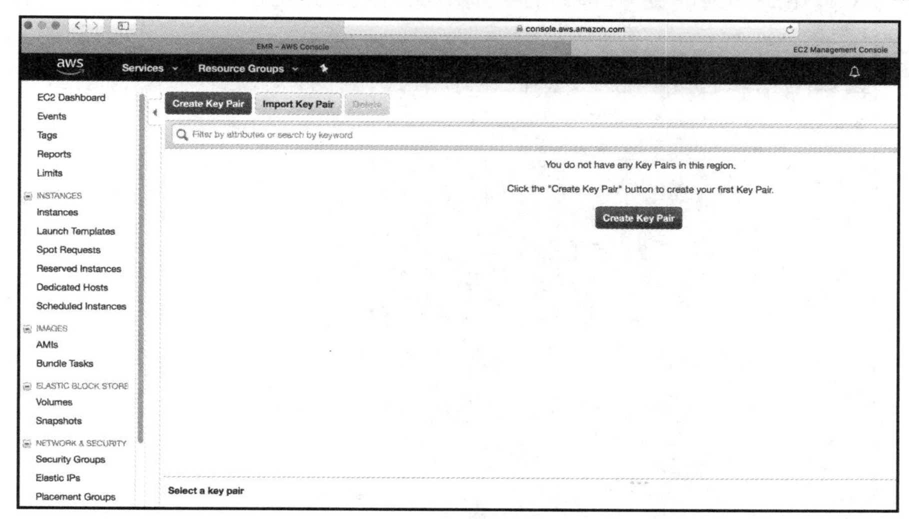

图 12.8

键对的命名方式如图 12.9 所示。

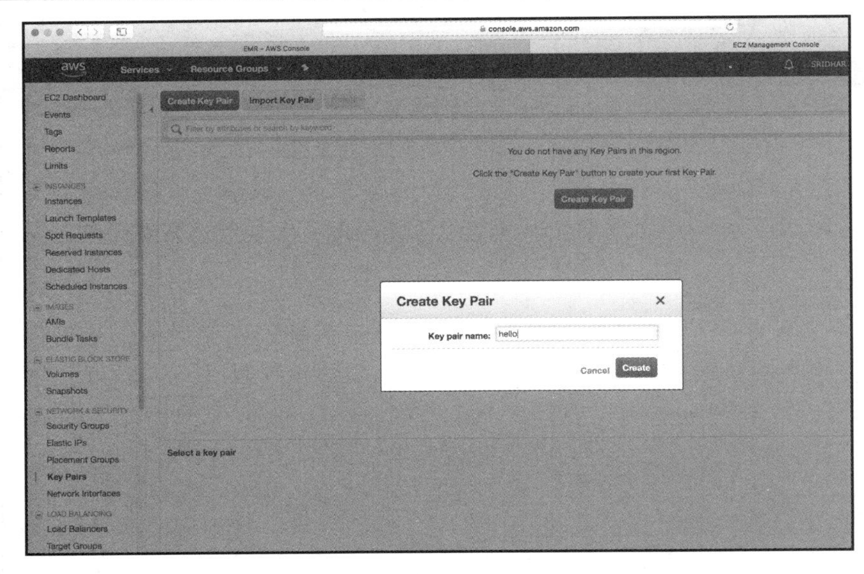

图 12.9

注意：

确保复制当前键对，后续过程中将无法再次执行这一操作。

图 12.10 显示了当前的键对，可对其加以保存以供后续操作使用。

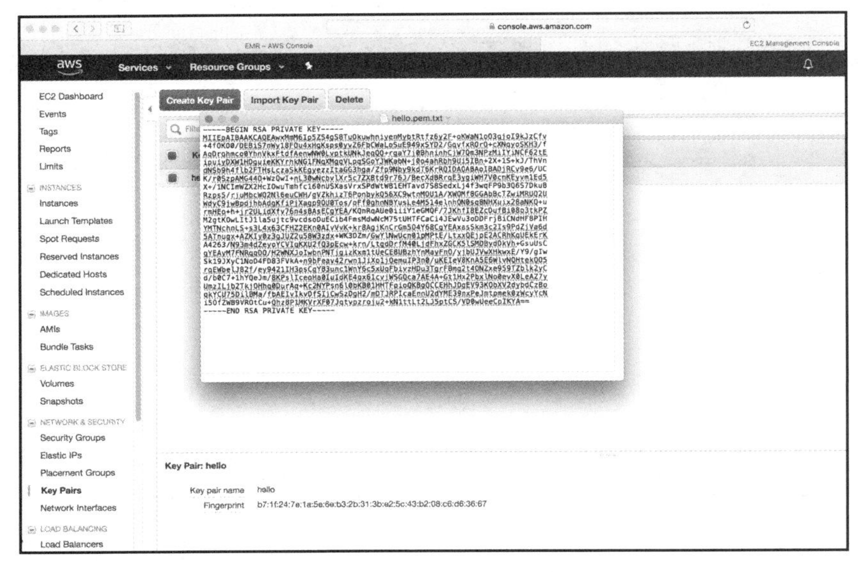

图 12.10

接下来，可利用刚生成的键对执行进一步操作，如图 12.11 所示。

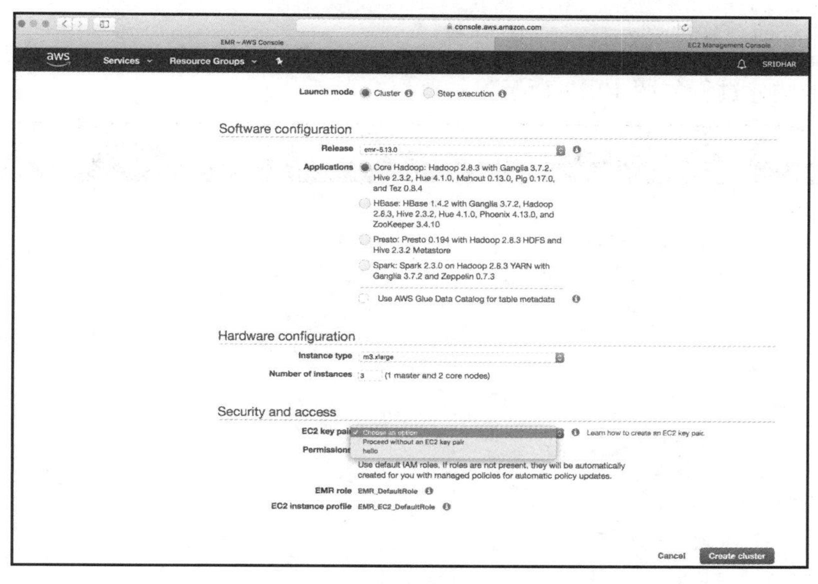

图 12.11

一旦选择了键对后，即可创建一个集群，如图 12.12 所示。

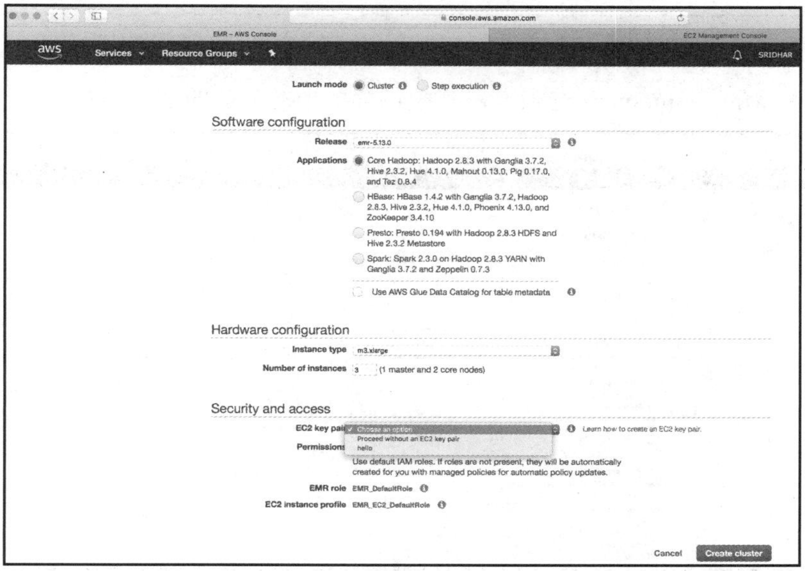

图 12.12

注意：

EMR 集群的创建过程大约需要 10 分钟。

EMR 的构建过程如图 12.13 所示。

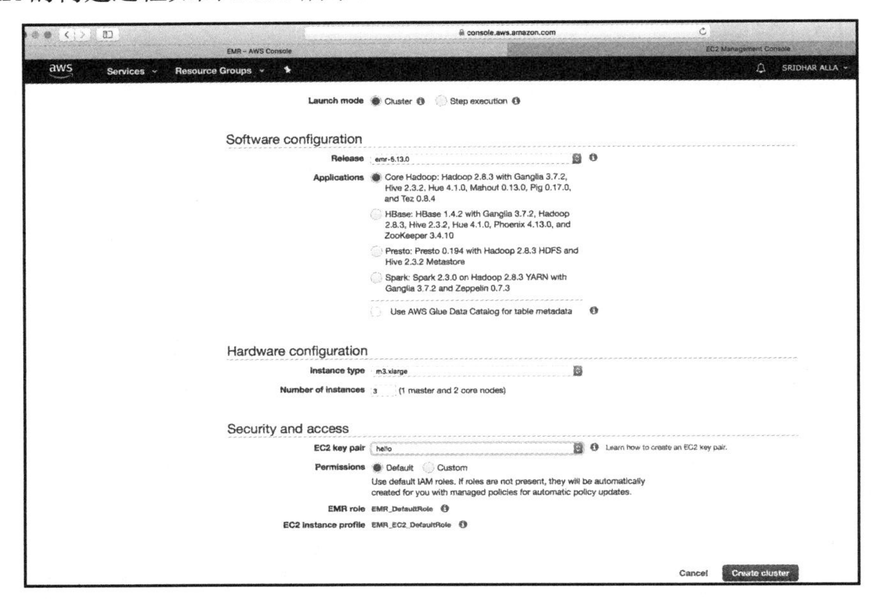

图 12.13

图 12.14 显示了 Summary 选项卡，其中包含了当前集群的细节信息。

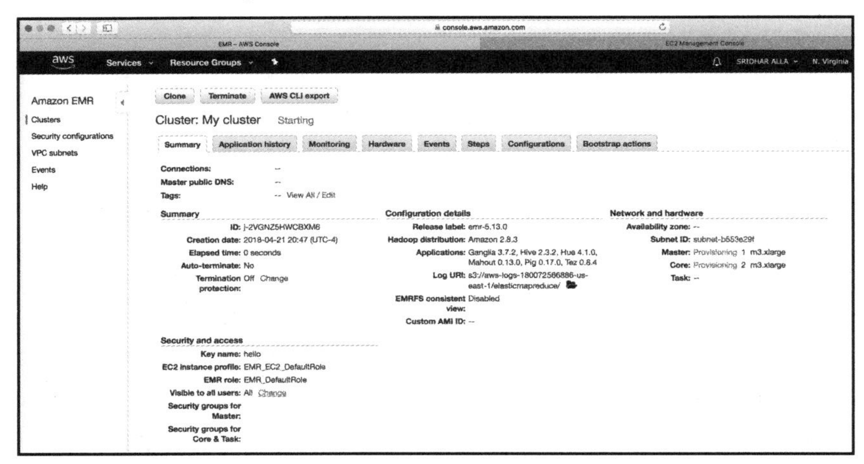

图 12.14

图 12.15 显示了 Hardware 选项卡，其中包含了集群的硬件信息。

Clone　Terminate　AWS CLI export

Cluster: My cluster　Waiting　Cluster ready after last step completed.

Summary　Application history　Monitoring　Hardware　Events　Steps　Configurations　Bootstrap actions

Add task instance group

Instance groups

Filter: Filter instance groups ...　2 instance groups (all loaded)

ID	Status	Node type & name	Instance type	Instance count	Purchasing optio
ig-2DZQGYXDH4GOP	Running	CORE Core Instance Group	m3.xlarge 8 vCore, 15 GiB memory, 80 SSD GB storage EBS Storage: none	2 Instances Resize	On-demand
ig-1DEJ243T1LBCK	Running	MASTER Master Instance Group	m3.xlarge 8 vCore, 15 GiB memory, 80 SSD GB storage EBS Storage: none	1 Instances	On-demand

图 12.15

图 12.16 显示了 Events 选项卡，其中包含了集群的事件。

Clone　Terminate　AWS CLI export

Cluster: My cluster　Waiting　Cluster ready after last step completed.

Summary　Application history　Monitoring　Hardware　Events　Steps　Configurations　Bootstrap actions

Time	Event description	Source ID	Source type	Event type	Severity	Full date & time
Apr 21 09:03 PM	Amazon EMR cluster j-2VGNZ5HWCBXM6 (My cluster) finished running all pending steps at 2018-04-22 01:02 UTC.	j-2VGNZ5HWCBXM6	Cluster	Cluster State Change	INFO	April 21, 2018 at 09:03:38 PM (UT
Apr 21 09:03 PM	Step s-1KJJF52B8QYQU (Setup hadoop debugging) in Amazon EMR cluster j-2VGNZ5HWCBXM6 (My cluster) completed execution at 2018-04-22 01:02 UTC. The step started running at 2018-04-22 01:02 UTC and took 0 minutes to complete.	s-1KJJF52B8QYQU	Step	Step State Change	INFO	April 21, 2018 at 09:03:36 PM (UT
Apr 21 09:02 PM	Amazon EMR cluster j-2VGNZ5HWCBXM6 (My cluster) began running steps at 2018-04-22 01:01 UTC.	j-2VGNZ5HWCBXM6	Cluster	Cluster State Change	INFO	April 21, 2018 at 09:02:49 PM (UT
Apr 21 09:02 PM	Step s-1KJJF52B8QYQU (Setup hadoop debugging) in Amazon EMR cluster j-2VGNZ5HWCBXM6 (My cluster) started running at 2018-04-22 01:02 UTC.	s-1KJJF52B8QYQU	Step	Step State Change	INFO	April 21, 2018 at 09:02:49 PM (UT
Apr 21 08:47 PM	Amazon EMR cluster j-2VGNZ5HWCBXM6 (My cluster) was requested at 2018-04-22 00:47 UTC and is being created.	j-2VGNZ5HWCBXM6	Cluster	Cluster State Change	INFO	April 21, 2018 at 08:47:27 PM (UT

图 12.16

考虑到安全设置，目前用户尚无法访问 EMR 集群。因此，需要开启相关端口，进而可从外部进行访问。随后，可查看 EMR 集群的 HDFS 和 YARN 服务。

注意：

在实际操作过程中，不应使用这一类不安全的 EMR 集群，此类操作仅供理解 EMR 所用。

图 12.17 显示了当前集群的 Security Groups（位于 EC2 Dashboard 中）。

编辑这两个安全组，并允许来自源 0.0.0.0/0 的所有 TCP 通信，如图 12.18 所示。

下面考查 EMR Master IP 地址（公开），即 http://EMR_MASTER_IP:8088/cluster，并以此访问 YARN 服务。

图 12.19 显示了资源管理器。

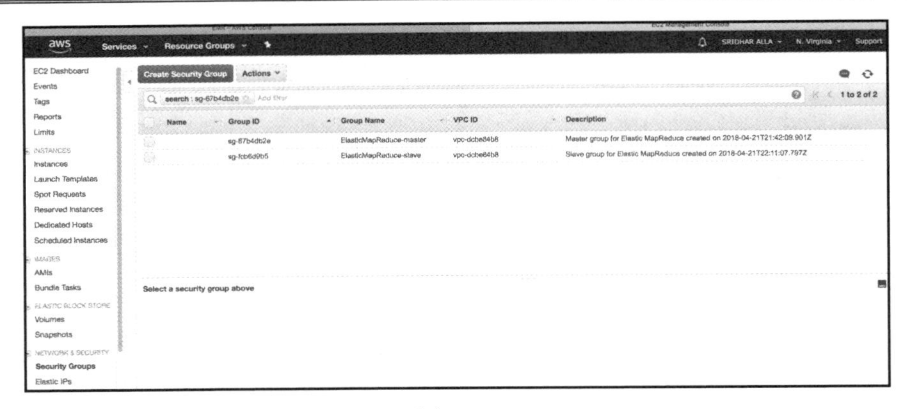

图 12.17

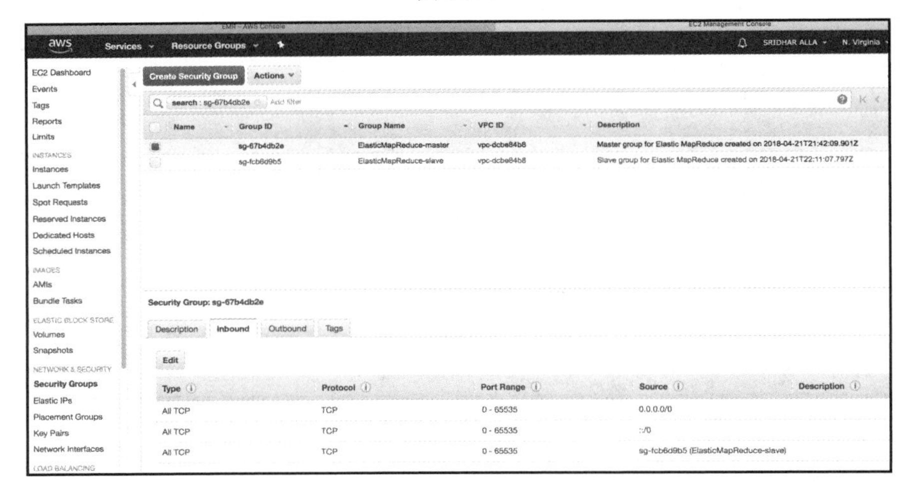

图 12.18

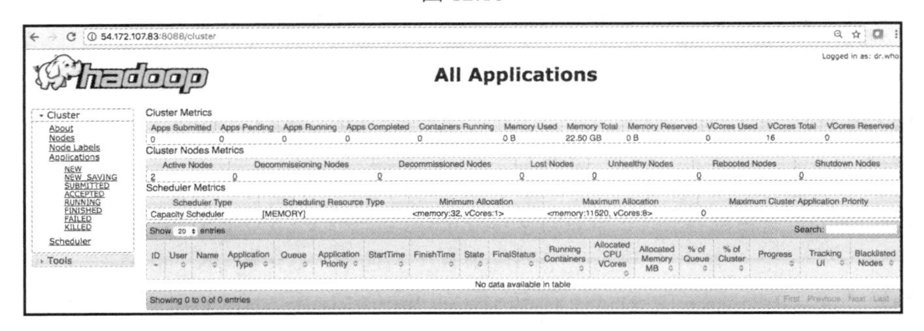

图 12.19

图 12.20 显示了资源管理器队列。

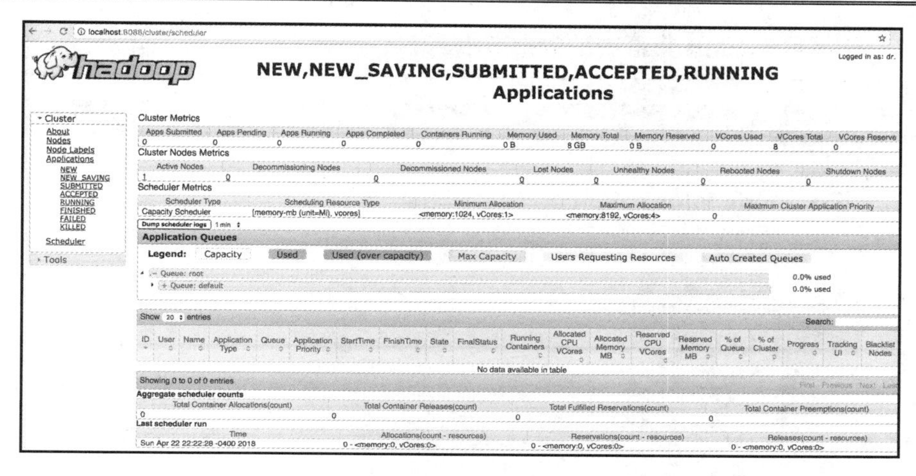

图 12.20

HDFS 也可通过相同的 IP 地址予以访问，即 http://<EMR-MASTER-IP>:50070。图 12.21 显示了 HDFS 入口界面。

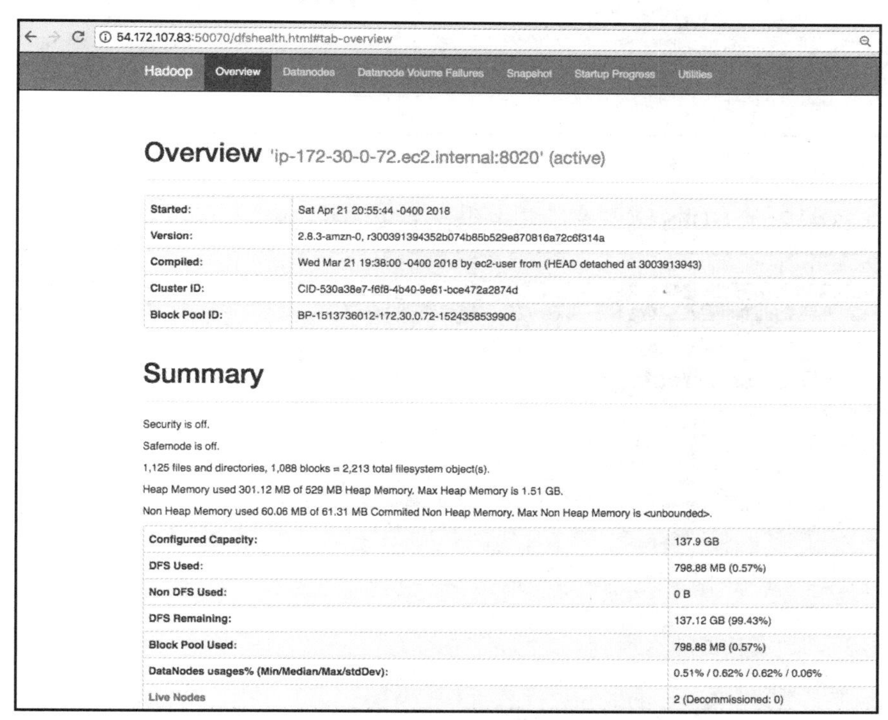

图 12.21

图 12.22 显示了 EMR 集群中的数据节点。

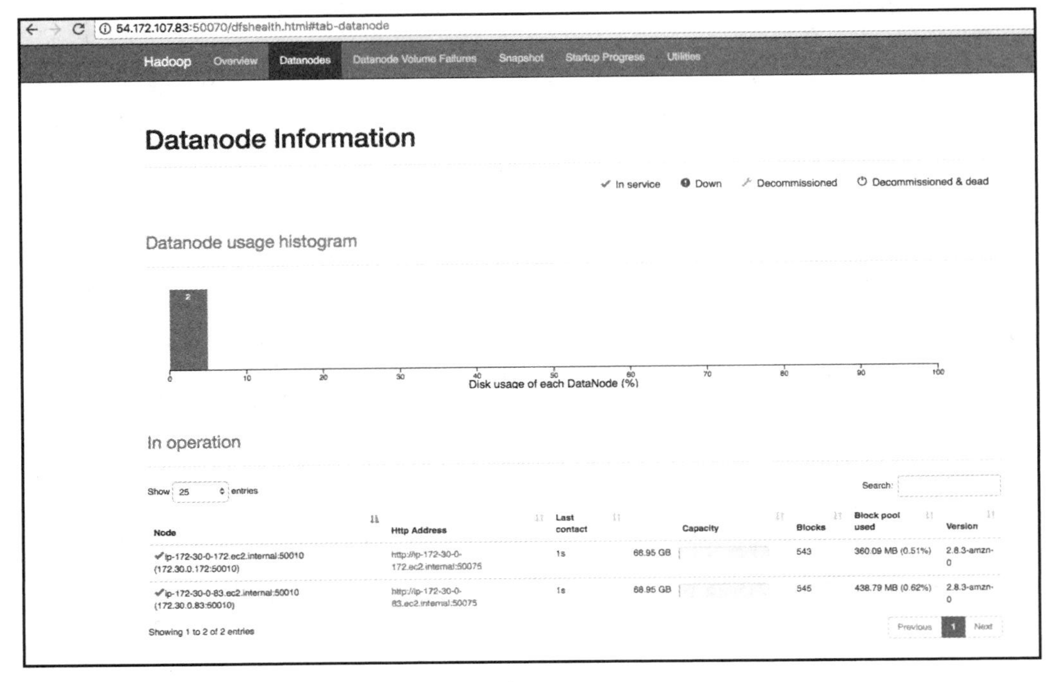

图 12.22

图 12.23 显示了 HDFS 浏览器，其中包含了文件系统中的目录和文件。

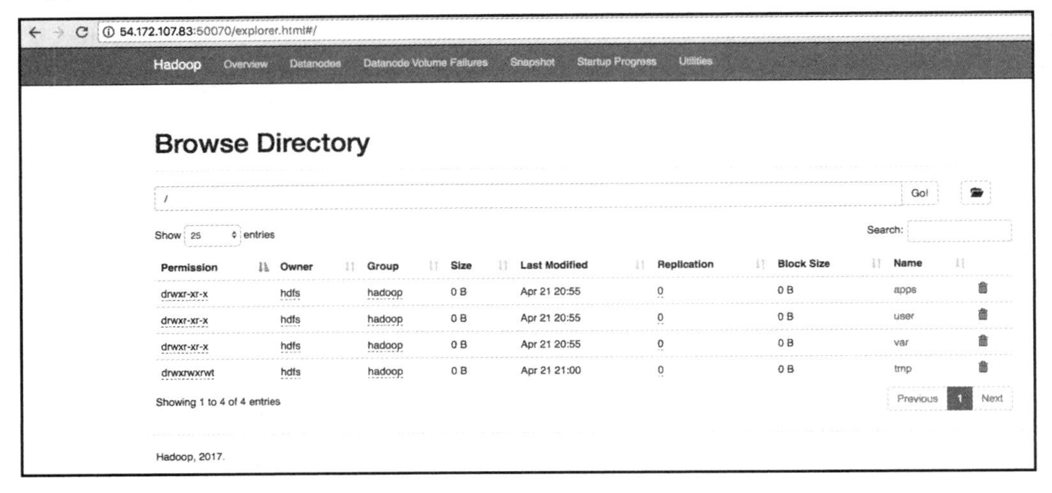

图 12.23

至此，我们已经演示了如何在 AWS 中构建 EMR 集群。

注意：

此时，应确保结束 EMR 集群的工作状态。

12.9　本章小结

针对云计算需求，本章主要讨论了与 AWS 云计算供应商相关的内容。